Proceedings of the Third International Conference on Plant Pathogenic Bacteria

Proceedings of the Third International Conference on Plant Pathogenic Bacteria, Wageningen, 14-21 April 1971

Editor: H. P. Maas Geesteranus

University of Toronto Press

ISBN 0-8020-1914-5

First published 1972 in Canada and the United States by University of Toronto Press, Toronto and Buffalo

Only for sale in Canada and the United States of America

Printed in the Netherlands by Zuid-Nederlandsche Drukkerij N.V., 's-Hertogenbosch

Organizing Committee

Drs H. P. Maas Geesteranus (Chairman)
Miss Dr E. Billing
Dr D. W. Dye
Dr D. C. Hildebrand
Dr I. Lazar

Executive Committee

Preface

The Third International Conference on Plant Pathogenic Bacteria was organized for those scientists who are working with plant pathogenic bacteria. Their interests include microbiology, biochemistry, physiology, pathology or taxonomy. Consequently the programme of this meeting included a variety of subjects.

One of the aims of the conference was to provide general information on current work, so that participants working in one discipline could learn of recent development in others. To achieve this, some sessions were devoted to reviews, combined in symposia on special topics prepared by specialists.

In addition, papers were read dealing with results of personal research and giving up-to-date information on current work. Besides there was the opportunity to present short papers on problems or negative results met with during investigations. These contributions were then discussed, either in a plenary session, or in small groups of people particularly interested in the subject. These discussions were intended to stimulate the exchange of knowledge and ideas but, as they were often long and detailed, it was impossible to summarize them. Therefore in this book only the papers that were presented are published.

The reader will also find in these Proceedings long review articles, reports of results and short contributions dealing only with the description of problems.

All together, the contents will give a good idea of the main subjects with which the phytobacteriologists are concerned at present.

The Editor

Contents

Symposium IV: Mycoplasmas

Symposium V: Enzymes and Toxins

Bacteria in buds and on leaf surfaces[1]

C. Leben and A. de Lange[2]

Department of Plant Pathology, Ohio Agricultural Research and Development Center, Wooster, Ohio 44691.

In addition to the existing terms 'rhizosphere' and 'phyllosphere', the bud as a source of the epiphytic microflora may be called the 'gemmisphere'. Micro-organisms have been reported to winter on or in normal dormant buds of some perennial plants. In spring these organisms may be distributed over plant parts. Growing buds seem to offer an especially favourable environment for micro-organisms. Bacteria are the most numerous of the microflora in active buds and the first colonizers of young tissues. Free moisture is essential for bacterial motility and is probably necessary for cell multiplication both on bud parts and leaves.

In the seedling bud, gemmisphere bacteria can originate in or on the seed, or they may enter the bud from the soil as the plant emerges. The bud flora may be supplemented by small animals, dust, rain-splashed soil or water drippings and splashings from the same or other plants. The bacterial population may be reduced by parasites, by the washing of rain and dew water or by a decrease in free water. Usually the same possibilities exist for the epiphytic microflora on leaves. The cycling of some gemmisphere and phyllosphere bacteria from seed to seed through the flower bud may take place.

Antagonistic and other relations among the epiphytic bacteria may well induce preferential multiplication of particular species. Evidence indicates that most gemmisphere bacteria die quickly on exposed, dry leaves; however, those that do remain viable may multiply quickly when the leaf is wet for some time and the weather favourable. Recent work suggests that some leaf-spotting bacterial pathogens may colonize normal, growing buds of healthy or diseased plants – the bud may thus be of epidemiological significance for diseases incited by these organisms.

1. Abstract of two papers published in: Ecology of leaf surface micro-organisms. T. F. Preece & C. H. Dickinson, eds, 1971. Academic Press, London & New York.

2. Present address: Phytopathologisch Laboratorium 'Willie Commelin Scholten', Baarn, The Netherlands.

Relationships between pathogenic and non-pathogenic bacterial inhabitants of aerial plant surfaces[1]

L. N. Gibbins

Department of Microbiology, University of Guelph, Guelph, Ontario, Canada

Abstract

Interrelationships between pathogenic and non-pathogenic bacterial inhabitants of plant surfaces are considered in three categories: 1. direct physiological interactions between epiphytes on plant surfaces, with little apparent (or passive) participation of the support plant; 2. physiological interactions in which the support plant plays a significant role (this category usually involves interactions within the plant); 3. genetic relationships between pathogens and non-pathogens, and the possible transformation of organisms from one group to another as the result of environmental circumstances. These categories are discussed with reference to appropriate examples, and particular mention is made of the non-pathogenic epiphyte *Erwinia herbicola*, and its inter-relationships, particularly with the 'fire-blight' pathogen, *Erwinia amylovora.*

Introduction

When the biological components of ecosystems are discussed, it is usually in terms of 'interactions' between the organisms, and the effect of these interactions on the whole system. This is probably quite satisfactory when trying to relate the activities of organisms from different kingdoms, or even from different classes, but as the ecosystem becomes less diverse in terms of the type of biological forms comprising it, then so does the term 'interaction' and its implications become inadequate. That this is so can perhaps be readily appreciated by considering, for example, the genus *Pseudomonas*, which contains many phytopathogens, but which also includes a wide and ubiquitous spectrum of soil-borne and other saprophytes. Given this situation, the question arises as to whether or not the vast populations of *Pseudomonas* spp. serve as sources from which new races or species of phytopathogen may arise. In other

1. It is a pleasure to acknowledge the collaboration of Dr A. K. Chatterjee, Dr W. J. Vail, Prof. J. A. Carpenter, and Mrs S. E. Cook in the experimental work. The work was supported financially by the National Research Council of Canada, and by the Ontario Department of Agriculture and Food. Table 1 and Figure 2 are reproduced by permission of the National Research Council of Canada from the Canadian Journal of Microbiology 15: 640–642 (1969). Table 2 is here reproduced by permission of the American Society for Microbiology, from the Journal of Bacteriology 105: 107–112 (1971).

words, the question of the 'relatedness' of the members of the ecosystem is of significance, in addition to their interactions. Consequently, in the present context, it is more appropriate to consider 'relationships' rather than 'interactions', and to be concerned with both the phenotypic expression of the component organisms, and the effect of environment on the various genotypes.

Although it has long been known that plants harbour a wide diversity of microbiological epiphytes (Ruinen, 1961; Leben, 1965; Last & Deighton, 1965), we know only enough about the relationships entered into by these organisms to be aware of their complexity. However, the implications of these relationships, from the standpoint of disease etiology, of plant and epiphyte nutrition, and of ecology generally, are being realised and extended as our knowledge of the identities and activities of the component members gradually accumulates. Within the bacteriological purview, it is now possible to recognize that the relationships between pathogenic and non-pathogenic bacteria, and the support plant, can be grouped, for convenience of discussion, into three categories.

1. There are direct physiological interactions between epiphytes on the plant surface, with passive participation of the plant.
2. Physiological interactions between epiphytes have been documented in which the support plant plays a very significant part, usually by virtue of the fact that these interactions occur actually within the plant. It is on this category that our interest will lie, as it is here that consideration of pathogen-plant interactions, and the possible effects of co-habiting non-pathogens on these interactions, and on the manifestation of disease, are of great importance.
3. There is evidence (Gibbins *et al.*, unpublished work) which suggests that, in some instances, the relationships between pathogens and 'recognised' non-pathogens may be close enough for interchange to occur under the appropriate conditions.

It should be emphasised, before embarking on an expansion of these categories, that they should not be regarded necessarily either as being mutually exclusive, or as being rigidly circumscribed, and we must avoid allowing our penchant for the construction of 'pigeon-holes' to obscure our view of the biological options available to ecosystems.

Ephiphyte inter-relationships with the passive participation of the plant

In 1965, Last & Deighton remarked that '. . . little is known by direct experimentation of the nutrition of leaf saprophytes . . .'. They were, however, able to delineate four characteristics of the bacteria saprophytic on leaves, namely, that they are not exacting nutritionally, that they are widely distributed, that they rarely occur in the soil to the same extent as on leaves, and that they are frequently pigmented. These characteristics suggest an innate ecological versatility with respect to aerial phytosystems in these organisms.

It is unfortunately true that relatively little new information has become available between 1965 and the present time. However, extensive studies have been made, by Libbert and his colleagues (Wichner & Libbert, 1968; Libbert *et al.*, 1969; Libbert &

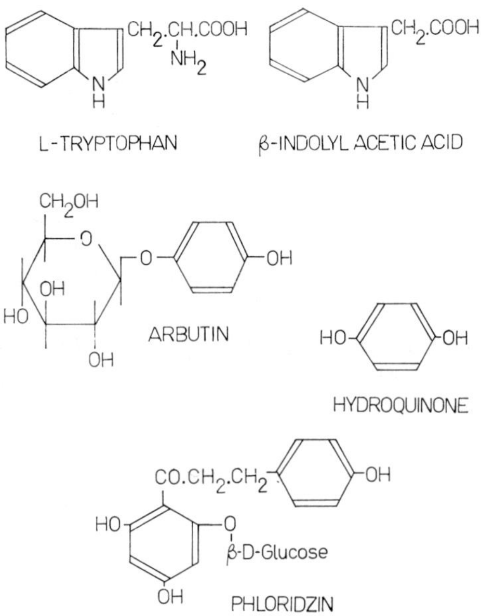

Fig. 1. Structural formulae of some aromatic compounds of significance in phytopathological situations.

Risch, 1969), on the physiology of bacteria epiphytic on *Pisum sativum*. The particular topic of interest was the synthesis, by the epiphytic organisms, of the well characterised plant hormone, β-indolylacetic acid (IAA), (Fig. 1). The precursor utilised in their *in vitro* experiments was L-tryptophan (Wichner & Libbert, 1968). The amount of IAA extractable from plants with a normal epiphytic flora was significantly greater than from sterile plants (Libbert *et al.*, 1969), but how these two observations are linked is not yet known. It was also demonstrated (Libbert & Risch, 1969) that some bacterial members of the flora could degrade IAA, and it is clear that physiological interactions will occur among the components of this ecosystem, as far as IAA is concerned. Libbert & Risch have suggested that the synthetic aspects of this association are of some significance to the plant. It may be conjectured that, as IAA is of great significance in many phytopathological as well as non-pathological situations (see, for example, Sequeira, 1965), the synthetic activities of non-pathogenic epiphytes in this regard may eventually be shown to have important repercussions with respect to the activities of certain bacterial and other phytopathogens.

Interactions between epiphytic pathogens and non-pathogens within the plant milieu

The documented interactions between pathogenic and non-pathogenic bacteria which normally live epiphytically on the aerial parts of plants, were observed either *in vitro*, in pure-culture experiments, or *in vivo*, in artifically produced disease situations. Evidence from both sources can lead to hypotheses as to the nature of the physiological relationships between the components of a natural ecosystem.

Many of the experiments performed so far have involved organisms associated with the 'fire-blight' syndrome, a serious and unpredictable disease of apples and pears. These studies have sought to define the relationships between the causal organism, *Erwinia amylovora*, the host plant, and an epiphytic non-pathogenic yellow-pigmented bacterium, which can almost always be isolated in relatively large numbers and often in almost pure culture from 'fire-blight' lesions. This organisms is now known as *Erwinia herbicola* (Geilinger) Dye (Bradbury, 1970a). The repeated isolation of strains of this organism from 'fire-blight' sources has raised the possibility that it may be making a significant contribution to the syndrome, and evidence is accumulating which suggest that this possibility may be substantiated.

The two plants of greatest economic significance which are susceptible to 'fire-blight' are the apple and pear trees. The predominant β-glucosides of the tissue of these two trees are arbutin (the β-D-glucoside of hydroquinone) and phloridzin (the β-D-glucoside of phloretin), respectively (Williams, 1960; Fig. 1). Hydroquinone is an inhibitor of the oxidative metabolism of *E. amylovora* (Chatterjee *et al.*, 1969), and probably of other biological oxidation systems, and can readily be liberated from arbutin by the action of β-glucosidase. It has been established that, when pear tissue is damaged, the plant enzyme and its substrate are brought together resulting in the formation of the toxic aglucone (Hildebrand & Schroth, 1964; Schroth & Hildebrand, 1965; Hildebrand *et al.*, 1969). Many strains of *E. herbicola* are capable of bringing about this same conversion, and they exhibit a remarkable resistance to the presence in the medium of very high concentrations of hydroquinone (Chatterjee *et al.*, 1969). *E. amylovora*, by contrast, has only a relatively feeble β-glucosidase activity (Table 1), and is very sensitive to hydroquinone (Fig. 2). On the basis of these observations, it has been postulated that, in the 'fire-blight' disease of pear trees, the non-pathogenic epiphyte, *E. herbicola*, may be making a contribution to the resistance of the host to

Table 1. The accumulation of hydroquinone in arbutin broth by isolates of *Erwinia herbicola* and *E. amylovora*.

Erwinia herbicola					*Erwinia amylovora*				
Isolate	O.D. 620 mμ		μg hydro-quinone/ml		Isolate	O.D. 620 mμ		μg hydro-quinone/ml	
	24 h	48 h	24 h	48 h		24 h	48 h	24 h	48 h
Y46	0.444	0.261	1033	1182	NCPPB 595	0.390	0.190	25	104
Y57	0.386	0.260	1282	1286	R43	0.431	0.391	80	130
Y69	0.296	0.425	433	136	P48	—	—	145	190
Y74	0.449	0.346	1103	974					
Y84-1	0.525	0.681	1723	1086					
Y111	0.444	0.318	1171	1129					
Y163	0.470	0.321	1176	1090					
Y178	0.519	0.359	978	970					
Y185	0.432	0.396	1087	1005					

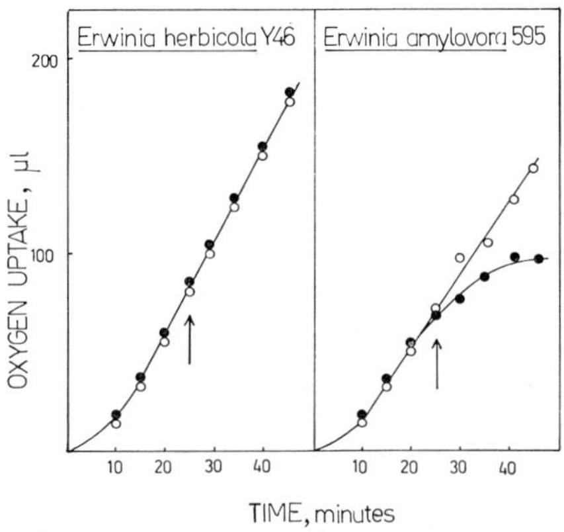

Fig. 2. Effect of hydroquinone on oxygen consumption by *Erwinia amylovora* NCPPB 595 and *Erwinia herbicola* Y46 with D-glucose as substrate. The twin side-armed Warburg flasks contained cell suspension, 80 µmoles D-glucose, and 600 µmoles KH_2PO_4 buffer, pH 6.5, in the main compartment. Hydroquinone (3000 µg) was tipped from the second side-arm at 25 minutes, as indicated by the arrows. ● = hydroquinone solution added; ○ = glass distilled water added. It should be noted that there was twice the number of cells in the *E. amylovora* experiment as in that involving *E. herbicola*.

invasions by the pathogenic *E. amylovora* (Chatterjee *et al.*, 1969). Whether or not the activities just described actually are enhanced by the presence of *E. herbicola* in the lesions has yet to be experimentally demonstrated, but the basic capability of strains of this species to act in this manner, *in vitro*, has certainly been demonstrated.

In an analogous situation, phloridzin is degraded, *in vitro*, by some *E. herbicola* strains to yield the aglucone phloretin, which is subsequently cleaved to yield phloroglucinol and p-hydroxyphenylpropionic acid (Chatterjee & Gibbins, 1969). These two compounds accumulate in large quantities in the experimental systems, and are apparently not degraded further by this organism. *E. herbicola* seems to suffer little ill-effect from being in contact with them. Of the products of phloridzin degradation, phloretin was found to be inhibitory towards the strains of *E. amylovora* tested (Chatterjee, 1970) but the pathogens are not as sensitive to this compound as they are to hydroquinone. This observation, together with the fact that phloretin is only sparingly soluble in aqueous media, perhaps mitigates against this compound being as effective a component of defence mechanisms in apple trees as hydroquinone apparently is in pear tissue.

The 'fire-blight' context has been the subject of other experiments designed to observe, in a more direct manner, the influence of epiphytic bacteria of the *E. herbicola* group on the disease syndrome. Goodman studied the effect of inoculating apple tissue with mixed cultures of *E. amylovora* and a yellow *Erwinia*-like isolate obtained from healthy apple buds. He found (a) that mixed cultures of the two organisms, when used as inoculum, were significantly less virulent then the pathogen alone (Goodman, 1965), and (b) that the pre-inoculation of the plant with the yellow bacterium had a protective effect when the plants were subsequently subjected to the pathogen (similar protection was effected by pre-inoculation with an avirulent strain of *E. amylovora* or with *Pseudomonas tabaci*, but not with *Xanthomonas pruni*; Goodman, 1967).

In an earlier study, Crosse (1965) demonstrated that saprophytic bacteria inhabiting the leaf surface of cherry trees, and which had affinities with the genus *Erwinia*, could effectively reduce the severity of leaf-scar infection of cherry by the canker pathogen,

Pseudomonas mors-prunorum. Attempts were made to extend this observation to the biological control of the disease, and cell suspensions of the antagonist were sprayed in orchard trials. Unfortunately, the populations of saprophytes fell rapidly to the equilibrium level within a few days, and the use of this approach to control the disease did not appear feasible at that time (Crosse, 1965). As *E. herbicola* is so widely distributed, it is probably that its involvement in other phytopathological situations will eventually be demonstrated.

Examples of synergistic interactions between aerial epiphytes in disease syndromes are not yet known, but an example of two soil-borne bacteria, namely *Pseudomonas caryophylli* and a *Corynebacterium* sp., acting this way has recently been documented (Brathwaite & Dickey, 1970a, 1970b, 1971).

Genetic relatedness between bacterial pathogens and non-pathogens

Bacterial phytopathogens are classified essentially in five genera: *Agrobacterium*, *Corynebacterium*, *Erwinia*, *Pseudomonas*, and *Xanthomonas*. Exceptions to this generalisation include the somewhat vaguely classified *Aplanobacter populi*, which causes bacterial canker of poplars (see Bradbury, 1970b), and, if the definition of pathogenicity is to be stretched somewhat, those strains of the genus *Rhizobium* which result in the formation of ineffective nodules or tumor-like structures (MacGregor & Alexander, 1971) on the roots of leguminous plants.

In classical terms, the intra- and inter-generic relationships among bacteria are recognised by definition, are extensively documented, and can perhaps be taken as read. The implication that the relationships between organisms within a taxon are essentially static has been prevalent for too long, but in recent years the advent of the computer and of numerical taxonomy has tended to eliminate the 'pigeon-hole' complex from which many biologists have suffered, in favour of a more general appreciation of the concept of the continuum of biological forms. A continuum implies, of course, very close genetic relatedness between adjacent components, and the achieving of the continuum in biology also implies genetic mixing on a continuing basis. It is this concept that is of importance to the plant pathologist, as he concerns himself with such fundamental problems as strain variation, the occurrence of biological races, the frequently observed unstable nature of virulence in phytopathogens *in vitro*, and the origin of pathogenicity itself.

Consideration of the origin of pathogenicity and of relationships between pathogens and non-pathogens invokes three important questions. What physiological events are associated with the occurrence of pathogenicity? Are pathogens representative of a class of organisms which achieved this capacity in the dim recesses of the evolutionary process? Are pathogens continually arising from non-pathogenic forms even at this moment?

The first question is amenable to experimentation, and the second only to conjecture. Investigation of the third question has so far been restricted to the analysis of circumstantial evidence pertaining to the occurrence and distribution of one phytopathogenic bacterium, namely *Pseudomonas solanacearum* (Buddenhagen & Kelman, 1964), and,

Table 2. Effect of incubation temperature on production of white variants of *Erwinia herbicola* Y46 grown in yeast beef broth shake cultures.

Incubation temperature (°C)	Viable, counts per ml		Percentage white variants
	yellow	white	
25	7.5×10^9	1.0×10^6*	0.01*
29	7.2×10^9	1.0×10^6*	0.01*
34	3.0×10^9	1.5×10^8	4.7
37	4.1×10^7	4.0×10^6	8.8
42	7.7×10^2	10*	0.1*

* These figures represent the lower limit of detection of white variants under the conditions of the experiment.

so far, it has not been possible to make direct observations of an experimental system.

Occasionally, however, we are given the suggestion, in the laboratory, that it may be possible, in the future, to study the transition to pathogenicity in a controlled and reproducible system. For example, recent studies in the Guelph laboratory have demonstrated that the genotypes of some strains of *Erwinia herbicola* can be changed, apparently irreversibly, simply by incubating cultures of the appropriate strains at temperatures slightly above those optimal for growth (Table 2) (Chatterjee & Gibbins, 1971). The variants produced are non-pigmented and auxotrophic, exhibiting an absolute requirement for thiamine not exhibited by the parent strain. Serological investigations indicate that the variants have a closer relationship with strains of the pathogenic *E. amylovora* than has the parent strain from which they were derived. As demonstrated in Fig. 3A, *Erwinia herbicola* Y46 has little serological identity with the strains of *E. amylovora* tested. However, two of the white variants of *E. herbicola* Y46, namely Y46-24 and Y46-70, show a relatively strong cross-reaction with the *E. amylovora* strains (Figs 3B and 3C, respectively). Current comparative taxonomic studies in the Guelph laboratory are designed to describe the relatedness of strains of *E. herbicola*, the white variants, and strains of *E. amylovora* in some detail. The question of whether or not the variants are phytopathogenic is also currently under active investigation.

The significance of this observation is the demonstration of the conversion of a non-pathogen, under mild, naturally attainable conditions (34 °C = 93.2 °F), to variants having an apparently closer relationship to pathogens then the strain from which they arose. It is quite conceivable that such a transformation could occur under field conditions: van der Zwet (1969) has suggested that the yellow *Erwinia*-like organisms associated with 'fire-blight' may be a phase in the life-cycle of the pathogenic *E. amylovora*. DNA base ratio studies (De Ley, 1968) demonstrate the close relationship between the yellow organisms and *E. amylovora*, and this, together with the observations reported here (Chatterjee & Gibbins, 1971), may possible be construed as

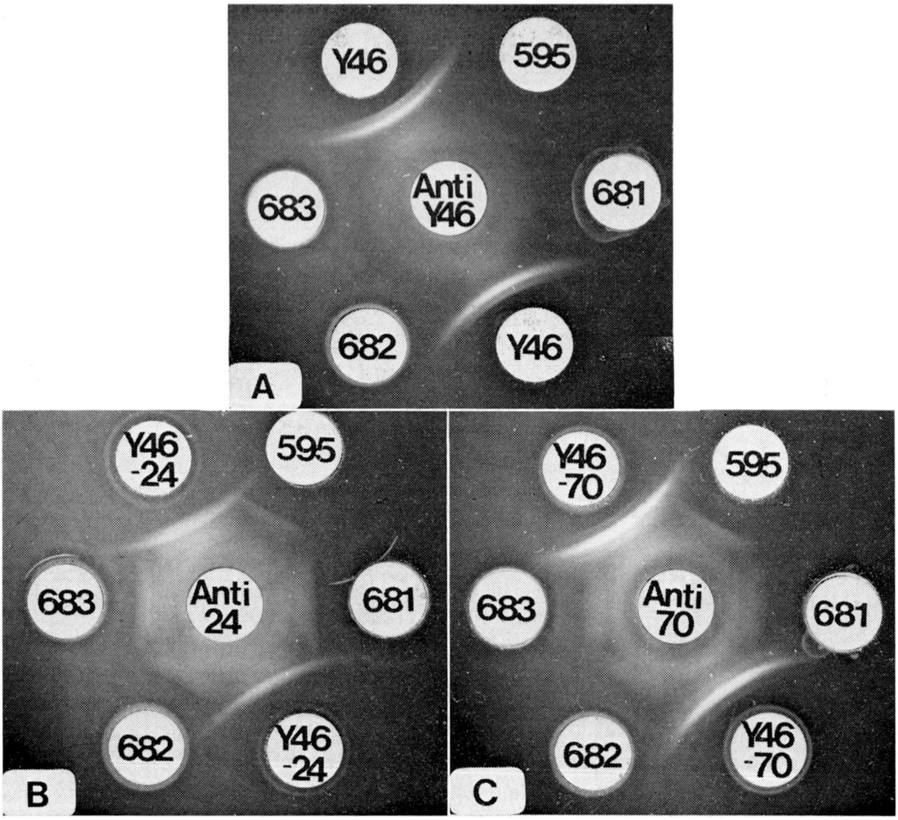

Fig. 3. Gel diffusion tests showing the serological relationships of *Erwinia herbicola* Y46, and two white variants, Y46-24 and Y46-70, with strains 595, 681, 682 and 683 of *Erwinia amylovora*. Antisera were prepared in rabbits in response to administration of whole, live cells of the appropriate bacterial strain. The test antigens used in these experiments consisted of thick suspensions of whole, live bacterial cells. The tests were carried out in 1.0% agar, containing 0.85% NaCl, and 0.01% methiolate, and incubation was performed at 25 °C for 3–7 days.

supporting van der Zwet's suggestion. However, the demonstrated involvement of these yellow organisms in syndromes caused by unrelated bacteria (Crosse, 1965) indicates that the life-cycle hypothesis, even if true, does not give a full explanation of the role of these organisms in phytopathology.

Concluding remarks

Two things will have become apparent during this brief, and necessarily circumscribed, discussion. Firstly, the amount of documented experimental work on the physiological interactions of bacterial epiphytes is very limited, and the number of contexts from which we have available data is small. Secondly, there may appear to have been an undue emphasis on the species *Erwinia herbicola*. This emphasis is justified on several counts, some of which have already been discussed. In addition, some strains of *E. herbicola* are lysogenic, and some will produce at least one bacteriocin without inducement (Gibbins *et al.*, unpublished work). The ecological ramifications of these properties are not clear but can be presumed to be significant. It is already known that at least one strain of *Xanthomonas vesicatoria* and two strains of *Aerobacter aerogenes* are sensitive to the bacteriocin produced by *E. herbicola* Y46 (Gibbins *et al.*, unpublished work). Of more general importance is the occurrence of strains of *E. herbicola* (or *Erwinia* spp.) in clinical situations (Gilardi *et al.*, 1970), and the role of these organisms in pathogenic processes of animals, including man, is currently a matter of some conjecture. In view of its ubiquitous nature, its apparent association with diseases of both plants and animals, and its interesting physiological and genetic characteristics, *E. herbicola* is worthy of considerable attention.

Bacteriology has made exciting advances during the past two decades, and the study of these tiny islands of life has led to the uncovering of an extraordinary array of principles, concepts, and processes with fundamental and far-reaching implications. We must not, however, allow ourselves to be deluded into believing that all the significant discoveries in biology and bacteriology have been made, and that all that remains for us is a scientific 'mopping-up' operation. The most cursory examination of microbial ecology is enough to point to the shallowness of our knowledge of the mechanisms and significance of biological interactions and relationships. It may be, perhaps, that soon we will see the eclipse of the single, pure-culture approach to experimental microbiology, in favour of mixed populations of micro-organisms, and of controlled ecosystems involving components from more than one biological kingdom. To a very large extent, future progress in phytopathology lies in such a transition.

References

Bradbury, J. F., 1970a. *Erwinia herbicola*. C.M.I. descriptions of pathogenic fungi and bacteria, No. 232.

Bradbury, J. F., 1970b. *Aplanobacter populi*. C.M.I. descriptions of pathogenic fungi and bacteria, No. 231.

Brathwaite, C. W. D. & R. S. Dickey, 1970a. Production of basal soft rot symptoms and maceration of carnation tissue by *Pseudomonas caryophylli* and *Corynebacterium* species. Phytopathology 60: 1040–1045.

Brathwaite, C. W. D. & R. S. Dickey, 1970b. Synergism between *Pseudomonas caryophylli* and a species of *Corynebacterium*. Phytopathology 60: 1046–1051.

Brathwaite, C. W. D. & R. S. Dickey, 1971. Permeability alterations in detached carnation leaf tissue inoculated with *Pseudomonas caryophylli* and *Corynebacterium* spp. Phytopathology 61: 317–321.

Buddenhagen, I. & A. Kelman, 1964. Biological and physiological aspects of bacterial wilt caused by *Pseudomonas solanacearum*. A. Rev. Phytopath. 2: 203–230.

Chatterjee, A. K., 1970. *Erwinia herbicola*: the degradation of phloridzin, and some observations on non-pigmented variants. Ph. D. thesis, University of Guelph, Ontario, Canada.

Chatterjee, A. K., L. N. Gibbins & J. A. Carpenter, 1969. Some observations on the physiology of *Erwinia herbicola* and its possible implication as a factor antagonistic to *Erwinia amylovora* in the 'fire-blight' syndrome. Can. J. Microbiol. 15: 640–642.

Chatterjee, A. J. & L. N. Gibbins, 1969. Metabolism of phloridzin by *Erwinia herbicola*: nature of the degradation products and the purification and properties of phloretin hydrolase. J. Bacteriol. 100: 594–600.

Chatterjee, A. K. & L. N. Gibbins, 1971. Induction of non-pigmented variants of *Erwinia herbicola* by incubation at supra-optimal temperatures. J. Bacteriol. 105: 107–112.

Crosse, J. E., 1965. Bacterial canker of stone fruits. VI. Inhibition of leaf-scar infection of cherry by saprophytic bacterium from the leaf surfaces. Ann. appl. Biol. 56: 149–160.

Crosse, J. E., 1966. Epidemiological relationships of the pseudomonad pathogens of deciduous fruit trees. A. Rev. Phytopath. 4: 291–310.

Crosse, J. E., 1968. The importance and problems of determining relationships among plant-pathogenic bacteria. Phytopathology 58: 1203–1206.

Ley, J. De, 1968. DNA base composition of yellow *Erwinia* strains. Antonie van Leeuwenhoek 34: 257–262.

Gilardi, G. L., E. Bottone & M. Birnbaum, 1970. Unusual fermentative, Gram-negative bacilli isolated from clinical speciments. I. Characterisation of *Erwinia* strains of the 'lathyri-herbicola' group. Appl. Microbiol. 20: 151–155.

Goodman, R. N., 1965. *In vitro* and *in vivo* interactions between components of mixed bacterial cultures isolated from apple buds. Phytopathology 55: 217–221.

Goodman, R. N., 1967. Protection of apple stem tissue against *Erwinia amylovora* infection by avirulent strains and three other bacterial species. Phytopathology 57: 59–63.

Hildebrand, D. C. & M. N. Schroth, 1964. Antibiotic activity of pear leaves against *Erwinia amylovora* and its relation to β-glucosidase. Phytopathology 54: 59–63.

Hildebrand, D. C., C. C. Powell, Jr. & M. N. Schroth, 1969. Fire-blight resistance in *Pyrus*: localisation of arbutin and β-glucosidase. Phytopathology 59: 1534–1539.

Klement, Z. & R. N. Goodman, 1967. The hypersensitive reaction to infection by bacterial plant pathogens. A. Rev. Phytopathol. 5: 17–44.

Last, F. T. & F. C. Deighton, 1965. The non-parasitic microflora on the surfaces of living leaves. Trans. Br. mycol. Soc. 48: 83–99.

Leben, C., 1965. Epiphytic micro-organisms in relation to plant disease. A. Rev. Phytopath. 3: 209–230.

Libbert, E., W. Kaiser & R. Kunert, 1969. Interactions between plants and epiphytic bacteria regarding their auxin metabolism. VI. The influence of the epiphytic bacteria on the content of extractable auxin in the plant. Physiologia Pl. 22: 432–439.

Libbert, E. & H. Risch, 1969. Interactions between plants and epiphytic bacteria regarding their auxin metabolism. V. Isolation and identification of the IAA-producing and destroying bacteria from pea plants. Physiologia Pl. 22: 51–58.

MacGregor, A. N. & M. Alexander, 1971. Formation of tumor-like structures on legume roots by *Rhizobium*. J. Bact. 105: 728–732.

Ruinen, J., 1961. The phyllosphere. I. An ecologically neglected milieu. Pl. Soil 15: 81–109.

Schroth, M. N. & D. C. Hildebrand, 1965. β-Glucosidase in *Erwinia amylovora* and *Pseudomonas syringae*. Phytopathology 55: 31–33.

Sequeira, L., 1965. Origin of indole acetic acid in tobacco plants infected by *Pseudomonas solanacearum*. Phytopathology 55: 1232–1236.

Wichner, S. & E. Libbert, 1968. Interactions between plants and epiphytic bacteria regarding their auxin metabolism. II. Influence of IAA-producing epiphytic bacteria on short-term IAA production from tryptophan in plant homogenates. Physiologia Pl. 21: 500–509.

Williams, A. H., 1960. The distribution of phenolic compounds in apple and pear trees. In J. B. Pridham, ed.: Phenolics in plants in health and disease. Pergamon Press, New York.

Zwet, T. van der, 1969. Study of 'fire-blight' cankers and associated bacteria in pear. Phytopathology 59: 607–613.

Interaction between bacteria and bacteriophages on plant surfaces and in plant tissues

E. L. Civerolo

Plant Science Research Division, Agricultural Research Service, U.S. Department of Agriculture, Beltsville, Maryland 20705

Abstract

The ecological role of phages of bacteria associated with plants in nature has not yet been clearly established. Poor nodulation of several leguminous plants has been attributed to lysis of *Rhizobium* species by phages in root nodules and soil. In several cases phages of phytopathogenic bacteria have been associated with naturally infected and diseased tissues. Experimentally, phages can reduce infection of various tissues by several phytopathogenic bacteria; presumably these phages interact with their bacterial hosts at infection sites on plant surfaces or within plant tissue. In relation to plant disease, phages may be effective only in protecting tissue from infection rather than by eradicating the pathogen.

At high temperatures, *Xanthomonas phaseoli* phages attack *Xanthomonas* and *Pseudomonas* species which are resistant at lower temperatures. Therefore, in addition to other factors, halo blight and brown spot diseases of bean, caused by *P. phaseolicola* and *P. syringae*, respectively, may be more prevalent at low temperatures because of resistance of these bacteria to phage at or below 22 °C.

Retardation of crown gall development after absorption of *Agrobacterium tumefaciens* phage PB2-A by tomato plants was attributed to blockage of the bacterial attachment site or to reduction of viable bacterial cells at the wound site.

Introduction

Bacteriophages active against bacteria associated with plant tissue have been known since 1923. However, several years earlier a bacteriolytic enzyme produced by *Rhizobium radicicola* was described (Laird, 1936). It was stated that bacteria in the nodules of leguminous plants were lysed due to accumulation of these bacteriolytic enzymes. These enzymes may not have been bacteriophages as known today, but their effects were similar.

Phages have been detected in soil, irrigation water, diseased tissues, and bacterial cultures using enrichment techniques. Interactions between bacteria and phages on plant surfaces and within plant tissues have been studied indirectly, primarily through their effects on plant disease development but also by studying phage-bacterium interactions *in vitro*.

Phages of root nodule bacteria were detected in soil and in plant tissue (Demolon & Dunez, 1935; Fred *et al.*, 1932; Laird, 1931; Vandecaveye & Katznelson, 1936), but the significance of phage in the symbiotic relationship between nodule bacteria and leguminous plants has not been elucidated. *Rhizobia* phages present in soil or plant tissues, multiplying at the expense of their bacterial hosts, may be responsible for poor nodulation of certain legumes, for low nitrogen fixation capability of certain strains of bacteria, and for reduced yields of old alfalfa stands. However, there is no conclusive evidence that phages interfere with symbiosis between *Rhizobia* spp. and host plants.

The relationship of phages of phytopathogenic bacteria to plant diseases has not been fully defined. For epidemiological studies, it is necessary to understand the interactions and ecological relationships between phages and their bacterial hosts, and the effects of these relationships on the plant host. Decline or disappearance, in nature and under experimental conditions, of various phytopathogenic bacteria has suggested that phages prevent or reduce infection of plants by reducing the number of bacterial host cells.

A variety of micro-organisms, along with viruses (including phages), may form relatively stable communities on plant surfaces. Predator micro-organisms, as components of the microflora of natural habitats, may function as homoeostatic regulators correcting microbial population imbalances caused by minor ecological disturbances (Mitchell, 1967). Phages, and possibly other microbial viruses with the ability to survive outside their host in spite of their obligate parasitism, may influence the composition of the microflora on plant surfaces. Extracellular viruses, however, may be susceptible to inactivation or destruction by some antiviral component of the indigenous microbial community (Mitchell, 1971).

Agrobacterium

Several studies have suggested that at least part of the tumor-inducing principle or process initiated by *Agrobacterium tumefaciens* may be, or at least involve, a bacteriophage or an infectious nucleic acid carried and transferred by the bacterium.

Electron microscopy has not conclusively demonstrated the role of phage-bacterium interactions in the etiology of crown gall disease. Intracellular phage was observed in ultra-thin sections of virulent *A. tumefaciens*, strain B6m, within pea seedling tissue (Kurkdjian *et al.*, 1968). Phage particles were detected in pea seedling tissue 28 hours after inoculation with a virulent or an avirulent clone of *A. tumefaciens*, strain 806 (Kurkdjian, 1970). However, no phage particles were observed in wounds infected by the attenuated strain A66 (Kurkdjian, 1970). Phage particles were also observed in tumor tissue from various plant species cultured for several years *in vitro* under sterile conditions; it was suggested that these phages may function in tumor transformation (Tourneur & Morel, 1970). No phage or virus-like particles were observed in sunflower crown gall tissue resulting from inoculation with *A. tumefaciens*, strain B6, lysogenic for phage omega (Manocha, 1970).

Phage PS8, active against *A. tumefaciens*, strain 806, was detected in homogenates of cultures of S8 crown gall tumor tissue originating from a sunflower plant infected with strain B6 and subsequently propagated *in vitro* by serial transfers. PS8 is morphologically similar to PB6, a temperate phage from *A. tumefaciens*, strain B6, contains DNA as does PB6, is inactivated by PB6 antisera, and exhibits one step growth similar to PB6. PS8 may have originated from phage introduced into plant cells by the bacterium used to incite the initial tumorous growth (Parsons & Beardsley, 1968). Although the persistence of phage for several years in plant cells grown *in vitro* may reflect the presence of the host bacterium in some undetected form, it is possible that phage DNA replication and formation of phage particles may occur in plant cells (Parsons & Beardsley, 1968; Sander 1964, 1967).

UV light and mitomycin C induce prophage release in *A. tumefaciens*, suggesting that these agents enhance tumor-inducing ability by inducing replication of a carcinogenic phage (Heberlein & Lippincott, 1965, 1967a, 1967b). However, there is no evidence for prophage induction at wound sites. Furthermore, chlorpromazine treatment, which also enhances the tumor-inducing ability of *A. tumefaciens*, did not promote release of temperate phage from 2 lysogenic strains of *A. tumefaciens* under conditions which support prophage induction by mitomycin C (Heberlein, 1970). Temperate phages may vary in their susceptibility to chlorpromazine induction, and conditions used to demonstrate induction may be significantly different at wound sites. Nevertheless, the data are consistent with the conclusion that chlorpromazine enhances the tumor-inducing ability of *A. tumefaciens* by a process other than phage induction (Heberlein, 1970).

Tumor initiating ability of *A. tumefaciens* is temperature sensitive (Lippincott & Lippincott, 1965; Stonier *et al.*, 1966). Temperature also influences the synthesis of phage LV-1 carried in *A. tumefaciens*, strain V-1. The prophage state is favored at 35 °C while phage synthesis is induced at 42 °C (Brunner & Pootjes, 1969). Furthermore, the pattern of phage release by this lysogenic strain parallels changes in tumor production by other tumorgenic strains (Brunner & Pootjes, 1969).

Free particles of phages omega and $PB2_1$ from the lysogenic strains B6 and B2, respectively, do not initiate tumor production in susceptible tissues (Klein & Beardsley, 1957; Stonier *et al.*, 1967). The ability of *A. tumefaciens* strain V-1, when cured of phage LV-1, to produce tumors on pinto bean leaves is similar to that of the original lysogenic strain (Brunner & Pootjes, 1969). Therefore, *A. tumefaciens* phages omega, $PB2_1$ and LV-1 do not seem to be the tumor-inducing principle. Phages PB6, PA6 and PS8 did not induce tumors on *Kalanchoë* plants (Roussaux *et al.*, 1968). Other undetected prophage(s) may be related to the tumor-inducing principle, since many *A. tumefaciens* strains are lysogenic (Zimmerer *et al.*, 1966). Phage particles may be associated with the tumor-inducing principle, or synthesized under conditions which favor tumor-inducing principle production (Brunner & Pootjes, 1969).

Recent molecular hybridization studies indicate that DNA extracted from crown-gall tissue possesses sequences common with DNA from virulent *A. tumefaciens* (Quetier *et al.*, 1969; Schilperoort *et al.*, 1967; Stroun *et al.*, 1969; Srivastava & Chadha, 1970). These same sequences do not exist in DNA from corresponding healthy tissue. These data suggest that cellular transformation may involve integration of sequences of bacterial and/or phage genome into the genome of the host plant. Preparations of DNA from *A. tumefaciens* phage PS8 may cause tumors on pinto bean primary leaves, on stems of *Helianthus annuus*, and on stems of *Nicotiana tabacum* when virulent bacteria are excluded (Leff & Beardsley, 1970). This report has not been confirmed.

Interaction between *A. tumefaciens* strains and phage in plant tissue has been described by Stonier *et al.* (1967) and by Boyd *et al.* (1971). $PB2_1$, applied to previously inoculated tobacco tissue *in vitro*, did not eliminate bacteria from tumor tissue. Either the bacteria, adapted to the environment provided by host tissue, were no longer susceptible to phage infection or could not support phage replication, or because location of the bacteria within the tissue made them relatively inaccessible to phage (Stonier *et al.*, 1967). The former explanation is less plausible, because presence of phage under conditions which ensure mass lysis, either by introducing phage (10^8–10^9 plaque forming units) into the wounds first or by premixing phage with bacteria, completely inhibited tumor initiation. Phage introduced immediately after the bacteria inhibited tumor initiation only slightly or not at all.

Intact tomato plants immersed in crude lysates containing 10^9–10^{10} plaque-forming units/ml of phage PB2-A of *A. tumefaciens*, strain B2-A, absorbed and translocated phage particles within 3 h (Boyd *et al.*, 1971). Phage was detected 168 h after phage absorption in leaves, stems and roots, and in stems and roots after 336 h. Surface sterilization of leaves, stems, and roots with 95% ethanol and 50% $CaOCl_2$ presumably inactivated phage particles on external tissue surfaces. There is no evidence that this phage was readily inactivated within tomato plants.

Increases in diameter and weight of tomato stem galls were significantly reduced when wounded plants absorbed phage-containing lysate 12 h before bacterial inoculation, but not when plants absorbed phage for only 3 or 6 h (Boyd *et al.*, 1971). Gall size was not reduced when plants were consecutively wounded and inoculated, either before or after phage treatment. Phage particles localized at the wound sites following absorption may block the site of, or prevent, bacterial attachment in the wound, and indirectly interfere with gall induction (Boyd *et al.*, 1971). The mechanism of gall retardation after treatment with phage-containing lysate was attributed to a reduced number of *A. tumefaciens* by cell lysis in the plant tissue, since the weight of stem galls formed on *Datura* plants may be a function of the number of viable bacteria introduced into the wound (Boyd *et al.*, 1971).

Corynebacterium

Specific phages for *Corynebacterium flaccumfaciens* and *Xanthomonas phaseoli* var. *fuscans* have been associated with heavily infected bean seeds (Klement, 1957; Klement & Lovas, 1960). *C. poinsettiae* was also susceptible to the phage for *C. flaccumfaciens* (Klement & Lovas, 1960). Two *C. insidosum* phages were isolated from a mixture of roots of wilted diseased alfalfa plants and soil adhering to the roots (Cook & Katznelson, 1960). The nature of the origin of *Corynebacterium* phages in diseased tissue is not known, nor is relationship of *Corynebacterium*-phage interaction to disease known.

Erwinia

Phages of *Erwinia carotovora* and *E. atroseptica* were isolated from soft-rotted carrots and soil, respectively (Coons & Kotila, 1925). Occurrence of these phages in nature was believed to be widespread and play an important role in the decline of their phytobacterial hosts in the soil, although no data are given. When these phages were spread on slices of susceptible plant tissue (carrot and potato), infection by *E. carotovora* and *E. astroseptica*, respectively, was prevented.

Phages (EN), specific for *Erwinia nigrifluens*, were associated with the bark canker disease of Persian walnut (Zeitoun & Wilson, 1969). Phage was isolated in a 'free state' from diseased walnut bark, from the black liquid exuded from infected bark, and from soil at the base of affected trees. No phage was isolated from healthy bark, or from soil at the base of trees free of bark canker. Lysogeny in *E. nigrifluens* was not demonstrated, but phage was liberated from *E. nigrifluens* cultures by UV-irradiation and chloroform treatment. Since the phage was capable of lysing the same *E. nigrifluens* isolate from which it was obtained, it was presumably a virulent phage introduced and co-existing in infected tissue with the bacterium (Zeitoun & Wilson, 1969).

E. rubrifaciens may occur in the same tree with *E. nigrifluens* (Zeitoun & Wilson, 1969). However, no phages capable of lysing *E. rubrifaciens* were found in or on diseased trees, soil, or irrigation water. Nevertheless, concentrated EN phage suspensions significantly inhibited growth of *E. rubrifaciens* on solid media (Zeitoun & Wilson, 1969). Samples of *E. rubrifaciens*, from agar in which growth had been arrested by a concentrated EN phage suspension, contained only a few rod-shaped bacterial cells, but numerous spherical bodies. The effect was similar to the phenomena of 'lysis from without' due to disorganization of the cell wall or 'abortive infection' (Zeitoun & Wilson, 1969). In the case of *E. rubrifaciens*, the affected cells become spherical and the density of heads of the attaching phage do not change. This indicates that phage DNA is not injected into the bacterial cell. Whether a type of abortive infection of *E. rubrifaciens* by EN phage occurs on plant surfaces or within plant tissue is not known.

Fire-blight diseased material contains Gram-negative, non-pathogenic, non-fluorescent bacteria related to *E. amylovora* by phage sensitivity (Goodman, 1965). However, nature of the relationships among components of mixed bacterial cultures from fire-blight diseased and apparently healthy tissue is not entirely clear. Yellowish-white mucoid isolates from apple buds were mixtures of at least two bacterial forms, a white virulent phage-sensitive *E. amylovora* and a yellow avirulent phage-sensitive bacterium (Goodman, 1965). The yellow component is bacteriostatic to the virulent white form of *E. amylovora in vitro*. Reduction of fire blight disease intensity in host tissue in summer accompanies an increase in prevalence of the yellow bacterium over the white virulent form of *E. amylovora* (Goodman, 1965). However, the manner in which the white virulent form becomes predominant in the host in spring is unknown, and there is no evidence that phage has a role in this phenomenon.

Several phages, 1507, 1508, P3, P5 and PP, lysed many Gram-negative, pathogenic and nonpathogenic bacterial isolates from fireblight diseased trees (Hendry *et al.*, 1967). *E. amylovora* isolates were distinguished by pigmentation, biochemical tests, serology and pathogenicity. Among 167 'typical' white isolates tested 113 were pathogenic, whereas of 12 grey 'atypical' isolates 5 were pathogenic. Fourteen isolates of a 'white miscellaneous group and 44 of a 'yellow' group were all non-pathogenic.

Although these Gram-negative, non-pathogenic, non-fluorescent bacteria are related to *E. amylovora* by phage sensitivity, the data do not suggest a role for these polyvirulent phages in affecting the bacterial population in fire blight diseased trees.

Pseudomonas

Field observations on occurrence of tobacco wildfire and blackfire diseases indicated there was no infection when an abundance of inoculum was present, with favorable weather conditions and occasional water congestion of the leaves (Fulton, 1950). Since it was also difficult to isolate *P. tabaci*, presence of a phage in wildfire-diseased tobacco was suspected. Phages which lysed both *P. tabaci* and *P. angulatum* were detected in filtered extracts of some diseased leaves, suggesting that phage may reduce the amount of wildfire on tobacco by lysing *P. tabaci* cells, thereby reducing the amount of inoculum.

Presence of phage on leaf surfaces was shown to reduce the amount of infection by *P. tabaci* (Fulton, 1950). In field tests, the percentage of diseased plants was reduced 63% when both phages 1 and 2 were sprayed on tobacco seedlings. When each phage was applied separately the percentage of infected plants was nearly equal to the control, but the number of wildfire lesions per plant ranged from 0 to 3, compared to 5 to 20 on control plants. The protective effect was demonstrated in greenhouse tests wherein the percentage of diseased plants was reduced 83% and 51% by phages 1 and 2, respectively. However, these data are not sufficient to indicate whether the phage-bacterium interaction is a factor limiting the amount of disease under natural conditions. It was suggested that light secondary infection following heavy primary infection, despite favorable conditions, may be due to the effectiveness of phage in reducing numbers of bacteria massed together in lesions (Fulton, 1950).

Presence of phage-resistant strains of bacteria in nature will affect any phage-bacterium interactions and may limit any effect phage has on disease. Data regarding occurrence and relative distribution of susceptible and phage-resistant phytopathogenic pseudomonad and xanthomonads are available (Baigent, 1965; Billing, 1963, 1970; Crosse & Garrett, 1961, 1963; Dye *et al.*, 1964; Fulton, 1950; Stolp & Starr, 1964; Stolp *et al.*, 1965). *P. tabaci* and *P. angulatum* isolates, resistant to phage 1 and 2, were encountered infrequently in field-infected tissue, suggesting that phage does not eliminate a significant proportion of bacteria or that phage-resistant variants are less pathogenic under natural conditions (Fulton, 1950). A polyvalent bacteriophage, which occurred in wildfire-infected tobacco leaves, remained active for 15 days on healthy 'Makhorka' tobacco leaves and reduced wildfire infection by 50% (Novikova, 1940).

Strains of *P. mors-prunorum* from plum and cherry, exhibiting degrees of pathogenic adaptation to their original host, were differentiated by typing against phages A1, A2, A3, A7 and A32 *(P. mors-prunorum)*, S3 and S8 (*P. syringae*; Crosse & Garrett, 1961). Plum isolates of *P. mors-prunorum* differed from cherry isolates in failing to react at the routine test dilutions with phages A2 and A7, giving isolated plaques or no reaction with phage A3, and being less sensitive to phage A32. They resembled *P. syringae* in typing pattern, except that they were more sensitive to phage A1 and less sensitive to *P. syringae* phages S3 and S8. Furthermore the plum isolates differed biochemically from *P. syringae*. The difference between plum and cherry isolates may reflect different phage-bacteria relationships under natural conditions. Cherry isolates may have failed to acquire resistance to phages A2, A7 and related phages because they are absent on cherry, or the phage types may be a result of selection by the host plants (Crosse & Garrett, 1961, 1963). It was also suggested that there is an association between phage sensitivity and specific virulence factors in *P. mors-prunorum* (Crosse & Garrett, 1961, 1963).

Sensitivity of *P. syringae* and *P. savastoni* isolates to 20 phages was also associated with host origin (Billing, 1970). There was no association between phage sensitivity patterns and host of origin with isolates of *P. coronofaciens*, *P. pisi* and *P. tomato* (Billing, 1970).

Xanthomonas

Phage active against *Xanthomonas malvacearum* was isolated from Nile River water and soil filtrates from fields of blight-infected cotton which had been overflowed (Rosberg & Parrack, 1955). Incidence of bacterial blight disease was reduced in fields that had been overflowed by the Nile. This effect was attributed to reduction of *X. malvacearum* in soil by phage carried in river water. Phage effective against *X. citri* was readily isolated from soil under infected citrus trees, but in only one case from infected leaves (Matsumoto & Okabe, 1937). Nevertheless, it was suggested that this phage may reduce the number of citrus canker organisms in soil.

Relationships of environmental factors to phage-bacteria interactions on plant hosts may exist. Strains of fluorescent phytopathogenic pseudomonads, *P. phaseolicola* and *P. syringae* are susceptible to several phages originally isolated for *X. phaseoli* strains (Vidaver & Schuster, 1969). Near optimum plating efficiency for several *X. phaseoli* phages on *X. phaseoli* occurs over a wide temperature range from about 13°–28°C. However, sensitivity of *P. phaseolicola* and *P. syringae* to the same *X. phaseoli* phages occurs in a narrow temperature range. Optimum EOP for phages on susceptible *P. phaseolicola* hosts occurs at 24° or 28°C and optimum EOP of phages of *P. syringae* at 28° or 34°C. *P. phaseolicola* and *P. syringae* are not susceptible to these phages below 20°C. Thus, in nature during hot weather, *X. phaseoli* may attack not only *Xanthomonas* species but also *Pseudomonas* species that are phage-resistant at lower temperatures. Although other factors may affect disease prevalence, the halo blight and brown spot diseases of bean caused by *P. phaseolicola* and *P. syringae*, respectively, may be more prevalent at or below 22°C because of phage resistance (Vidaver & Schuster, 1969).

Some correlation has been reported between general phage typing patterns and plant habitat of several *Xanthomonas* species (Stolp & Starr, 1964). These data may help to explain the ecological significance of *Xanthomonas* phages. Certain phage typing patterns of about 365 *Xanthomonas* cultures reflected specific plant habitats, whereas other patterns were shared by isolates from various plants (Stolp & Starr, 1964).

Virulence of *X. phaseoli* strains was associated with sensitivity to 4 virulent phages. Phage subtypes of *X. phaseoli* prevalent in an infected bean crop were most virulent and subtypes, which were only occasionally found in the bean crop, were least virulent (Sutton & Wallen, 1967). These data are very limited but may reflect the ecological significance of *X. phaseoli* strains and their phages to plant hosts.

Failure to detect *X. phaseoli* in soil from infected fields and to isolate *X. phaseoli* phages without enrichment suggested low probability of contact between phage and susceptible *X. phaseoli* (Sutton & Wallen, 1967). It was concluded that conditions for phage to have a significant ecological effect on the frequency of occurrence of *X. phaseoli* strains are unlikely (Sutton & Wallen, 1967).

Bacterial leaf blight disease of rice, caused by *X. oryzae* is forecasted indirectly by measuring phage titer of rice fields, irrigation canals and rivers (Okabe & Goto, 1963; Kuo *et al.*, 1967). These phages are produced from their host bacteria in the rhizosphere of *Leersia oryzoides* and on leaves of the rice plants (Okabe & Goto, 1963). This method of disease forecasting may be questionable as it may not be very sensitive or reliable.

Phage Xf of *X. oryzae* is filamentous and flexible in shape and contains single-stranded DNA with the 4 usual bases (Kuo *et ai.*, 1967, 1968). Compared to other *X. oryzae* phages, Xf has a wider host range. Xf phage does not affect the growth rate of infected cultures, that is, it does not lyse host cells (Kuo *et al.*, 1969). Infection by Xf phage does not affect lysis of the host *in vitro* by *X. oryzae* phage XP12. No specific ecological significance was attributed to phage Xf. Presumably Xf-infected *X. oryzae*

Table 1. Inhibition of bacterial spot disease development by phage applied to 'Sunhigh' peach seedling foliage 1 and 24 h before inoculation with *Xanthomonas pruni*.

Treatment[1]	Av. % infected leaves	Av. disease rating per infected leaf[2]	Disease index per plant[3]	% disease reduction
X. pruni-control	55	3.3	182	
Phage first, then *X. pruni* 1 h later	28	1.9	53	71
X. pruni-control	45	2.5	113	
Phage first, then *X. pruni* 24 h later	27	1.8	49	57

1. Values are averages from 3 experiments. *X. pruni* inocula contained 2–4 $\times$ 10^8 colony-forming units (c.f.u.)/ml. Titers of phage-containing crude lysates were 0.3–2.3 $\times$ 10^{10} plaque-forming units (p.f.u.)/ml.
2. Disease rating: 0 = no symptoms, 1 = 0–3%, 2 = 3–6%, 3 = 6–12%, 4 = 12–25%, 5 = 25–50%, 6 = more than 50% of leaf surface exhibiting symptoms.
3. Disease index = Av. % infected leaves/plant $\times$ Av. disease rating/infected leaf.

All phage controls: rating 0.

Data adapted from Civerolo & Keil, 1969.

Table 2. Bacterial spot disease development of 'Sunhigh' peach seedlings foliage treated with phage 1 and 24 h after inoculation with *Xanthomonas pruni*.

Treatment[1]	Av. % infected leaves	Av. disease rating per infected leaf	Disease index per plant	% disease reduction
X. pruni-control	58	3.5	203	
X. pruni first, then phage 1 h later	51	3.1	158	22
X. pruni first, then phage 24 h later	67	3.8	255	

1. *X. pruni* inoculum contained approximately 3×10^8 c.f.u./ml. Titer of phage-containing lysate was 2×10^{10} p.f.u./ml.
All phage controls: rating 0.
Data adapted from Civerolo & Keil, 1969.

cells associated with host plant tissue would continue to grow and divide, and be susceptible to lysis by other *X. oryzae* phages. Effect of Xf infection on pathogenicity of *X. oryzae* was not indicated.

Phage present on peach foliage or within intercellular spaces can reduce infection by *X. pruni* under experimental conditions (Civerolo, 1970; Civerolo & Keil, 1969; Table 1). Phage is stable for at least 24 h on peach leaves in absence of host bacteria and is still effective in protecting foliage against *X. pruni* infection. The effect is primarily prophylactic rather than eradicative, since application of high titer phage suspensions after inoculation with *X. pruni* does not significantly reduce infection (Table 2). Failure of phages to eradicate *X. pruni* even when applied as soon as 1 h after inoculation may be due to *X. pruni* cells rapidly occupying sites within intercellular spaces where they are inaccessible to phage, to lack of sufficient infectious phage particles in intercellular spaces at presumed infection sites to ensure contact with bacterial cells, or to presence of phage-resistant but virulent cells (Civerolo, 1970; Civerolo & Keil, 1969; Stonier *et al.*, 1967). There is no evidence that *X. pruni* cells in the environment of the intercellular spaces of peach leaves are resistant to phage infection or that phage cannot replicate within such cells. *X. pruni* cells are probably restricted to the intercellular spaces of infected leaves (Feliciano & Daines, 1970).

The relative roles of two *X. pruni* phages, Xp3-A and Xp3-I, in bacterial spot disease development were evaluated (Civerolo, 1970). Growth characteristics of Xp3-A and Xp3-I in their homologous bacterial host are given in Table 3. Xp3-A and Xp3-I particles adsorb rapidly to *X. pruni* cells. Xp3-I has a longer latent period than Xp3-A. Furthermore, infective Xp3-A particles are released immediately after being synthesized, while newly synthesized Xp3-I particles are released slowly from infected cells. The average burst sizes ranged from 42–49 for Xp3-A and 176–296 for Xp3-I.

Xp3-I-infected cells were only slightly more effective in reducing bacterial spot disease than Xp3-A-infected cells (Tables 4, 5). The difference could be because a

Table 3. Growth characteristics of two *Xanthomonas pruni* phages.

Characteristic[1]	Phage	
	Xp3-A	Xp3-I
Plaque morphology	large, round, clear, with narrow halo	small, irregular shaped, turbid
Adsorption rate constant K	$0.2–1.2 \times 10^{-9}$ ml/min	$1.3–3.3 \times 10^{-9}$ ml/min
Latent period	30–45 min	60–75 min
Rise period	30 min	130 min
Eclipse period	0–15 min	15–30 min
Burst size	42–49 p.f.u./cell	176–296 p.f.u./cell

1. Host *Xanthomonas pruni*; peach-'Coronet' isolate.
Data adapted from Civerolo, 1970.

Table 4. Effect of Xp3-A pruniphage on bacterial spot disease *(Xanthomonas pruni)* of peach foliage.

Treatment	Av. % infected leaves	Av. disease rating/infected leaf	Disease index/ plant	% disease reduction
X. pruni-control				
appr. 2×10^8 c.f.u./ml	77	2.0	154	
Xp3-A-containing crude lysate				
9×10^9 p.f.u./ml	46	1.3	60	61
Xp3-A-infected *X. pruni* control			34	
p.f.u./c.f.u. = 16	31	1.1		
Xp3-A-infected *X. pruni*				
p.f.u./c.f.u. = 16	61	1.7	104	32
Terramycin, 100 ppm	9	0.8	7	96

Data adapted from Civerolo, 1970.

larger number of phage particles were released by Xp3-I-infected cells on the leaf surface or within the intercellular spaces. There was greater protection against *X. pruni* infection when foliage was treated with free Xp3-A particles (8.6×10^9 p.f.u./ml) than when treated with free Xp3-I particles (1.0×10^{11} p.f.u./ml). *In vitro* growth characteristics of Xp3-A and Xp3-I suggest that the time following adsorption and subsequent release of progeny phage particles from bacterial cells infected by Xp3-I within the intercellular spaces may be sufficient for enough cells to escape phage attack and initiate infection. In contrast, phage exhibiting a growth pattern like that of Xp3-A may rapidly infect most of the invading *X. pruni* cells, release progeny phage rapidly, and check establishment of *X. pruni* infection.

Table 5. Effect of Xp3-I pruniphage on bacterial spot disease *(Xanthomonas pruni)* of peach foliage.

Treatment	Av. % infected leaves	Av. disease rating per infected leaf	Disease index per plant	% disease reduction
X. pruni-control about 2 × 10^8 c.f.u./ml	59	2.2	130	
Xp3-I-containing crude lysate 1 × 10^{11} p.f.u./ml	54	1.3	70	46
Xp3-I-infected *X. pruni* control p.f.u./c.f.u. = 57	8	1.1	8.8	
Xp3-I-infected *X. pruni* p.f.u./c.f.u. = 57	55	1.3	72	45
Terramycin (100 ppm)	5	0.7	4	97

Data adapted from Civerolo, 1970.

References

Baigent, Nancy L., J. DeVay & M. Starr, 1963. Bacteriophages of *Pseudomonas syringae*. N.Z. Jl. Sci. 6: 75–100.

Billing, Eve, 1963. The value of phage sensitivity tests for the identification of phytopathogenic *Pseudomonas* spp. J. appl. Bact. 26: 193–210.

Billing, Eve, 1970. Further studies on the phage sensitivity and the determination of phytopathogenic *Pseudomonas* spp. J. appl. Bact. 33: 478–491.

Boyd, R. J., A. C. Hildebrant & O. N. Allen, 1971. Retardation of crown gall enlargement after bacteriophage treatment. Pl. Dis. Reptr 55: 145–148.

Brunner, N., & C. F. Pootjes, 1969. Bacteriophage release in a lysogenic strain of *Agrobacterium tumefaciens*. J. Virol. 3: 181–186.

Civerolo, E. L., 1970. Comparative relationships between two *Xanthomonas pruni* bacteriophages and their bacterial host. Phytopathology 60: 1385–1388.

Civerolo, E. L. & H. L. Keil, 1969. Inhibition of bacterial spot of peach foliage by *Xanthomonas pruni* bacteriophage. Phytopathology 59: 1966–1967.

Cook, F. D. & H. Katznelson, 1960. Isolation of bacteriophages for the detection of *Corynebacterium insidiosum*, agent of bacterial wilt of alfalfa. Can. J. Microbiol. 6: 121–125.

Coons, G. H. & J. E. Kotila, 1925. The transmissible lytic principle (bacteriophage) in relation to plant pathogens. Phytopathology 15: 257–370.

Crosse, J. & Constance M. E. Garrett, 1961. Relationship between phage type and host plant in *Pseudomonas mors-prunorum* Wermald. Nature, Lond. 192: 379–380.

Crosse, J. E. & Constance M. E. Garrett, 1963. Studies on bacteriophage of *Pseudomonas mors-prunorum*, *P. syringae* and related organisms. J. appl. Bact. 26: 159–177.

Demolon, A. & A. Dunez, 1935. Recherches sur le role du bacteriophage dans le fatigue des luzernières. Annls agron. 5: 357–370.

Feliciano, A. & R. H. Daines, 1970. Factors influencing ingress of *Xanthomonas pruni* through peach leaf scars and subsequent development of spring cankers. Phytopathology 60: 1720–1726.

Fred, E. B., I. L. Baldwin & E. McCoy, 1932. Root nodule bacteria and leguminous plants. Univ. Wis. Stud. Sci. 5.
Fulton, R. W., 1950. Bacteriophages attacking *Pseudomonas tabaci* and *P. angulatum*. Phytopathology 40: 936–949.
Goodman, R. N., 1965. *In vitro* and *in vivo* interactions between components of mixed bacterial cultures isolated from apple buds. Phytopathology 55: 217–221.
Heberlein, G. T., 1970. Enhancement of *Agrobacterium tumefaciens* infectivity by chloropromazine hydrochloride. Infect. Immun. 2: 468–473.
Heberlein, G. T. & J. A. Lippincott, 1965. Photoreversible ultraviolet enhancement of infectivity in *Agrobacterium tumefaciens*. J. Bact. 89: 1511–1514.
Heberlein, G. T. & J. A. Lippincott, 1967a. Ultraviolet-induced changes in the infectivity of *Agrobacterium tumefaciens*. J. Bact. 93: 1246–1253.
Heberlein, G. T. & J. A. Lippincott, 1967b. Enhancement of *Agrobacterium tumefaciens* infectivity by mitomycin C. J. Bact. 94: 1470–1474.
Hendry, A. T., J. A. Carpenter & E. H. Garrad, 1967. Bacteriophage studies of isolates from fire blight sources. Can. J. Microbiol. 13: 1357–1364.
Klein, R. M. & R. E. Beardsley, 1957. On the role of omega bacteriophage in formation of crown-gall tumor cells. Am. Nat. 91: 330–331.
Klement, Z., 1957. Two new bacteriophages for bacterial pathogens of bean. Nature, Lond. 180: 41–42.
Klement, Z. & B. Lovas, 1960. Biological and morphological characterization of the phage for *Xanthomonas phaseoli* var. *fuscans*. Phytopath. Z. 37: 321–329.
Kuo, T-T., T-C. Huang, R-Y. Wu & C-M. Yang, 1967. Characterization of three bacteriophages of *Xanthomonas oryzae* (Uyeda et Ishiyama) Dowson. Bot. Bull. Acad. sin., Shanghai 8: 246–254.
Kuo, T-T., T-C. Huang, R-Y. Wu & T-Y. Chow, 1969. A filamentous bacteriophage from *Xanthomonas oryzae*. Virology 39: 548–555.
Kurkdjian, A., 1970. Observations sur la présence des phages dans les plaies infectées par différentes souches d'*Agrobacterium tumefaciens* (Smith et Town) Conn. Annls Inst. Pasteur, Paris 118: 690–695.
Kurkdjian, A., R. Beardsley & P. Manigault, 1968. Bacteriophages d'*Agrobacterium tumefaciens*. II. Souche B6m lysogène dans les tissues de *Pisum sativum* L., et effet de la glycine sur la production des phages. Annls Inst. Pasteur, Paris 114: 555–561.
Laird, D. G., 1932. Bacteriophage and the root nodule bacteria. Arch. Mikrobiol. 3: 159–193.
Leff, Judith & R. E. Beardsley, 1970. Action tumorigène de l'acide nucléique d'un bactériophage présent dans les cultures de tissu tumoral de Tournesol (*Helianthus annuus*). C.r. hebd. Séanc. Acad. Sci., Paris 270: 2505–2507.
Manocha, M. S., 1970. Fine structure of sunflower crown gall tissue. Can. J. Bot. 48: 1455–1458.
Matsumoto, T. & N. Okabe, 1937. Preliminary note on the bacteriophage for *Bacterium citri* (Hasse) Doidge. Agriculture Hort., Tokyo 12: 2055–2059.
Mitchell, R., 1971. Role of predators in the reversal of imbalances in microbial ecosystems. Nature, Lond. 230: 257–258.
Novikova, N. S., 1940. Bacteriophage prophylaxis against bacterial wildfire in Makhorka tobacco. Mӯkrobiol. Zh. 7: 127–140. Abstract in Biol. Abstr. 16 (1942) 1249.
Okabe, N. & M. Goto, 1963. Bacteriophages of plant pathogens. A. Rev. Phytopath. 1: 397–418.
Parsons, C. L. & R. E. Beardsley, 1968. Bacteriophage activity in homogenates of crown gall tissue. J. Virol. 2: 651.
Quetier, F., T. Huguet & Etienne Guille, 1969. Induction of crown gall: Partial homology between tumor-cell DNA, bacterial DNA and the G+C-rich DNA of stressed normal cells. Biochem. biophys. Res. Commun. 34: 128–133.
Rosberg, D. W. & A. L. Parrack, 1955. The electron microscopy of a bacteriophage attacking *Xanthomonas malvacearum*. Phytopathology 45: 49–51.

Roussaux, J., A. Kurkdjian & R. E. Beardsley, 1968. Bacteriophages d'*Agrobacterium tumefaciens* (Smith et Town.) Conn. I. Isolement et caractères. Annls Inst. Pasteur, Paris 114: 237–247.
Sander, Evamarie, 1964. Evidence of the synthesis of a DNA phage in leaves of tobacco plants. Virology 24: 545–551.
Sander, Evamarie, 1967. Alteration of fd phage in tobacco leaves. Virology 33: 121–130.
Schilperoort, R. A., H. Veldstra, S. O. Warnaar, G. Mulder & J. A. Cohen, 1967. Formation of complexes between DNA isolated from tobacco crown gall tumors and RNA complementary to *A. tumefaciens* DNA. Biochim. biophys. Acta 145: 523–525.
Srivastava, B. I. S. & K. C. Chadha, 1970. Liberation of *Agrobacterium tumefaciens* DNA from the crown gall tumor cell DNA by shearing. Biochem. biophys. Res. Commun. 40: 968–972.
Stolp, H. & M. P. Starr, 1964. Bacteriophage reactions and speciation of phytopathogenic xanthomonas. Phytopath. Z. 51: 442–478.
Stolp, H., M. P. Starr & Nancy L. Baigent, 1965. Problems in speciation of phytopathogenic pseudomonads and xanthomonads. Ann. Rev. Phytopath. 3: 231–264.
Stonier, T., R. E. Beardsley, L. Parsons & J. McSharry, 1966. *Agrobacterium tumefaciens* Conn. III. Effect of thermal shock on bacteria in relation to tumor-inducing ability. J. Bact. 91: 266–269.
Stonier, T., J. McSharry & T. Speitel, 1967. *Agrobacterium tumefaciens* Conn. IV. Bacteriophage $PB2_1$ and its inhibitory effect on tumor induction. J. Virol. 1: 268–273.
Stroun, M., P. Gahan & S. Sarid, 1969. *Agrobacterium tumefaciens* RNA in non-tumerous tomato cells. Biochem. biophys. Res. Commun. 37: 652–657.
Sutton, M. D. & V. R. Wallen, 1967. Phage types of *Xanthomonas phaseoli* isolated from beans. Can. J. Bot. 45: 267–280.
Tourneur, J. & G. Morel, 1970. Sur la présence de phages dans les tissus de 'Crown-gall' cultivés *in vitro*. C.r. hebd. Séanc. Acad. Sci., Paris 270: 2810–2812.
Vandecaveye, S. C. & H. Katznelson, 1936. Bacteriophage as related to the root nodule bacteria of alfalfa. J. Bact. 31: 465–477.
Vidaver, Anne K. & M. L. Schuster, 1960. Characterization of *Xanthomonas phaseoli* bacteriophages. J. Virol. 4: 300–308.
Zeitsoun, F. M. & Z. E. Wilson, 1969. The relation of bacteriophage to the walnut tree pathogens, *Erwinia nigrifluens* and *Erwinia rubrifaciens*. Phytopathology 59: 756–761.
Zimmerer, R. P. ,R. H. Hamilton & C. F. Pootjes, 1966. Isolation and morphology of temperate *Agrobacterium tumefaciens* bacteriophage. J. Bact. 92: 746–750.

The significance of the vegetation for the survival of plant pathogenic bacteria

M. Goto

Faculty of Agriculture, Shizuoka University, Iwata, Shizuoka, Japan

Abstract

Due to the low efficiency of the applied techniques, the results from earlier investigations on the survival of plant pathogenic bacteria in association with vegetations generally referred solely to populations with more than 10^4 to 10^5 cells per gram. Populations with normally lower surviving numbers were considered non-dependent and the possibility of their survival outside the host was ignored. The data obtained with *Xanthomonas citri*, previously considered to be confined to its host, suggest however that many other bacterial pathogens may survive at rather low population densities, either in association with non-host plants or in the soil.

To design effective control measures, it is often necessary to know where the plant pathogenic bacteria survive in nature during cold or dry seasons, as part of their life cycle. Such reservoirs or hiding places of plant pathogenic bacteria are found in soils, seeds, infected plant debris, living plant tissues (either growing or in a dormant state), alternate hosts, weeds, or in some cases even in insects. Of course, each pathogen prefers or even needs its own reservoirs. Sometimes soil-inhabiting pathogenic bacteria are the major inoculum source (the so-called soil-borne diseases). In certain other bacteria the seed is the inoculum source (the so-called seed-borne diseases). In bacterial diseases of orchard crops, dormant plant tissues usually play the most important role. In addition the way of carry-over is by no means always exclusive, particularly with pathogens mainly occurring in plant tissues where the surviving stages present in the soil are often either overlooked, underestimated, or disregarded. The tendency is to consider the non-host phase in the life-cycle of these pathogens as transient, not at all contributing to the propagation of the disease. More careful observations of bacterial plant diseases have recently made clear, that new data are needed for an adequate explanation of the sporadic nature of epidemics caused by bacteria in the field.

Since Valleau *et al.* (1944) have demonstrated that virulent *Pseudomonas tabaci* and *P. angulata* are present in the soil and apparently persist there, increasing attention has been paid to the importance of survival of the pathogen as a saprophyte in soils, in the rhizosphere of the host as well as of other plants. My own investigations on the overwintering of *Xanthomonas citri* in association with plants unrelated to *Citrus* spp., or in soils where no citrus trees were grown, strongly suggest that our theories on the life cycle of the bacterium under natural conditions must be revised.

This review will discuss the significance of the vegetation on the saprophytic survival of the various plant pathogenic bacteria in nature, and the necessary techniques for their detection. Though the epiphytic phase in the life cycle of pathogenic bacteria also plays an important role in their survival, this topic will not be touched upon here. It will be treated in another place.

Efficiency of the techniques used for detecting bacteria

To study saprophytical survival of plant pathogenic bacteria, adequate techniques are needed. They must reveal low population densities of pathogenic bacteria under natural conditions. Different species may need different techniques so that it will not always be possible to compare the results, especially with respect to their population levels. As only for a few pathogenic bacteria efficient selective media have been developed, most experiments have been performed with host plants as the selecting agent of pathogens from mixed microbial populations. Again it is not possible to apply one and the same technique in all cases: it will be different for each disease and depends on such factors as the mode of infection, the properties of plant tissues to be inoculated, or the type of disease. Only a careful study of the minimum level on which a species can be detected can lead to results.

Selective media To investigate saprophytic survival of pathogenic bacteria, efficient selective media are needed. Their usefulness depends mainly on the average recovery percentage of the involved pathogenic bacteria at low population levels in a mixture with quite high population densities of saprophytes. It is also important that all naturally occurring strains of this bacterium can grow well on the media. Attempts to develop such media for plant pathogenic bacteria have been unsuccessful, however, because they were insufficiently efficient as compared with such techniques as inoculation or immunofluorescent methods. Because the microflora on root surfaces is more complex than that in soil, even the media that will select from natural soils are not necessarily efficient for selection from the rhizospheres or from soil samples that contain intact rootlets. Therefore, the suitability of selective media from various types of soil, vegetation or other localities greatly different in their microflora must be considered. The requirements seem to be partly fulfilled in *Agrobacterium* and some selective media have been developed for it, independently, by Schroth *et al.* (1965) and by Kado & Heskett (1970). With a known number of cells added to soil, the first-mentioned authors recovered 38% of them, the latter 48%. Furthermore, Schroth *et al.* found that this recovery percentage could be obtained at a level as low as 10^2 cells/g soil. They used their medium with success for a survey of the distribution of the *Agrobacterium* in California soils.

A selective medium for *Erwinia amylovora* was developed by Miller & Schroth (1970) and successfully used for the assay of the microflora on pear trees.

A modified Drigalski's medium was used for ecological studies of *Erwinia* soft-rot bacteria (Tsuyama, 1962) and *Pseudomonas solanacearum* (Okabe, 1969) in Japan. The application of this medium had certain limitations, however, because the lowest detectable population levels were about 10^4 cells/g soil. In addition, it could not depress the growth of crystal violet tolerant, gram negative saprophytic soil bacteria, so that the development of the colonies of the pathogenic bacteria at lower population densities was obscured or hindered and the bacterial population below this level could not be detected on this medium.

Spraying inocula over leaves or pouring inocula over water-soaked leaves To detect *Corynebacterium michiganense* surviving in soils, Strider (1967) used the spray inoculation method in which the soil suspension was sprayed with an atomizer on 3 to 4 weeks old tomato seedlings in the greenhouse. Layne (1968) developed as a bioassay for *C. michiganense* the tomato leaf rub inoculation method (with the use of carborundum) in which infection was obtained at population densities above 8×10^5 cells/ml.

The method used by Valleau *et al.* (1944) and Diachun & Valleau (1946) in their studies on overwintering of *Pseudomonas tabaci*, *P. angulata*, *P. phaseolicola*, *Xanthomonas vesicatoria*, and *X. phaseoli* var. *sojense* consisted of pouring soil suspensions, in which the pathogenic bacteria were expected to be present, over water-soaked undersurfaces of leaves of host plants. Essentially this technique is a stomatal inoculation method in which the minimum inoculation doses are considered to be around 10^5 cells/ml. The technique is superior to Strider's, since water congesting substantially increases the rate of stomatal infection. In the case of citrus canker, for instance, the stomatal infection did not occur on unwounded, mature leaves, whereas a considerable number of stomatal infections could be induced when artificial water congestion was applied prior to spraying bacterial suspensions on unwounded undersurfaces of the leaves (Goto, unpublished).

Carrot slice inoculation When slices of such plants as carrots or potatoes were used (Tsuyama, 1962), the minimum detection level of soft rot *Erwinia* from soil was 10^3 cells/g soil. Togashi & Sakamoto (1967) obtained, however, the development of soft rot on carrot slices inoculated with a mixture of 8.9 cells of the pathogen and 5.6×10^5 cells of soil bacteria.

Partial-vacuum technique Boosalis (1950, 1952) devised a partial-vacuum technique for water-congesting the host tissues and applied it in inoculating seedlings of cereal crops and grasses with *Xanthomonas translucens* to determine the mode of overwintering of the bacterium. Because this technique forces water into leaf tissues through the stomata by negative pressure (about 0.008 mm mercury for 1 to 5 minutes), the amount of inoculum introduced into the tissues must be larger than that by Valleau's technique, and the minimum detection level could be lower, although no reported experimental data confirm this.

Leaf infiltration method Although the applicability may be limited to certain kinds of host plants, the infiltration technique would be one of the most efficient of the methods using the host plant as a selective agent. This technique was applied independently by Kennedy (1969) and Goto *et al.* (1970) for the detection of *Pseudo-*

monas glycinea in soybean and *Xanthomonas citri* from soils and weeds, respectively. In this technique, the inoculum is infiltrated into the mesophyll of the leaves with a syringe. In citrus leaves, water could be injected, irrespective of plant species, at an average rate of 7 $\mu l/cm^2$. Since a single cell of the bacterium injected into the mesophyll could successfully multiply in the tissues and finally develops a lesion, infection was apparent with a population of about 10^2 cells/ml of *X. citri*. When the technique was applied to open systems which contained *X. citri* together with various soil micro-organisms, the pathogen could selectively grow in the living leaf tissues and form lesions after a latent period of one to three weeks, depending on the inoculum load. The recovery percentage was approximately 10 at bacterial populations below 10^2 cells/ml soil suspension (5 g soil/liter water). To lower the detection levels, the samples were fractionated and concentrated by centrifugation. One hundred ml of soil suspension was first centrifuged at 2,000 rpm (450 *G*) for 20 minutes to remove coarse soil particles and impurities, and the supernatant was again centrifuged at 6,000 rpm (4000 *G*) for 20 minutes to concentrate the pathogenic bacteria. The sediments were resuspended into one ml of 1% peptone-sucrose solution and injected. Following this procedure, the detection level became 1 to 10 cells of *X. citri* per ml of the original samples which were prepared by suspending at most 25 g of field soil into 100 ml of sterilized water. The number of lesions could be increased by adding peptone and sucrose for enrichment to the supernatant of the first centrifugation at the ratio of 1 mg/ml, and incubating at 28°C for 5 hours or overnight at 20°C before the second centrifugation. The amount of the nutrients to be added, the incubation temperature, and the incubation time were very important because the harmful metabolites produced by the saprophytes quickly accumulated in the mixture. The ordinary clay, as well as the saprophytic soil micro-organisms, showed no injurious effects to the citrus leaves, even at the maximum concentration which mechanically allowed the injection into mesophyll. However, the clay of a particular acidic red soil collected at the western part of Shizuoka Prefecture, Japan, turned leaf colour yellow and seemed to reduce the activity of the leaves.

Multineedle inoculation Multineedle inoculation, which was used in the studies on overwintering of *Xanthomonas citri* (Goto *et al.*, 1970), made it possible to detect this bacterium at a concentration around 10^2 cells/ml when a diluted pure culture was used. This technique, in which 500 pricks are given with a bundle of 50 fine needles, may be useful when infiltration is impossible, such as in samples of dense clay. However, with this method the latent period at low doses is usually longer than with the infiltration method.

To detect *X. oryzae* surviving in the saprophytic phase, Isaka (1964, 1968, 1970) used the multineedle inoculation method combined with the microscopical examination of bacterial exudation from the inoculated leaves. In bacterial blight of rice, the infiltration method has not yet been applied, although Hagborg (1970) has devised a technique for infiltrating the leaves of cereals. Because the disease is one of the typical vascular diseases, the pathogenic bacteria should be introduced into the vascular system to obtain effective infection and to detect the bacteria later on. In Isaka's method

(1964), young rice leaves were twice or three times punctured with a bundle of 100 fine needles so that at least some 200 pricks were given at one place of a leaf. Five days later, all inoculated leaves were microscopically examined for bacterial ooze from the vascular bundles at a distance of 1 to 5 cm from the puncture. The detection level of this technique was 1.4×10^2 cells/ml and said to be the most efficient among the various methods used in the ecological researches on bacterial leaf blight of rice. For instance, the detection level was above 10^4 cells/ml when the symptom readings were made on the inoculated leaves (Mizukami, 1961).

Growing susceptible plants on the infested soil In the earlier studies on citrus canker, disinfected citrus seeds were germinated on the soils infested with *Xanthomonas citri* to examine the survival ability of the bacteria in soil (Doidge, 1916; Kawakami, 1921; Fulton, 1920; Lee, 1920). To detect *X. citri* under natural conditions, however, this method seems to be less effective. Citrus seeds could germinate without infection in soils carrying fairly high numbers of bacterial cells which could easily be detected with the leaf infiltration technique (Goto, unpublished).

Pseudomonas solanacearum could be detected by planting the susceptible tomato plants in the soils to be examined. Infection of unwounded tomato roots, however, did not occur unless the number of bacterial cells reached 5×10^4 per ml of nutrient solution in hydroponic units (Kelman & Sequeira, 1965). This population level was confirmed by infection of wounded tomato roots grown in soil (Jenkins *et al.*, 1967; Okabe, 1969).

Phage technique The phage technique, first used by Katznelson & Sutton (1951) and Katznelson *et al.* (1954) to detect bacterial infection in bean seeds, was combined by Wakimoto (1954) with the process of bacterial enrichment either in a medium or through host plants. This method was extensively applied by Japanese workers to study the overwintering of *Xanthomonas oryzae* in nature (Tagami *et al.*, 1963, 1964; Yoshimura, 1963). However, the population level of the host bacteria or the contaminants present in the samples played a critical role for the successful application of the method (Goto, 1969; Goto *et al.*, 1970). In the experiments on the phage-bacteria interaction of *X. oryzae* and *X. citri*, the phage population started to rise only when the bacterial population reached a level of 10^5 cells/ml, regardless of the phage population tested below 2×10^5 particles/ml. Therefore, the poorer the bacterial inoculum, the longer the lag phase of phage growth, and the efficiency of the phage technique was restricted by the population of contaminants in the samples to be tested. It was difficult to detect *X. oryzae* or *X. citri* at a population level below 10^4/ml from samples of rhizospheres or soils containing high concentrations of saprophytic microorganisms.

A quantitative method has also been proposed by Wakimoto & Yoshii (1955) and Wakimoto (1959) to determine the number of the living cells of *X. oryzae* in soils or other materials. Although this technique has been used to some extent for the ecological study of bacterial blight of rice in Japan (Tagami *et al.*, 1963, 1964), it has some limitations in evaluating the efficiency of the method. Although it seems to be satisfactory to estimate the relative bacterium population under natural conditions, the lowest detection level has not been fully investigated. The reliability of this technique has been evaluated only in the closed system (Wakimoto & Yoshii 1955) and it is at least open to question whether every infected bacterial cell can liberate the same number of phage particles when mixed with a high population of contaminants as in the closed system.

Technique with the use of antisera Jenkins *et al.* (1967) developed a serological technique to detect *Pseudomonas solanacearum* in soil, which was said to be more quick and reliable than the other methods. They obtained positive precipitin reactions in agar diffusion plates at 2.5×10^4 bacterial cells/ml soil after centrifugation of the soil filtrates to concentrate bacterial density, whereas 2.5×10^6 cells/ml could be detected by direct isolation when the filtrate was centrifuged, and 2.5×10^4 cells/ml by an inoculation method which needed 21 days for confirmative reading.

The immunofluorescent staining method can be used to detect the pathogenic bacteria. In *Erwinia* soft-rot bacteria, the microculture impression method (bacteria grown for 6 to 8 hours at 30 °C) was superior to the plating method with Drigalski's medium (Kikumoto & Sakamoto, 1967, 1969). *E. aroideae* could be detected with this technique from soil suspensions containing 52 cells of the pathogen and 1.5×10^7 cells of soil micro-organisms per ml. Since this bacterium is extremely heterologous in its serological properties (Okabe & Goto, 1956; Goto & Okabe, 1958), the technique can only be used in experiments with isolates of known serological properties. The same objection holds for other pathogens in applying serological techniques to assay natural populations.

Survival of the plant pathogenic bacteria on vegetation

Saprophytic survival on susceptible host plants Extensive ecological researches on bacterial leaf blight of rice in Japan have revealed three species of wild grasses to be infected by *Xanthomonas oryzae* under natural conditions; they play an important role as infection sources for the next crop. They are *Leersia oryzoides* and its var. *japonica*, and *Zizania latifolia* (Goto *et al.*, 1953; Inoue *et al.*, 1957; Tagami *et al.*, 1963, 1964; Yoshimura, 1963; Yoshimura *et al.*, 1956). The first two are particularly important for the overwintering stage of the pathogen since they grow well on river-banks and along irrigation canals, ditches and footpaths in the paddy fields in areas where the disease occurs (Mizukami & Wakimoto, 1969). The phage technique showed that the bacteria overwinter as saprophytes on roots and in the rhizospheres of these grasses. The bacterial population on *L. oryzoides* var. *japonica* decreased rapidly in winter, but rose again at the beginning of spring. The bacteria multiplied vigorously as soon as the aerial parts of the plants developed, although these parts did not always develop disease symptoms. The population on subterranean stems and roots also increased in spring.

In Minnesota it was demonstrated that *Xanthomonas translucens* could overwinter on such hosts as winter wheat, winter rye, quack grass and brome grass (Boosalis, 1952).

The *Erwinia* soft rot bacteria have been shown to live in soil, especially in the rhizosphere, on root surfaces and on leaf surfaces of susceptible host plants (Tsuyama, 1962). According to Kikumoto & Sakamoto (1969), the rhizosphere of cruciferous plants stimulated the bacterial population. The *Erwinia* was shown to be a member of the specific rhizosphere microflora for such plants as chinese cabbage *(Brassica chinensis)* and *Allium tuberosum*, and living for a long time in fields in such a saprophytic phase. Nevertheless they preferred certain growing stages of the host plants, as on young plants the population of the pathogen seemed to decrease because of intense microbial competition around the roots.

Saprophytic survival on non-host plants The saprophytic survival of plant pathogenic bacteria on root surfaces or in rhizospheres of susceptible non-host plants was first revealed by Valleau *et al.* (1944). At present we know that many other pathogenic bacteria are able to live and overwinter in association with roots of non-host plants, although it is often not yet clear how the non-host phase is important as an actual inoculum source for epidemics.

Valleau *et al.* (1944), in their extensive survey, have shown that *Pseudomonas angulata* overwinters in close association with the roots of several winter-cover crops and weeds: crimson clover, vetch, barley, wheat, rye, henbit, and chickweed. Likewise, the survival of *P. tabaci* in fields where wildfire has occurred in the previous crop is closely associated with roots of rye and wheat. It is very remarkable that the bacteria could be detected from the roots of cover-crop plants in the second winter after the tobacco crop. These bacteria survive as minute colonies on the surface of the rootlets.

Xanthomonas vesicatoria is another bacterium for which Valleau *et al.* have revealed the existence in association with roots of wheat under natural conditions, but they could not prove that *Pseudomonas phaseolicola* and *Xanthomonas phaseoli* var. *sojense*, for which the host seed is assumed to be the major reservoir of the inoculum, have a similar saprophytic stage. Smith (1962) obtained similar results with *X. malvacearum*, and detected it in 14 out of 161 types of weed collected from blighted fields (48 grass species, 11 unidentified grasses, alfalfa, an *Amaranthus* sp., grain sorghum and 113 other weeds) by inoculating young cotton seedlings. However, only samples collected in summer and fall gave positive results, those collected in the winter or from blight-free fields never did.

Kikumoto & Sakamoto (1969) revealed that the growth of soft-rot *Erwinia* was selectively stimulated in the rhizospheres of the following weeds which are common in fields of chinese cabbage: *Agrostis perennans*, *Portulaca oleracea*, *Sonchus oleraceus*, *Chenopodium album*, and *Commelina communis*. Such an effect was not observed in seven leguminous crop plants: red bean, kidney bean, soybean, pea, sweet pea, broad bean and red clover. They also demonstrated the survival of *E. aroideae* at lower populations in the rhizospheres of raddish, wheat, oat, red bean, tomato, and sponge-gourd, but the pathogen could be detected only by the immunofluorescent staining method and could not be isolated on Drigalski's medium.

Using the multineedle inoculation method, Isaka (1969) demonstrated that in paddy fields *Xanthomonas oryzae* was able to overwinter in association with various weeds. On out of 44 weeds growing in paddy fields in early spring (17 grasses and 27 other plant species) he detected the surviving pathogen on 11 grasses and 16 other, not closely allied weeds. The bacterium occurred more frequently on aboveground parts than on underground parts. Among the grasses the following species carried high populations of the pathogen: *Phalaris arundinacea*, *Alopecurus fulvus*, *Leersia oryzoides* var. *japonica*, *Zizania latifolia*, *Miscanthus sinensis* and *Bromus unioloides*. In particular *Phalaris arundinacea* was thought to serve as a most important reservoir in causing disease epidemics, more than *Leersia oryzoides* var. *japonica*. Carryover on *Trifolium repens*, *Astragalus sinicus*, *Rumex japonicus* and *Artemisia vulgaris* var. *indica* suggested that these weeds may also be important as inoculum sources. *Xanthomonas oryzae* was also isolated from *Vicia faba*, *Brassica chinensis*, *Cyperus japonica*, *Rumex acetosa* and *Nelumbo nucifera* by Wakimoto (1956) and by Mizukami (1961).

Calystegia japonica is a common weed found in the citrus groves in Japan. The saprophytic survival of *Xanthomonas citri* on the surface of the rhizomes of this plant and in unsterilized soil, both of which were artificially infested with the bacterium, was examined monthly during winter by the infiltration method by Goto (1970). Outdoors, the bacterium survived for 6 months on the rhizomes and for 5 months in the soil, but in the greenhouse only for 3 months in the soil and for 4 months on the surface of rhizomes. The difference is probably due to the lower activity of soil micro-organisms at low temperatures.

Xanthomonas citri could frequently be detected in soils, citrus roots, and various weeds collected in citrus groves before the new shoots developed in spring (Goto, 1970). The involved weeds were: *Alopecurus fulvus*, *Agrostis perennans*, *Artemisia vulgaris*, *Avena sativa*, *Bromus japonicus*, *Calystegia japonica*, *Capsella bursa-pastoris*, *Cerastium holosteoides*, *Eragrostis curvula*, *Erigeron linifolius*, *Gnaphalium multiceps*, *Poa annua*, *Senecio vulgaris*, *Stellaria media*, *Stellaria uliginosa*, *Vetiveria zizanioides*, and *Zoysia japonica*. None of these species, however, were susceptible to the bacterium in inoculation tests, so that it was presumed that they occurred only on the surface of the weeds and not in the tissue. Based on the infiltration technique, these populations contained as a rule less than 10^3 cells/g sample, rarely over 10^4 cells/g. The infestations seemed to be due to the transport by splashing rain from the lesions on the citrus leaves, because the soils and the weeds around healthy citrus plants were usually free from the bacterium. As the number of bacterial cells appearing in rain water from the lesions of Unshu plants were negligible after late December, the bacteria detected on the weeds in spring must has been arrived there before late fall and survived until the next spring. We also found that *X. citri* can survive on dry grass straw used as mulch, and on dead dry weeds for, a long time, in quite high numbers.

However, the data recently obtained on *Zoysia japonica* strongly suggest that *X. citri* on weeds may not always represent simple infestation in which the bacterial population gradually declines, but that it may survive in close association with roots of weeds (Goto & Ohta, 1971). We could detect the bacterium from the rhizomes and roots of this grass almost all the year round, though at low numbers. Recovery experiments from artificially infested grass confirmed this. Further, the majority of the bacterial cultures isolated from this grass were different in their phage sensitivity patterns, and their ability to utilize mannitol, from those occurring in the neighbouring citrus grove, indicting that the bacteria from these two sources were surviving independently of each other. Earlier workers have suggested that the citrus canker pathogen may survive from one season to the next in the soil, because of the appearance of diseased sprouts from the roots of diseased trees which are burnt (Wolf, 1916; Stirling, 1915), and because the infection occured on the newly planted trees in the fields where citrus trees had been cut the year before (Doidge, 1916).

Later experiments with artificially infested soils have indicated that *X. citri* survives only a week or so in natural field soils (Fulton, 1920; Lee, 1920; Kawakami, 1921; Loucks, 1930). The discrepancy between these field observations and the results of the recovery experiments may be due to the poor techniques used in the latter. Since the disease is said to have been eradicated from citrus areas in the United States, however, it may be implied that the pathogen does not possess a sufficient saprophytic ability to retain high population levels in the non-host phase for a long time, or that the infection is hardly established through the pathogen in this phase.

From the standpoint of the epidemiology of citrus canker, Unshu plants are significantly different from the other susceptible citrus plants in that development of lesions on twigs is very poor, and if they occur, they quickly heal by the formation of a callous layer. The causal organism in diseased tissue also shows rather quick decrease in numbers. Decline of the living pathogen in the lesions of spring leaves is much faster in Unshu than in Natsudaidai. Thus, the lesions of stems and spring leaves of Unshu trees are only of minor importance as a source of infection at the time of the epidemic in the following spring: the role of these lesions in epidemiology is negligible.

Summer shoots hardly develop on adult Unshu trees in full bearing, but do on younger trees. This may explain why the disease is frequent and severe on younger Unshu trees, but rarely so on adult ones. In spite of such general patterns of disease occurrence, epidemics may develop on adult Unshu plants once in several years or in a decade. We have, however, no satisfactory explanation of the mechanism of such sporadic epidemics. The pathogen saprophytically living on weeds might play a role in gradually building up the inoculum in the fields under favorable environmental conditions, although no investigations have been conducted along this line. The pathogen in the non-host phase may possibly be able to serve as the main source of infection in the nurseries, where the dense foliages of young seedlings grow close to the soil surface.

The effect of soil from the rhizosphere of weeds on the growth of *Pseudomonas solanacearum* was studied by Okabe (1969). The populations of the pathogen in artificially infested natural soils collected from weed rhizospheres varied considerably, depending on the kind of plant. They were very high in *Capsella* soil (either fresh or air-dried), very poor on *Stellaria* and *Poa* soils, and high on *Equisetum* soil only when it was air-dried before infestation. This suggests that differences in quality and quantity of nutrients secreted from the roots of these plants can influence the pathogen or, by stimulating or depressing the microflora, depress or stimulate the pathogen. Smith (1944) has also demonstrated that with *Pseudomonas solanacearum* differences in resistance of the plants grown in rotation affected the population of the bacterium to different degrees, presumably due to biological effects of the crop on soil microbs.

Conclusion

Because of the low efficiency of the applied techniques, the results obtained from the past investigations on the survival of plant pathogenic bacteria in association with the vegetation generally refer only to population densities above 10^4 to 10^5 cells per gram. Therefore, with bacteria whose normal survival populations are below this level, we have assumed that they survive in nature on their hosts and we have ignored the possibility of their survival on non-hosts. The data now obtained with *Xanthomonas citri*, one of the bacteria previously considered to be confined to its host, suggests that many other bacterial pathogens may survive at rather low population densities, either in association with non-host plants, or in the soil. Since most pathogenic bacteria can survive in sterilized soil for a long time, their population levels in the non-host phase have to be dynamic and result from the interaction between the pathogen and the resident soil microflora. The interaction must be intense when the population of a pathogen is high, and only pathogens with a high saprophytic ability may overcome the competition with the soil bacteria and maintain relatively high populations in the non-host phase. The populations of pathogens with poor saprophytic ability will quickly decline. However, during the latter process, the interaction may become less intensive and allow the survival of the pathogen at lower populations levels by escaping the competition with the soil microflora. This may be due to heterogeneity, either of the soil components, or of the bacterial populations.

The population levels of the pathogens surviving in the non-host phase, therefore, represent their degree of saprophytic ability. More research is needed to obtain sufficient information on the interaction between the pathogens and soil microorganisms at different population levels to reveal the behaviour of the pathogens at the lower population densities under non-host conditions.

As yet there is no direct evidence that *Xanthomonas citri*, living at very low population densities, serves as a source for natural infection of citrus trees in the orchard, although field observations suggest this to be very likely.

As Buddenhagen (1965) has stated, it may be unrealistic to expect pathogens to occur in large populations in the soil. But, on the other hand, surviving at low populations should not be underestimated, because we cannot neglect the possibility that growth of pathogens is stimulated when the roots of the host plants grow into the micro-environment of the pathogens.

Although the long term presence of plant pathogenic bacteria in nature may be host-dependent, indications on saprophytic survival in association with the vegetation are increasing. Therefore, regardless of the population levels in the non-host phase, Valleau's statement that the parasitic phase of the life cycle of the plant pathogenic bacteria is more or less accidental and is probably not essential for their perpetuation, would be applicable to many other plant pathogenic bacteria.

References

Anderson, E. S., 1957. The relation of bacteriophages to bacterial ecology. In: Microbial Ecology. 7th Symp. Soc. Gen. Microbiol. Cambridge.

Baines, R. C., 1947. Perennial nightshade, a host for *Corynebacterium michiganense*. Phytopathology 37: 359 (abstr.).

Baldwin, C. H. & R. N. Goodman, 1963. Prevalence of *Erwinia amylovora* in apple buds as detected by phage typing. Phytopathology 53: 1299–1303.

Basu, P. K., 1970. Temperature, an important factor determining survival of *Corynebacterium michiganense* in soil. Phytopathology 60: 825–827.

Boosalis, M. G., 1950. A partial-vacuum technique for inoculating seedlings with bacteria and fungi. Phytopathology 40: 2–3.

Boosalis, M. G., 1952. The epidemiology of *Xanthomonas translucens* (J. J. et R.) Dowson on cereals and grasses. Phytopathology 42: 387–395.

Buddenhagen, I. W., 1965. The relation of plant-pathogenic bacteria to the soil. Ecology of soil-borne plant pathogens, preclude to biological control. Univ. Calif. Press. p. 269–284.

Calavan, E. C., 1956. Citrus canker – a bacterial disease caused by *Xanthomonas citri*. Bull. Calif. Dep. Agric. 45: 259–262.

Cameron, H. R., 1970. Pseudomonas content of cherry trees. Phytopathology 60: 1343–1346.

Davis, J. R., H. English & J. E. Devay, 1969. Some factors related to the development of bacterial canker in peach. Phytopathology 59: 10–11 (abstr.).

Diachun, S. & W. D. Valleau, 1946. Growth and overwintering of *Xanthomonas vesicatoria* in association with wheat roots. Phytopathology 36: 277–280.

Doidge, E. M., 1916. The origin and cause of citrus canker in South Africa. Sci. Bull. Dep. Agric. For. Un. S. Afr. 8.

Feliciano, A. & R. H. Daines, 1970. Factors influencing ingress of *Xanthomonas pruni* through peach leaf scares, tissues invaded, and subsequent development of spring canker. Phytopathology 60: 574 (abstr.).

Fulton, H. R., 1920. Decline of *Pseudomonas citri* in the soil. J. agric. Res. 19: 207–223.

Goto, K., R. Fukatsu & K. Ohata, 1953a. Natural occurrence of bacterial leaf blight of rice on weeds. Ann. phytopath. Soc. Japan 17: 154 (abstract).

Goto, K., R. Fukatsu & K. Ohata, 1953b. Relationship between the occurrence of bacterial leaf blight disease on rice and that on wild grasses in endemic areas. Pl. Prot., Tokyo 7: 365–368 (Japanese).

Goto, M., 1962. Studies on citrus canker. I. Bull. Fac. Agric. Shizuoka Univ. 12: 3–72 (Japanese, English summary).

Goto, M., 1969a. Ecology of phage-bacteria interaction of *Xanthomonas oryzae* (Uyeda et Ishiyama) Dowson. Bull. Fac. Agric. Shizuoka Univ. 19: 31–67 (Japanese, English summary).

Goto, M., 1969b. Studies on citrus canker in Japan. Proc. 1st internat. Citrus Symp. 3: 1251–1252.

Goto, M. 1970. Studies on citrus canker. III. Survival of *Xanthomonas citri* (Hasse) Dowson in soils and on the surface of weeds. Bull. Fac. Agric. Shizuoka Univ. 20: 21–29 (Japanese, English summary).

Goto, M. & K. Ohta, 1971. Saprophytic survival of *Xanthomonas citri* on *Zoysia japonica*. Ann. Phytopath. Soc. Japan (in press).

Goto, M. & N. Okabe, 1958. Studies on the strains of *Erwinia carotovora* (Jones) Holland. V. Antigenic structure of soma, their relation to biochemical properties, and heat inactivation of agglutinins. Bull. Fac. Agric. Shizuoka Univ. 8: 1–30 (Japanese, English summary).

Goto, M., S. Serizawa & M. Morita, 1970. Studies on citrus canker. II. Leaf infiltration technique for the detection of *Xanthomonas citri* (Hasse) Dowson, with special reference to the comparison with phage method. Bull. Fac. Agric. Shizuoka Univ. 20: 1–19 (Japanese, English summary).

Graham, J. H., 1953. Overwintering of three bacterial pathogens of soybean. Phytopathology 43: 189–192.
Hagborg, W. A. F., 1970. A device for injecting solutions and suspensions into thin leaves of plants. Can. J. Bot. 48: 1135–1136.
Inoue, Y., K. Goto & K. Ohata, 1957. Overwintering and mode of infection of leaf blight bacteria of riceplant. Bull. hort. Div. Tokai-Kinki agr. exp. Stn 4: 74–82 (Japanese, English summary).
Isaka, M., 1964. Microscopic observation of the bacterial exudation from inoculated rice leaves for detection of *Xanthomonas oryzae*. Proc. Ass. Pl. Prot. Kyushu 12: 26–30 (Japanese).
Isaka, M., 1968. Studies on bacterial leaf blight of riceplant. The relation between the phage population on the seedlings and carryover of the pathogen. Proc. Ass. Pl. Prot. Kyushu 16: 1–4 (Japanese).
Isaka, M., 1969. Studies on bacterial leaf blight of rice plant. On some grasses and weeds as carriers of the pathogen. Proc. Ass. Pl. Prot. Kyushu 17: 14–19 (Japanese).
Isaka, M., 1970. A new detection method for the pathogen of bacterial leaf blight of rice *(Xanthomonas oryzae)* by means of bacterial exudation. Ann. phytopath. Soc. Japan 36: 313–318 (Japanese, English summary).
Isaka, M. & M. Miyakoshi, 1967. A study on the detection of *Xanthomonas oryzae* at low population. Ann. phytopath. Soc. Japan 33: 110 (Japanese, English abstract).
Jenkins Jr., S. F., D. J. Morton & P. D. Dukes, 1965. Detection of *Pseudomonas solanacearum* in infested soils by serological techniques. Phytopathology 55: 1063 (abstract).
Jenkins Jr., S. F., D. J. Morton & P. D. Dukes, 1967. Comparison of techniques for detection of *Pseudomonas solanacearum* in artificially infected soils. Phytopathology 57: 25–27.
Kado, C. I. & M. G. Heskett, 1970. Selective media for isolation of *Agrobacterium*, *Corynebacterium*, *Erwinia*, *Pseudomonas* and *Xanthomonas*. Phytopathology 60: 969–976.
Katznelson, H. & M. D. Sutton, 1951. A rapid phage plaque count method for detection of bacteria as applied to the demonstration of internally born bacterial infections of seed. J. Bact. 61: 689–701.
Katznelson, H., M. D. Sutton & S. T. Bailey, 1954. The use of bacteriophage of *Xanthomonas phaseoli* in detecting infection in beans, with observations of its growth and morphology. Can. J. Microbiol. 1: 22–29.
Kawakami, K., 1921. Studies on citrus canker (Japanese).
Keil, H. L., B. C. Smale & R. A. Wilson, 1966. Role of injury and longevity of *Erwinia amylovora* in the epidemiology of fire blight of pear. Phytopathology 56: 464–465.
Keil, H. L. & T. Van der Zwet, 1969. Presence of *Erwinia amylovora* in apparently healthy pear and apple tissue. Phytopathology 59: 1035 (abstract).
Kelman, A., 1953. The bacterial wilt caused by *Pseudomonas solanacearum*. Techn. Bull. N. Carol. agric. Exp. Stn 99.
Kelman, A. & L. Sequeira, 1965. Root-to-root spread of *Pseudomonas solanacearum*. Phytopathology 55: 304–309.
Kenedy, B. W., 1969. Detection and distribution of *Pseudomonas glycinea* in soybean. Phytopathology 59: 1618–1619.
Kikumoto, T., 1968. Ecological studies on the soft-rot bacteria of vegetables. V. The variation of the rhizosphere microflora in relation to the development of chinese cabbage. Jubilee Pub. Commemotion 60th birthday Prof. M. Sakamoto: 355–365 (Japanese, English summary).
Kikumoto, T. & M. Sakamoto, 1967. Ecological studies on the soft-rot bacteria of vegetables. III. Application of immuno-fluorescent staining for the detection and counting of *Erwinia aroidea* in soil. Ann. phytopath. Soc. Japan 33: 181–186 (Japanese, English summary).
Kikumoto, T. & M. Sakamoto, 1969a. Ecological studies on the soft-rot bacteria of vegetables. VI. Influence of the development of various plants on the survival of *Erwinia aroidea* added to soil. Ann. phytopath. Soc. Japan 35: 29–35 (Japanese, English summary).

Kikumoto, T. & M. Sakamoto, 1969b. Ecological studies on the soft-rot bacteria of vegetables. VII. The preferential stimulation of the soft-rot bacteria in the rhizosphere of crop plants and weeds. Ann. phytopath. Soc. Japan 35: 36–40 (Japanese, English summary).

Koizumi, M., 1969a. Ecological studies on citrus canker caused by *Xanthomonas citri*. II. Factors influencing multiplication of Cpl phages. Bull. hort. Res. Stn Japan, Ser. B, 9: 117–127 (Japanese, English summary).

Koizumi, M. 1969b. Ecological studies on citrus canker caused by *Xanthomonas citri*. III. Seasonal changes in number of causal bacteria and its bacteriophage Cpl in rain water flowing down from the diseased trees. Bull. hort. Res. Stn Japan, Ser. B, 9: 129–144 (Japanese, English summary).

Koizumi, M., S. Yamamoto & S. Yamada, 1966. Ecological studies on citrus canker caused by *Xanthomonas citri*. I. Some ecological studies on bacteriophages of causal bacterium in lesions of various citrus leaves. Bull. hort. Res. Stn Japan, Ser. B, 5: 105–117.

Lange, A. & C. Leben, 1970. Colonization of cucumber buds by *Pseudomonas lachrymans* in relation to leaf symptoms. Phytopathology 60: 1865–1866.

Layne, R. E. C., 1968. A quantitative local lesion bioassay of *Corynebacterium michiganense*. Phytopathology 58: 534–535.

Leben, C., 1965. Epiphytic microorganisms in relation to plant disease. A. Rev. Phytopath. 3: 209–230.

Leben, C., 1969. Colonization of soybean buds by bacteria: observation with the scanning electron microscope. Can. J. Microbiol. 15: 319–320.

Leben, C., G. C. Daft & A. F. Schmitthenner, 1968. Bacterial blight of soybeans: population levels of *Pseudomonas glycinea* in relation to symptom development. Phytopathology 58: 1143–1146.

Leben, C., V. Rusch & A. F. Schmitthenner, 1968. The colonization of soybean buds by *Pseudomonas glycinea* and other bacteria. Phytopathology 58: 1677–1681.

Lee, H. A., 1920. Behavior of the citrus canker organism in the soil. J. agr. Res. 19: 189–206.

Loucks, K. W., 1930. Some physiological studies of *Phytomonas citri*. J. agr. Res. 41: 247–258.

Miller, T. D. & M. N. Schroth, 1970. Selective medium for the isolation of *Erwinia amylovora* and other *Erwinia* spp. Phytopathology 60: 1304 (abstr.).

Mizukami, T., 1961. Studies on the ecological properties of *Xanthomonas oryzae* (Uyeda et Ishiyama) Dowson, the causal organism of bacterial leaf blight of rice plant. Sci. Bull. Fac. Agr. Saga Univ. 13: 1–85 (Japanese, English summary).

Mizukami, T. & S. Wakimoto, 1969. Epidemiology and control of bacterial leaf blight of rice. Ann. Rev. Phytopath. 7: 51–72.

Obata, T., F. Tsuboi & S. Wakimoto, 1969. Studies on the detection of *Xanthomonas citri* by phage technique and the surface sterilization of Unshu orange for export to the United States. Res. Bull. Pl. Prot. Serv. Japan 6: 26–37.

Okabe, N., 1969. Population changes of *Pseudomonas solanacearum* and soil microorganisms in artificially infested natural field soils. Bull. Fac. Agric. Shizuoka Univ. 19: 1–29 (Japanese, English summary).

Okabe, N. & M. Goto, 1956. Studies on the strains of *Erwinia carotovora* (Jones) Holland. I. Antigenic structures of flagellae and their relation to pathogenicity and maltose fermentation. Bull. Fac. Agric. Shizuoka Univ. 6: 16–32 (Japanese, English summary).

Okabe, N. & M. Goto, 1963. Bacteriophages of plant pathogens. A. Rev. Phytopath. 1: 397–418.

Panagopoulos, C. G., 1966. Studies on the source of inoculum for blast and black pit of citrus. Prec. 1st Congr. Mediter. Phytopath. Union 340–345.

Rusch, V. & C. Leben, 1968. Epiphytic microflora: the balloon print isolation technique. Can. J. Microbiol. 14: 486–487.

Schaad, N. W. & E. E. Wilson, 1970. Survival of *Erwinia rubrifaciens* in soil. Phytopathology 60: 557–558.

Schroth, M. N., J. P. Thomson & D. C. Hildebrand, 1965. Isolation of *Agrobacterium tumefaciens-A. radiobacter* group from soil. Phytopathology 55: 645–647.

Smith, T. E., 1944. Control of bacterial wilt *(Bacterium solanacearum)* of tobacco as influenced by crop rotation and chemical treatment of the soil. Circ. U.S. Dep. Agric. 692.

Smith, T. E., 1962. A variant culture of *Xanthomonas malvacearum* obtained from weed roots. Phytopathology 52: 1313–1314.

Strider, D. L., 1967. Survival studies with the tomato bacterial canker organism. Phytopathology 57: 1067–1071.

Stirling, F., 1915. On the firing line before the citrus canker. Proc. Fla St. hort. Soc. 28: 85–89.

Tagami, Y., 1959. Relationship between the occurrence of bacterial leaf blight of rice and the populations of causal bacteria and its phages in rice growing seasons. Pl. Protect. Japan 13: 389–394 (Japanese).

Tagami, Y. & T. Mizukami, 1962. Historical review of the researches on bacterial leaf blight of rice caused by *Xanthomonas oryzae* (Uyeda et Ishiyama) Dowson. Spec. Rep. Pl. Dis. Insect Pests forecast. Serv. 10 (Japanese).

Tagami, Y., S. Kuhara, T. Kurita, H. Fujii, N. Sekiya, S. Yoshimura, T. Sato & B. Watanabe, 1963. Epidemiological studies on bacterial leaf blight of rice, *Xanthomonas oryzae* (Uyeda et Ishiyama) Dowson. I. The overwintering of the pathogen. Bull. Kyushu agric. Exp. Stn 9: 89–112 (Japanese, English summary).

Tagami, Y., S. Kuhara, T. Kurita, H. Fujii, N. Sekiya & T. Sato, 1964. Epidemiological studies on the bacterial leaf blight of rice, *Xanthomonas oryzae* (Uyeda et Ishiy.) Dowson. II. Successive change in the population of the pathogen in paddy fields during rice growing period. Bull. Kyushu agric. Exp. Stn 10: 23–50 (Japanese, English summary).

Togashi, J. & M. Sakamoto, 1967. Studies on the outbreak of the soft-rot disease of the chinese cabbage by *Erwinia aroideae* (Towns.) Holl. III. On the behavior of *Erwinia aroideae* in field soil. Bull. Inst. agric. Res. Tohoku Univ. 19: 65–81 (Japanese, English summary).

Tsuyama, H., 1962. Studies on the soft rot disease of chinese cabbage caused by *Erwinia aroideae* (Towns.) Holland. Bull. Inst. agric. Res. Tohoku Univ. 13: 221–345 (Japanese, English summary).

Valleau, W. D., E. M. Johnson & S. Diachun, 1944. Root infection of crop plants and weeds by tobacco leaf-spot bacteria. Phytopathology 34: 163–174.

Wakimoto, S., 1954. The determination of the presence of *Xanthomonas oryzae* by the phage technique. Sci. Bull. Fac. Agric. Kyushu Univ. 14: 495–498 (Japanese, English summary).

Wakimoto, S., 1955. Overwintering of *Xanthomonas oryzae* on unhulled grains of rice. Agriculture Hort., Tokyo 30: 1501 (Japanese).

Wakimoto, S., 1956a. Considerations on the overwintering and the infection mechanism of *Xanthomonas oryzae*. Pl. Protect. Japan 10: 421–424 (Japanese).

Wakimoto, S., 1956b. Overwintering of *Xanthomonas oryzae* in soil. Agriculture Hort., Tokyo, 31: 1413–1414 (Japanese).

Wakimoto, S., 1957a. A simple method for comparison of bacterial population in a large number of samples by phage technique. Ann. phytopath. Soc. Japan 22: 159–163 (Japanese, English summary).

Wakimoto, S., 1957b. Relationship between the root of some plants and the bacteria causing leaf blight of rice. Proc. Ass. Pl. Protect. Kyushu 3: 2–5 (Japanese, English summary).

Wakimoto, S., 1967. Some characteristics of citrus canker bacteria, *Xanthomonas citri* (Hasse) Dowson, and the related phages isolated from Japan. Ann. phytopath. Soc. Japan 33: 301–310.

Wakimoto, S. & H. Yoshii, 1955. Quantitative determination of the population of a bacteria by the phage technique. Sci. Bull. Fac. Agr. Kyushu Univ. 15: 161–169 (Japanese, English summary).

Wharton, A. L., 1967. Detection of infection by *Pseudomonas phaseolicola* (Burkh.) Dowson in white-seeded dwarf bean seed stocks. Ann. appl. Biol. 60: 305–312.

Winstead, N. N. & A. Kelman, 1952. Inoculation technique for evaluating resistance to *Pseudomonas solanacearum*. Phytopathology 42: 628–634.

Wolf, F. A., 1916. Citrus canker. J. agric. Res. 6: 69–100.

Yoshimura, S., 1960. Considerations on overwintering of *Xanthomonas oryzae*. Pl. Protect. Japan 14: 343–347 (Japanese).
Yoshimura, S., 1963. Diagnostic and ecological studies of rice bacterial leaf blight, caused by *Xanthomonas oryzae* (Uyeda et Ishiy.) Dowson. Bull. Hokuriki agric. Exp. Stn 5: 27–182 (Japanese, English summary).
Yoshimura, S., T. Morihashi & Y. Suzuki, 1956. Distribution of the important grasses for overwintering of *Xanthomonas oryzae* and their diseased situations in the field. Ann. phytopath. Soc. Japan 24: 6 (abstr. in Japanese).
Zwet, T. V., 1969. Study of fire blight cankers and associated bacteria in pear. Phytopathology 59: 607–613.
Zwet, T. V. & H. L. Keil, 1970. Relative susceptibility of succulent and woody tissue of Magness pear of infection by *Erwinia amylovora*. Phytopathology 60: 593.

Mechanism of disease initiation by Pseudomonas tabaci

H. A. J. Hoitink and O. E. Bradfute

Ohio Agricultural Research and Development Center, Wooster, Ohio 44691

Abstract

Stable gymnoplasts of *Pseudomonas tabaci* were prepared by washing cells in buffer, followed by resuspension in a heavy metal-free solution of sucrose, glucose, EDTA, lysozyme and HEPES buffer (pH 7.0). These cells converted to gymnoplasts within 5 min after incubation and were sensitive to osmotic shock. Electron microscopy of these gymnoplasts showed that the cell wall had been completely removed. Stabilized gymnoplasts incorporated ^{14}C-leucine into hot TCA precipitates at rates similar to whole cells, indicating that they were viable. Gymnoplasts did not induce a susceptible reaction in tobacco or a hypersensitive reaction in soybean, although they were detected in free-hand sections of injected leaf tissue for over 24 h.

A heat labile cell-free preparation that induced water-soaking in tobacco was extracted from *P. tabaci*. A similar fraction was obtained from isolates of *P. glycinea*, but this fraction induced necrosis. Small quantities of these toxins were extracted from culture filtrates. The activity was in the 25–55% $(NH_4)_2SO_4$ fraction. Pronase destroyed activity of both fractions.

Little is known about the role of L-forms, gymnoplasts (protoplasts), or spheroplasts in causing plant disease. For animal pathogens, however, Klieneberger (1938) demonstrated that injection with L-forms of *Streptobacillus moniliformis* did not result in disease. When these unstable L-forms reverted to bacterial forms *in vivo*, the animal succumbed to the bacterial infection, suggesting that the cell wall carried a virulence mechanism. Rubio-Huertos & Beltra (1962) found that glycine-induced spheroplasts of *Agrobacterium tumefaciens* caused disease. These spheroplasts, however, reverted to the original strain in the absence of glycine.

This paper presents data which show that stable gymnoplasts of *Pseudomonas tabaci* do not cause disease when injected into White Burley tobacco leaves. Furthermore, a cell-free extract of high molecular weight was extracted from *P. tabaci* which induced water-soaking in tobacco. A similar extract was obtained from *P. glycinea* which mimicked the hypersensitive reaction.

Preparation of gymnoplasts Gymnoplasts were prepared from washed cells of *P. tabaci*, strain Pt5, with a modified method of Costerton *et al.* (1967). A cell suspension (7.9×10^7 cells/ml) was incubated in a water bath with very slow rotary motion at 25° C for 5 min in a heavy metal-free solution of 0.05 M HEPES buffer (N-2-hydroxyethyl piperazine-N-2-ethane sulfonic acid), 0.5 M sucrose, 0.05 M glucose, 5×10^{-4} M Na-ethylene diamine tetraacetate (EDTA), and 250 μg lysozyme/ml (Worthington

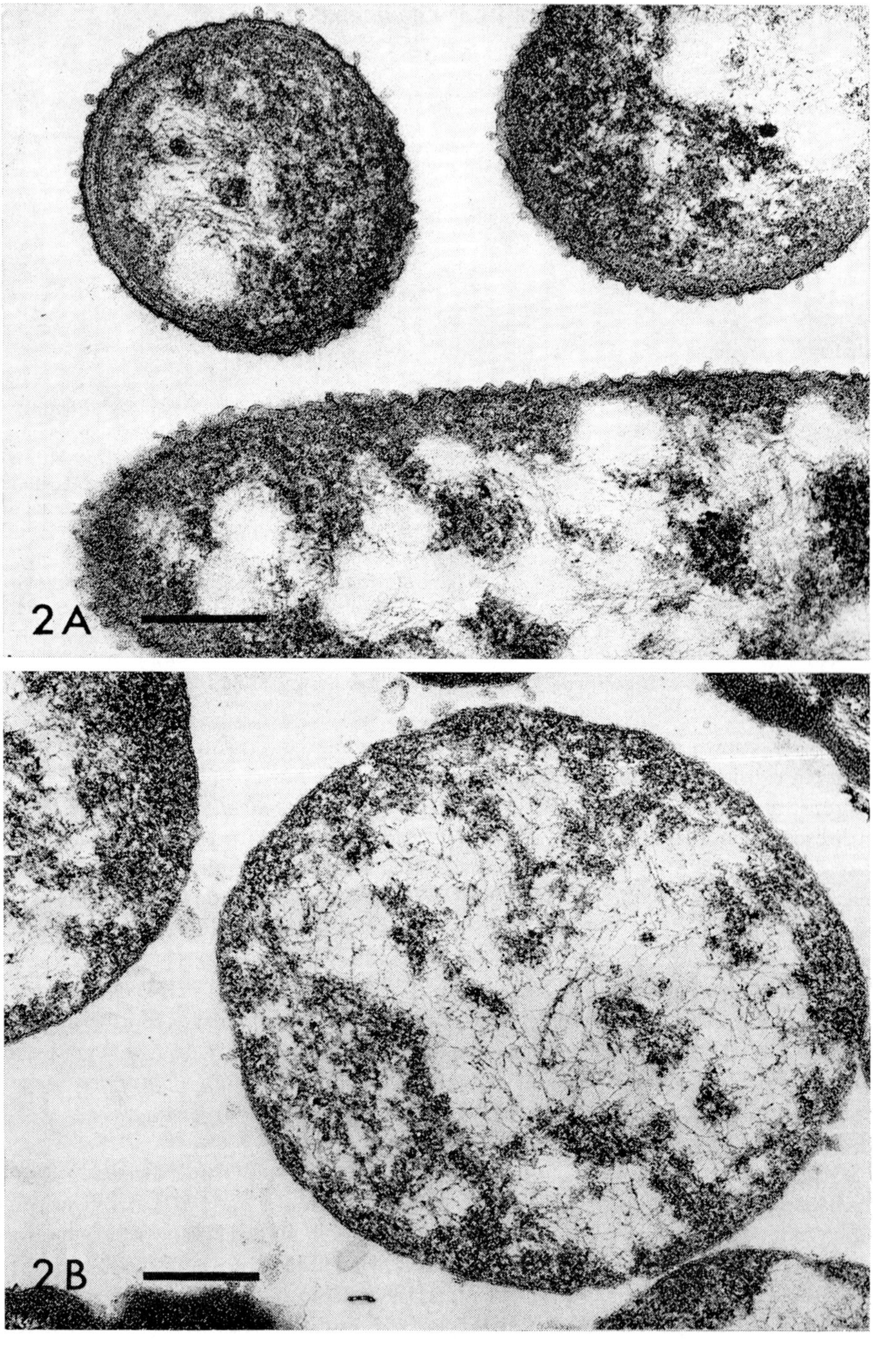
2A
2B

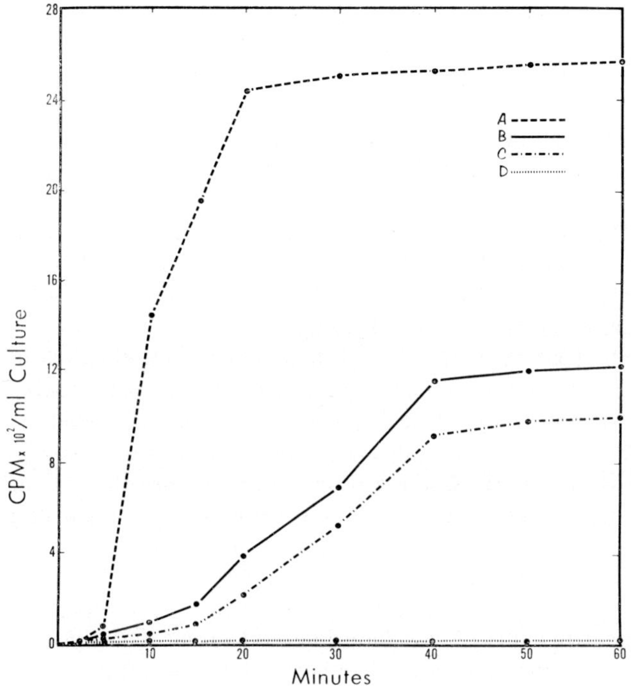

Fig. 2. Electron micrographs of *P. tabaci*. Sections fixed with aldehyde followed by OsO_4, embedded in arildite-epon, and post stained with uranyl acetate and lead citrate. A: whole cells; B: gymnoplasts. Marker represents 0.2 μm.

Biochemical Corp., Freehold, N. J.) (pH 7.0). Resulting gymnoplasts were centrifuged for 15 min at 6,000 *g* (4° C) and partially resuspended in 0.5 M sucrose.

These gymnoplasts were stable in 0.5 M sucrose for over 24 h but lysed in 0.35 M sucrose, indicating cell wall removal. Incorporation of ^{14}C-leucine by gymnoplasts into acid-precipitable material (Rodenberg *et al.*, 1968) was similar to whole cells (Fig. 1). Incorporation by cell-free extracts (lysed gymnoplasts) was nil.

Preparation of gymnoplasts was confirmed by electron microscopy of thin sections which showed complete removal of the cell wall from otherwise intact bacterial ultrastructure (Fig. 2).

Gymnoplasts when infiltrated into White Burley tobacco leaves (Klement, 1963) at concentrations from 10^3–10^9/ml did not cause disease, although they could be detected in free-hand sections of injected leaves for over 24 h.

Fig. 1. Incorporation of D-L-leucine 1-^{14}C into protein of whole cells, protoplasts and a cell-free system of *P. tabaci*. Reaction mixtures (20 ml) were 0.05 M glucose, 0.025 M $MgSO_4$, 0.05 M potassium phosphate, 0.05 M HEPES buffer (pH 7.0), and 0.015 μ Ci/ml D-L-leucine 1-^{14}C. Reactions were incubated at 25 °C and stopped by adding 2 ml of 15% trichloroacetic acid to 1 ml reaction mixture. A: cells (10^8/ml) in reaction mixture; B: cells (10^8/ml) in reaction and 0.5 M sucrose; C: gymnoplasts (10^8/ml) in 0.5 M sucrose; D: cell-free incorporation (lysed gymnoplasts, 10^8/ml).

Extraction of toxic cell-free preparations Cells of *P. tabaci* strains Ptl, 3 and 5 or several strains of *P. glycinea* were harvested by centrifugation (8,000 *g*, 0 °C) from 24 h broth cultures (27 °C) and resuspended into precooled (0 °C) 0.05 M potassium phosphate buffer and 0.1 % thioglycollate. The cold suspension was sonicated at 1/2 min intervals for a total of 4 min. The probe was chilled intermittantly in a mixture of NaCl and ice. The suspension was then centrifuged at 10,000 *g* for 30 min and supernates were recentrifuged at 18,000 *g* (30 min, 0° C). The latter supernate was brought to 25° C and treated with 0.01 M Na-azide (2 h). Toxins were next precipitated with $(NH_4)_2\ SO_4$ in the 25–55 % fraction. The precipitate was resuspended in 0.1 % thioglycollate in 0.05 M potassium phosphate buffer and dialyzed against phosphate buffer. The dialyzed solution was frozen and lyophylized.

Tobacco plants (Klement, 1963) were injected with toxin (10 mg/ml) dissolved in 0.005 M phosphate buffer. This cell-free extract of *P. tabaci* induced water-soaking within 5 h after injection. These leaves remained water-soaked for up to three days. Similar preparations of *P. glycinea* isolates induced necrosis but no water-soaking. Small quantities of these extracts were obtained from culture filtrates. Both toxic extracts were destroyed by heating at 60 °C for 10 min. Furthermore pronase (Rosselet *et al.*, 1969) destroyed activity, which agrees with properties described for a similar cell-free extract from *P. solanacearum* (Sequeira & Ainslie, 1969).

The foregoing data suggest that the disease inducing mechanism of *P. tabaci* is localized in the cell wall of the pathogen and does not occur as a soluble factor in the cytoplasm. Similar evidence is known for several animal pathogens (Guze, 1968).

References

Costerton, J. W., Cecil Forsberg, T. J. Matula, F. L. A. Buckmire & R. A. MacLeod, 1967. Nutrition and metabolism of marine bacteria. XVI. Formation of protoplasts, spheroplasts, and related forms from a gram-negative marine bacterium. J. Bacteriol. 94: 1764–1777.

Guze, L. B., Ed., 1968. Microbial protoplasts, spheroplasts and L-forms. The Williams and Wilkins Co., Baltimore.

Klement, Z., 1963. Rapid detection of the pathogenicity of phytopathogenic pseudomonads. Nature 199: 299–300.

Klieneberger, E., 1968. Pleura pneumonia-like organisms of diverse provenance: some results of an inquiry into methods of differentiation. J. Hyg. 38: 458–476.

Rodenberg, S., W. Steinberg, J. Piper, K. Nickerson, J. Vary, R. Epstein & H. O. Halverson, 1968. Relationship between protein and ribonucleic acid synthesis during outgrowth of spores of *Bacillus cereus*. J. Bacteriol. 96: 492–500.

Rosselet, J. P., S. G. Murphy, R. J. Adamski, M. J. Fletcher & B. J. Ludwig, 1969. Endotoxin-free biologically active component of *Escherichia coli*. J. Bacteriol. 98: 434–436.

Rubio-Huertos, M. & R. Beltra, 1962. Formas L fijas del *Agrobacterium tumefaciens* obtenidas por medio de glicocola. Microbiol. Españ. 219: 1–12.

Sequeira, L. & Virginia Ainslie, 1969. Bacterial cell-free preparations that induce or prevent the hypersensitive reaction in tobacco. Abstracts XI International Botanical Congress, Seattle, Washington.

Relationship of host metabolism to bacterial infection

H. R. Cameron

Department of Botany and Plant Pathology, Oregon State University, Corvallis, Oregon 97331, U.S.A.

Abstract

The presence of pathogenic bacteria in apparently healthy plant tissue suggests that host metabolism may be a critical factor in the initiation of bacterial infection.

Plant pathologists have directed most of their attention to the physiology, variability and mode of infection of the pathogen, but in many cases they have failed to consider adequately the variability of host cells and their influence on infection. There are numerous references to failure of inoculations, lack of expected symptom development and changes in host susceptibility.

In woody plants, the presence of pathogenic bacteria does not insure infection. Cameron (1962, 1970) and Dowler (1969) have reported the occurrence of *Pseudomonas syringae* in healthy sweet cherry *(Prunus avium)*, pear *(Pyrus communis)* and peach *(Prunus persica)*. Keil & van der Zwet (1969) have found pathogenic cells of *Erwinia amylovora* in susceptible symptomless pears (1969). Pathogenic isolates of *Agrobacterium tumifaciens* have been isolated from symptomless trees and Laben *et al.* (1970) have observed a resident phase of *P. syringae* on healthy bean leaves. These bacteria incited symptoms when inoculated into a similar host and therefore were capable of infection.

Injured tissues are considered to be more susceptible to infection than healthy tissues, but little is known about metabolic changes in wounded tissues. Research is needed on what effect environmental and cultural changes have on host metabolism and, in turn, what changes in host metabolism are involved in the infection process. When these are known, it may be possible to adjust horticultural practices to reduce host susceptibility.

Changes in the environment and cultural practices that influence the susceptibility of the host include: mineral and nutritional levels of the host, level or composition of plant growth substances, metabolic changes due to aging of tissue, and changes in metabolism resulting from physical damage of host cells.

Mineral nutrients usually do not effect host susceptibility unless the element is in excess or is at so low a level that the plant is deficient. The addition of nitrogen, phosphate or potash did not change host susceptibility to *P. syringae* (Wormald & Garner, 1938; Wilson, 1939), but increased susceptibility to *Erwinia amylovora*

(Thomas & Ark, 1939). In Oregon (Compton, 1960; Cameron, 1962) the addition of 1.5 kg nitrogen per tree, and in combination with boron, zinc, potassium and phosphate, failed to effect susceptibility. Excessive amounts of nitrogen may keep trees growing late in the fall, making them more susceptible to infection associated with cold injury. Keil & Wilson (1962) report an increased resistance of pear leaves to *E. amylovora* following the addition of aluminum sulfate to the soil.

Research at Eli Lilly Co. (unpublished) correlates the level of plant growth substances with the development of crown gall by *Agrobacterium tumifaciens*. English & Davis (1969) reported an increase in resistance after the application of gibberellic acid and both English & Davis (1969) and Thomas & Ark (1939) found that susceptibility was reduced by defoliation.

Young tissues are usually more susceptible to infection than mature tissues (Layne *et al.*, 1968; Cameron *et al.*, 1969; van der Zwet & Keil, 1970). Hydrogen ion level has also been associated with increased susceptibility following liming (Wormald & Garner, 1938) and a decrease in pH has been noted in maturing tissues (Challice, 1969).

English & Davis (1969) reported that peach seedlings subjected to 25–30 days at 6 °C were more susceptible to infection by *P. syringae* than trees held at 16 °C. Crosse & Panagopoulus (1968) and Cameron (1968) observed drastic increases in bacterial infections after spring frosts and Wilson (1939) suggested that temperature was frequently the limiting factor in infection and symptom development.

All of the above factors influence the metabolism of the host by changing the rate of certain biochemical reactions. These different host cell metabolites react with the pathogen. Hirai (1970) points out that both viruses and fungi are unable to proceed beyond the infection site without the proper environmental conditions. He suggests that this is due to an unsatisfactory physiological interaction between host and pathogen.

Numerous papers have been published on the possible correlation of phenols and quinones with disease resistance (Schroth & Hildebrand, 1964; Smale & Keil, 1966), but relatively little work has been done on the effect of environmental and cultural practices on the phenolic level of the host.

The effects of different nutrient levels on phenolic content is apparently unknown, but presumably a high carbon to nitrogen ratio would favor the production of phenolics. Boron deficiency causes an accumulation of phenolic acids (Lee & Aronoff, 1967), but no attempt has been made to associate these with disease resistance.

The metabolic bases for increased susceptibility following freezing are not easily sorted out. Olien (1967), summarizing the effects of freezing, stated 'The critical metabolic activity in a specific biological system depends on which aspect of the several simultaneously occurring patterns first causes a lethal effect.' Phenolic compounds are probably released from frozen cells as from damaged cells in general but other reactions occur simultaneously.

What affect do the metabolic activities of the host cell have on the infection process? This is a difficult problem, but it is undoubtedly a factor in the success or failure of the pathogen to infect and needs to be worked on by plant physiologists, biochemists and plant pathologists.

References

Cameron, H. R., 1962. Diseases of deciduous fruit trees incited by *Pseudomonas syringae* van Hall. Stn tech. Bull. Ore. agric. Exp. Stn 66.

Cameron, H. R., 1968. Symptoms of *Pseudomonas syringae* infection with respect to freezing. Rep. Ore. St. hort. Soc. 16: 77.

Cameron, H. R., M. N. Westwood & P. B. Lombard, 1969. Resistance of *Pyrus* species and cultivars to *Erwinia amylovora*. Phytopathology 59: 1813–1815.

Cameron, H. R., 1970. Pseudomonas content of cherry trees. Phytopathology 60: 1343–1346.

Compton, O. C., 1960. Some effects of fertilizers on brining quality of Royal Anne Cherries. Rep. Ore. St. hort. Soc. 52: 85.

Crosse, J. E. & C. G. Panagopoulos, 1968. Factors predisposing infection of pear blossoms by *Pseudomonas syringae*. First Int. Congr. Pl. Path. London p. 38 (Abstr.).

Challice, J. S., 1969. Studies on the nature, distribution and biosynthesis of phenolic compounds in *Pyrus* and *Malus*. Master's Thesis Bath Univ. of Tech., Bristol England.

Dowler, W. M., 1969. Bacteriophage labeling of *Pseudomonas syringae*. Phytopathology 59: 245–246.

English, H. & J. R. Davis, 1969. Effect of temperature and growth state on the susceptibility of *Prunus* spp. to *Pseudomonas syringae*. Phytopathology 59: 1025 (Abstr.).

Hirai, Tokuzo, 1970. Comparison of biochemical changes due to fungus infection versus virus infection. Phytopathology 60: 256–266.

Keil, H. L. & R. A. Wilson, 1962. Inhibition of *Erwinia amylovora* by tissue, extracts, and ask from resistant and susceptible pears. Phytopathology 52: 1219 (Abstr.).

Keil, H. L. & T. van der Zwet, 1969. Presence of *Erwinia amylovora* in apparently healthy pear and apple tissue. Phytopathology 59: 1035 (Abstr.).

Laben, C., M. N. Schroth & D. C. Hildebrand, 1970. Colonization and movement of *Pseudomonas syringae* on healthy bean seedlings. Phytopathology 60: 677–680.

Layne, R. E. C., Catherine H. Bailey & L. F. Hough, 1968. Efficiency of transmission of fire blight resistance in *Pyrus*. Can. J. Pl. Sci. 48: 231–243.

Lee, S. & S. Arnoff, 1967. Boron in plants: a biochemical role. Science 158: 798–799.

Olien, C. R., 1967. Freezing stresses and survival. A. Rev. Pl. Physiol. 18: 387–408.

Schroth, M. N. & D. C. Hildebrand, 1964. β-Glucosidase activity in *Erwinia amylovora* and *Pseudomonas syringae*. Phytopathology 54: 906 (Abstr.).

Smale, B. C. & H. L. Keil, 1966. A biochemical study of the intervarietal resistance of *Pyrus communis* to fireblight. Phytochemistry 5: 1113–1120.

Thomas, H. Earl & P. A. Ark, 1939. Some factors affecting the susceptibility of plants to fire blight. Hilgardia 12: 299–322.

Wilson, Edward E., 1939. Factors affecting development of the bacterial canker of stone fruits. Hilgardia 12: 259–298.

Wormald, H. & R. J. Garner, 1938. Manurial trial on nursery trees with reference to effect on plum bacterial canker. Rep. E. Malling Res. Stn 1937 Sect. III: 194–197.

Zwet, T. van der & H. L. Keil, 1970. Relative susceptibility of succulent and woody tissue of Macus pear to infection by *Erwinia amylovora*. Phytopathology 60: 593–595.

Changes in amino acids and electrophoretic characters of cytoplasmic components by some strains of Erwinia

Elvira Grou

Tr. Săvulescu Institute of Biology, Bucureşti, Romania

Abstract

Results are given of comparative analyses of cytoplasmic components of eleven *Erwinia* strains belonging to the 'carotovora', 'amylovora' and 'herbicola' groups.

The contents in proteins and nucleic acids, the absorption spectra in ultraviolet, the composition in amino acids and the electrophoretic characteristics were determined. The absorption in ultraviolet shows the presence of molecular complexes of protein-nucleic acids. The ratio basic amino acids to acidic amino acids gives values less than 1 for all analysed strains. The electrophoretic spectrum of proteins detected with amido black shows different numbers of bands in densitometric curves, but the table of Ef values nevertheless emphasizes many similarities between the strains.

Recently various proposals have been made to classify the *Erwinia* spp. into a few groups, or to distribute them over various genera of Enterobacteriaceae, Martinec & Kocur (1963) lumped the phenotypes they had examined into two species: *E. amylovora* and *E. carotovora*. Dye (1968), in a similar study, recommended a lumping into five species: *E. amylovora*, *E. herbicola*, *E. uredovorus*, *E. stewartii*, and *E. carotovora*. Starr & Mandel (1969) compared the GC-DNA contents of different species of Erwinia but considered that no nomenclatural recommendations could be made merely on the basis of phenotypic data.

Parallel with other research on the different species of Erwinia, Grou & Lazar (1968) and Grou (1971) studied the physico-chemical characteristics of their cytoplasmic proteins. Absorption in ultraviolet, composition in amino acids and electrophoretic behaviour of extracts from various species were examined to accumulate sufficient new data for as complete a characterization as possible.

Materials and methods

The bacterial strains used in this study, now in the collection of the 'Tr. Savulescu' Institute of Biology, were received from Dr D. C. Graham (Agricultural Scientific Services, East Craigs, Edinburg, Scotland), R. A. Lelliott (National Collection of Plant Pathogenic Bacteria, Harpenden, England), C. M. Kocur (Czechoslovak Collection of Micro-organisms, C. E. Purkine University, Brno, Czechoslovakia) and from Dr D. C. Hildebrand (University of California, Berkeley, USA).

Table 1. Physico-chemical characteristics of cytoplasmic proteins of some *Erwinia* species.

		Protein %	RNA %	DNA %	UV absorption max./min.	Ab/Aa	Protein/ARN E280/E260	Number of bands
E. chrysanthemi	377	78.1	8.1	16.8	255–270/240	0.5	0.75	8
	517	63.1	9.1	3.8	260/240	—	0.64	3
	910	50.7	1.6	24.0	270/240	0.5	0.71	9
E. carotovora	1065	40.9	6.9	3	255–260/230–235	—	0.49	3
	402	40.1	4.1	3.4	255/230–235	—	0.46	5
	G-123	49.0	7.3	2.6	255/240	0.9	0.60	4
E. aroidea	1231	24.2	1.4	4.2	255–260/230–235	—	0.50	1 1
	119V	34.9	17.3	7.3	260/240–245	0.1	0.55	3
E. atroseptica	531	35.6	2.5	14.1	255/230	—	0.45	5
E. amylovora	1114	35.0	19.8	17.5	255–265/235	0.3	0.71	2
	Fb9	1.1	5.4	—	255–260/240	—	0.57	2
E. herbicola	2407	57.0	7.4	+	275–280/245–265	0.2	1.33	9
	2406	2.2	2	+	255–260/235–240	—	0.48	4
E. lathyri	G-144	18.6	14	—	255–265/235	—	0.58	2
	G-157	35.2	5.5	6.8	255–260/235	—	0.51	2

Ab = basic aminoacids
Aa = acid aminoacids

Table 2. EF values of some *Erwinia* species on polyacrylamide gel in tris-borat buffer.

		Classes									
		0.01	0.1	0.2	0.3	0.4	0.5	0.6	0.7	0.8	0.9
E. atroseptica	531		0.11		0.31		0.55	0.62	0.76		
E. carotovora	G-123					0.43	0.53	0.65		0.86	
	1065			0.22		0.40	0.53				
E. aroideae	119V				0.39		0.53			0.88	
E. chrysanthemi	377	0.06		0.20	0.37	0.42	0.50	0.60	0.77		0.98
	517			0.23		0.43			0.71		
	910	0.08	0.11	0.22	0.33	0.40	0.53	0.66	0.76	0.80	
E. amylovora	1114			0.28	0.31						
E. herbicola	2407	0.08	0.1	0.26	0.32	0.40	0.57	0.65	0.75		0.93
	2406				0.30	0.45	0.57		0.70		

Dry or fresh bacterial mass was extracted in 0.1 M phosphate buffer (pH 7), by mortar-grinding with quartz sand, cooled, and then centrifuged at 4,000 and 17,000 rev/min, dialysed against tap water and distilled for 24–28 h.

Proteins, ribose and DNA were estimated in extracts (Chargaff & Davidson, 1965; Weinberg, 1968). Amino acids were determined by paper-chromatography after hydrolysation of the extracts with 6 N HCl for 18 h at 110 °C (Hais & Macek, 1963) and ultraviolet spectra were obtained with an SF_4 spectrophotometer for 1 cm layer.

Electrophoresis was carried out on polyacrylamide gel with 6% Cyanogum in tris-borate buffer (pH 8.3) (Raymond, 1964). The cultures of the analysed bacteria are included in Tables 1 and 2.

Results and discussion

Amino acid analysis revealed small differences in the number and amount of components; small amounts of phenylalanine and tyrosine were likewise detectable by the analysis of these strains. But in other *Erwinia* strains, Grou (1971) showed their absence in the amino acid complement. The ratio basic amino acids to acidic amino acids was always less than 1, being between 0.2 and 0.9 (Table 1).

Absorption in ultraviolet In spectra absorption maxima and minima were similar for all strains, except for *E. chrysanthemi* 377 and 910 and for *E. herbicola* 2407.

But according to the ratio of protein and nucleic acid the extinction values between 280 and 260 μm showed that the species are less similar. Similarities have been recorded both between species, and between strains of one species. Table 1 shows that the ratio varies between 0.46 and 1.33. Therefore these nucleoproteins have the same composition.

Electrophoretic analysis First the electrophoretic plates were examined by counting

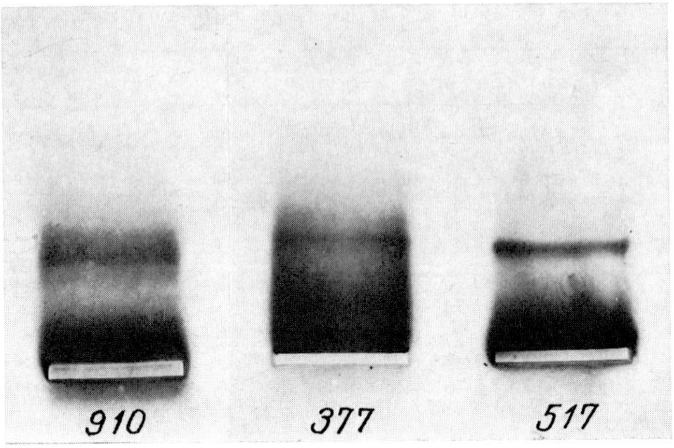

Fig. 1. Electrophoretic patterns of cytoplasmic proteins of some *Erwinia* species.

the bands and determining corresponding Ef values; then an integrator was used to make absorption measurements (Table 2).

The results showed that the *E. chrysanthemi* 377 and *E. herbicola* 2407 samples had most bands (8–9) while in the others the number was between 2 and 5. The distribution of bands varied strongly. Each species had its own spectrum (Fig. 1). For instance, though the strains *E. chrysanthemi* 575, *E. carotovora* 1065 and *E. aroideae* 119V had an equal number of bands, they differed in mobility of the bands. Densitometry also showed difference in the electrophoretical profiles, but were weaker due to the lack of sensitivity of the respective integrators.

Analysis of the results shows that in Erwinia strains cytoplasmac components contain different nucleoproteins. A common characteristic of these nucleoproteins is that in all analysed extracts the protein component is non-histonic. This is due to the preponderance of acidic amino acids over the basic ones, as well as to the small amounts of phenylanine and tyrosine. They likewise contain both RNA and DNA. Nevertheless they differ in the ratio protein/nucleic acids, both between the different species and between strains of the same species. In Chargaff's classification (1965) of nucleoproteins, they obviously belong to the third category 'of bacterial origin'.

The electrophoretic patterns obtained on polyacrylamide gel also revealed a strong heterogeneity among the analysed species. The present data, together with those from previous investigations (Fig. 3) concerning electrophoretic patterns of proteins from various Erwinia strains, illustrate this.

The complexity of fractions, as well as the absorption in ultraviolet characteristics recorded for *E. chrysanthemi* 910 and 377 and for *E. herbicola* 2407 show the presence of some nucleoproteins of a different structure. From current research we may find in some species of Erwinia certain molecular RNA-DNA-protein complexes similar to those isolated by Tongur from *E. coli.*

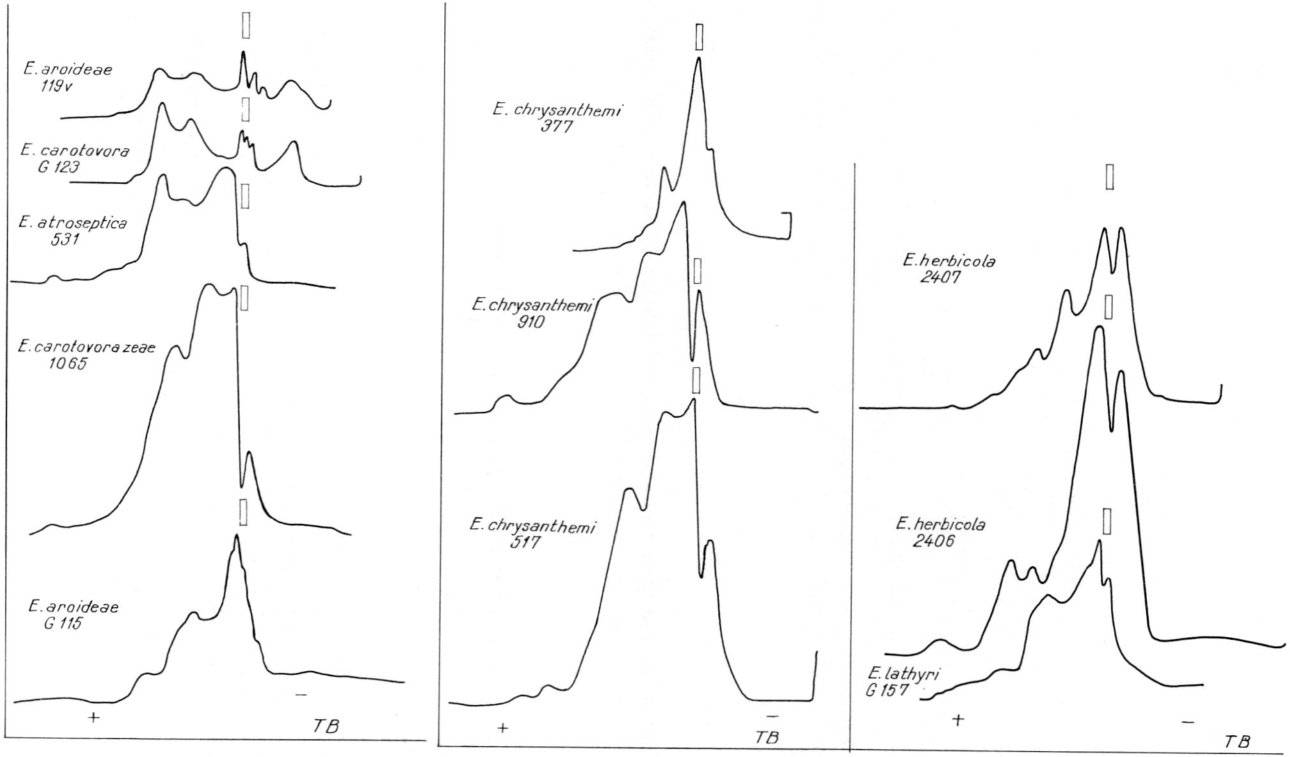

Fig. 2. Electrophoretic profiles of cytoplasmic proteins of the soft rot group of *Erwinia* (left and middle) and of the yellow group of *Erwinia* (right).

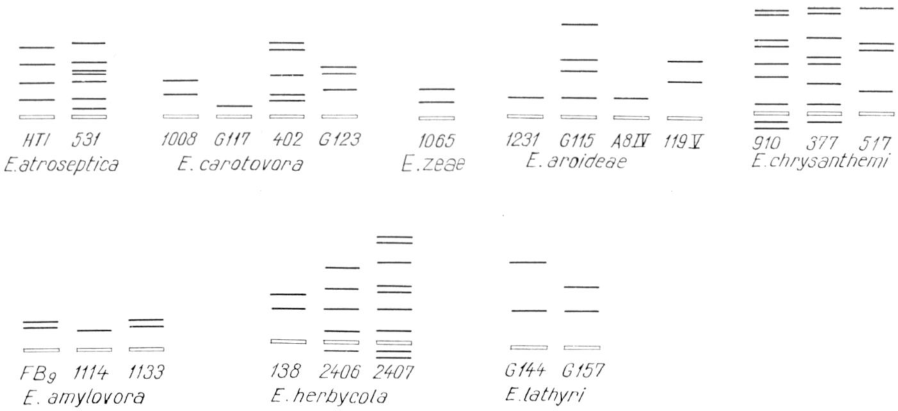

Fig. 3. Protein patterns of some *Erwinia* species.

By the action of certain tensioactive compounds upon the electrophoretic behaviour of these proteins we also intend to obtain a more complete characterization of these proteins.

We do not intend to use these characteristics for nomenclature until our research with other Enterobacteriaceae is complete.

References

Chargaff, E. & J. N. Davidson, 1965. The nucleic acids. Vol. 1. New York.

Dye, D. W., 1968. A taxonomic study of the genus *Erwinia*. 1st. Int. Congr. Pl. Path. (London) July 1968, Abstract of Papers: 52.

Grou, Elvira & I. Lazar, 1968. Proprietățile fizico-chimice și acțiunea toxică pe iepuri a extractelor nucleoproteice de la speciile de *Erwinia* care produc putreziri moi. Microbiologia 1: 183–188.

Hais, Y. M. & K. Macek, 1963. Paper chromatography. Prague.

Martinec, T. & M. Kocur, 1963. A taxonomic study of the genus *Erwinia*. Folia, Publ. Fac. nat. Sci. J. E. Purkynye Univ. (Biologia) 4: 1–163.

Raymond, B., 1964. Acrylamide gel electrophoresis. Ann. N.Y. Acad. Sci. 121: 350–365.

Starr, M. P. & M. Mandel, 1969. DNA base composition and taxonomy of phytopathogenic and other enterobacteria. J. gen. Microbiol. 56: 113–123.

Tongur, V. S., N. S. Wladytchenskaya & N. M. Kotchikina, 1968. A natural RNA-DNA complex in bacterial cells. J. mol. Biol. 33: 451–464.

Weinberg, R., 1968. An electrophoretic analysis of the isozymes of malate dehydrogenase in several different plants. Pl. Physiol. 43(4): 622–628.

The biological activity of phytotoxic bacterial glycopeptides

G. A. Strobel

Department of Botany and Microbiology, Montana State University, Bozeman, Montana 59715

The wilt-producing capacity of large molecular weight substances released by a plant pathogenic bacterium in its host has traditionally been attributed to a plugging action in the plant stem. Investigations on the mechanism of wilt production by toxic high molecular weight glycopeptides, produced by some plant pathogenic corynebacteria, have provided evidence for an alternative hypothesis.

Dye and ^3H_2O movement experiments in toxin treated plant stems, originally suggesting membrane damage, was caused by the glycopeptide. Electron micrographic studies on toxin treated and on naturally infected plant stems corroborated this hypothesis. Using ^{14}C labelled toxin it was possible to demonstrate a plant host-toxin complex. This experiment initially suggested that the toxic glycopeptide possessed an 'active site'. To evaluate the 'active site' hypothesis, the toxin was modified chemically. To evaluate the chemically modified toxin, the degree of wilt expressed by the test plant had to be quantified. This was done in a device that measures stem wilt. Effective blockage of the carboxyl group of 2-keto-3-deoxygluconic acid by methylation removed biological activity from the toxin of *Corynebacterium sepedonicum*. Although the molecular weight of toxin of *C. insidiosum* is 5×10^6, it too produces wilt in plant stems. Polysaccharides of similar molecular weight are not effective as stem-wilt-inducing substances. A summary of the information supporting the hypothesis of toxin-induced membrane damage is presented in the table.

Although support for a toxin effect on cellular membranes has been presented, it does not rule out the possibility that plugging in vascular tissues of infected plants occurs. Bacteria, alive and dead, plant host polymers, etc. could all be contributing factors to plug initiation which can often be seen in plants infected by these pathogens. Nevertheless, these water soluble toxins may also play a role in symptom production.

Studies on wilt toxins are not without problems, however. The most difficult is due to the observation that toxin wilted plant cuttings are capable of regaining their turgidity after placement into water, provided the wilting process has not proceeded too far. This observation, too, indicates that plugging is not a factor in wilt production, and it suggests that membrane damage initially caused by the toxin is repaired so that the cells may regain their turgidity. This aspect of the problem deserves further investigation.

Summary of evidence for membrane damage in *Corynebacterium* species induced by phytotoxic glycopeptides.

Experiment	Toxin source	Observation
Movement of dyes	*C. sepedonicum* *C. michiganense*	Dyes moved through toxin treated stems faster than through control stem.
Electron microscopy of plant stems	*C. michiganense* *C. sepedonicum*	Toxin treated stems showed membrane rupture and disorganization. No plugs were observed in the vascular tissue.
Micro-autoradiography of ^{3}H-toxin treated stems	*C. sepedonicum*	The labelling appeared associated with chloroplasts and other cytoplasmic membranes. Again no plugging was observed in the stem tissue.
Observation of a toxin-host complex	*C. sepedonicum*	An acid dissociable complex was isolated from toxin-^{14}C treated stem tissue. After dissociation the toxin could be recovered.
Movement of $^{3}H_2O$ through leaves	*C. sepedonicum*	Plant cuttings, saturated with $^{3}H_2O$ and then treated with toxin lost $^{3}H_2O$ faster than the controls.
Conduction of washings of plant leaves	*C. sepedonicum*	Cuttings treated with the toxin lost electrolytes at a more rapid rate than the controls.
Autoradiography of plant cuttings	*C. michiganense* *C. sepedonicum* *C. insidiosum*	^{14}C-toxin of each of these organisms when administered to plant cuttings were spreading throughout the plant stem; at the primary site of wilting tissue collapsed. Some labelling appeared at leaf margins.
Comparative studies with other polysaccharides	*C. michiganense* *C. insidiosum* *C. sepedonicum*	In all cases, using young tomato plants in the assay technique, polysaccharides bracketing the toxins in M.W. were not capable of stem wilt induction.
Active site studies	*C. sepedonicum*	Methyl esterification abolished activity.

For literature references: see p. 365.

The significance of catalase activity in Pseudomonas solanacearum

B. Digat

Institut National de la Recherche Agronomique, Station de Phytobactériologie, 49 Beaucouze, France

Abstract

Catalase activity of *Pseudomonas solanacearum* is mainly extracellular and located in the glycoproteic fraction of the bacterial slime. Avirulent mutants lack functional catalase. However, if a porphyrinic coenzyme is supplied, they can temporarily recover, in catalase activity as well as by becoming highly virulent.

It is suggested that bacterial catalase could take a leading part in the virulence and in the first stages of pathogenesis.

Introduction

Though pathogenesis of bacterial wilt caused by *Pseudomonas solanacearum* race 1 was intensively studied during the last years (see Maine, 1960; Buddenhagen & Kelman, 1964; Kelman & Cowling, 1965; Husain & Kelman, 1968), our understanding about the critical first stages of establishment of the pathogen is still elusive.

During the infection process, enzyme activities of both pathogen and host have been studied minutely (Sequeira, 1964; Husain & Kelman, 1968) and the results explain to some extent the course of pathogenesis, but not the first stages nor the virulence of the bacterium.

In causing wilting, the extracellular bacterial polysaccharides undoubtedly play an important role (Husain & Kelman, 1958), but the subtle mechanisms of virulence have remained in the dark.

In this study, emphasis is on the most important role of bacterial catalase during the process of establishment of the pathogen in the host.

Materials and methods

Two isolates of *P. solanacearum* from tomato were used in the experiments: a highly virulent one obtained in 1968 by the author in French Guiana (SEAC de Kourou) from the variety 'Floralou', and an avirulent stable mutant obtained by repeated subculture of a strain originating from a 'Floralou' isolate in the French West Indies (Basse-Pointe, Martinique, 1965).

The following media were used:

1. Basal medium agar: yeast extract 5 g, pancreatic peptone 5 g, bacto-agar 20 g,

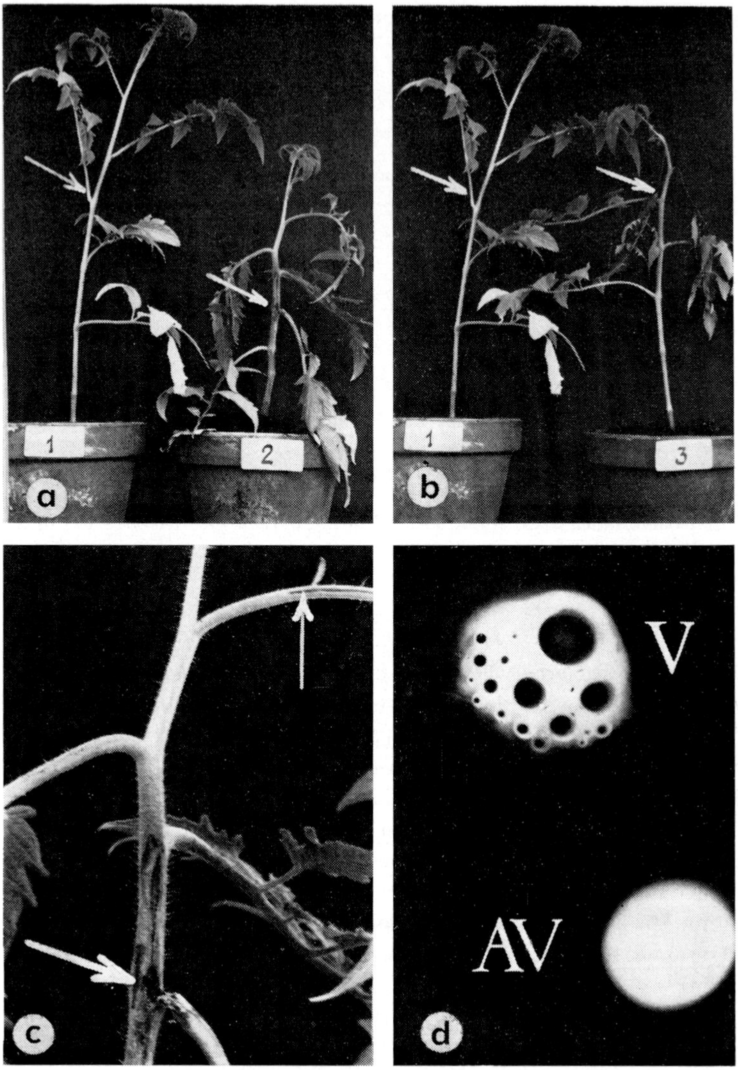

a, b: tomato plant reactions 6 days after inoculation by stem puncture technique with avirulent mutant (1), avirulent mutant grown on hemin (20 μg/ml) agar (2), and highly virulent strain (3); note difference between slow wilt in 2 and sudden wilt in 3;

c: typical symptoms in host plant 2 caused by the avirulent mutants, grown on hemin agar; note heavy necrosis of vascular system and diffusion of bacteria in the whole plant;

d: strong catalasic activity of highly virulent strain (V) as compared with the undetectable one of the avirulent mutant (AV).

dextrose 10 g, made up to 1000 ml with distilled water, pH adjusted to 6.8–7.0 (YPD).
2. Kelman medium to distinguish virulent colonies and avirulent mutants (Kelman, 1954), containing 2,3,5 triphenyl tetrazolium chloride (TTC+), or without dye (TTC−).
3. Heated blood agar (HB) from the Pasteur Institute in Paris.
4. Hemin agar (TTC+ or TTC−) medium with hemin chlorhydrate added from a stock solution (200 μg/ml) from the Pasteur Institute.

To estimate the hydrogen peroxide splitting activity (catalase activity), a stock solution of hydrogen peroxide was prepared by diluting H_2O_2 (20 volumes) with 0.01 M phosphate buffer, pH 6.5 (1 : 1).

The activity of the colonies was visually demonstrated by effervescence on addition of hydrogen peroxide (10 volumes) to the colony. Negative results were checked by placing the colony into H_2O_2.

To find the site of catalase activity, the bacteria were grown in Roux bottles at 30 °C on YPD-agar. After 48 h they were collected and suspended in sterile distilled water. By shaking vigorously, the bacterial slime was dissolved. Centrifuging (25.000 *g*) for 15 minutes at + 4 °C eliminated the bacterial cells. Then the pH was adjusted to 7.0 with 0.2 N NaOH and sufficient solid ammonium sulphate was added to achieve 95% saturation. The glycoprotein fraction was left for 12 h at +4 °C to precipitate. Afterwards it was removed by centrifuging and put into some distilled water. After 36 h of dialysis against distilled water, the viscous extracellular glycoprotein fraction was tested for its catalase activity.

The susceptible tomato varieties F_1 hybrid I.N.R.A. 63-4 and I.N.R.A. 63-5 were inoculated. The plants were grown individually in pots. The inoculations were made by the stem puncture technique described by Winstead & Kelman (1952). The plants were kept alternatingly in the dark at 25 °C for 10 h and in the light at 30 °C for 18 h.

Disease estimates were made eight days after inoculation.

Results

Intensity of catalase activity The catalase activity of the colonies, and of the bacterial suspensions of the highly virulent strain were high (Fig. d : V). For the avirulent mutant, catalase activity was very weak or undetectable (Fig. d : AV).

Site of catalase activity Catalase activity was for the most part extracellular and located in the slime. The glycoproteinic fraction of slime contained the highest rate of catalase.

Recovery of catalase activity and virulence The avirulent mutants growing on heated blood or hemin agar restored their catalase activity to a level as high as that of highly virulent bacteria.

The avirulent mutants were inoculated in tomato plants. All mutants grown on a medium of either heated blood or hemin were virulent, first causing foliar epinasty, after six days necrosis spreading along the xylem, and after eight days wilting (Figs a

The effect of different media on catalase activity and virulence of avirulent mutants.

Medium	Catalase activity	Virulence
HB	++++	yes
Hemin 20μg/ml	+++	yes
Hemin 5μg/ml	+++	yes
TTC+	0	no
TTC−	0	no
LPGA	0	no

++++: instantaneous and strong effervescence
+++ : instantaneous effervescence
0 : no effervescence

and c). However, the wilt caused by the 'neo-virulent' mutant was slower than that caused by the highly virulent strain (sudden wilt, see Fig. b).

Loss of catalase activity and loss of virulence After growing on heated blood or hemin agar, avirulent mutants were grown subsequently on TTC+, TTC− and YPG agar, and catalase activity quickly decreased.

In the same way, isolates from wilted plants immediately lost their catalase activity when grown on TTC or YPG agar, and they became avirulent when inoculated on tomato plants.

Discussion

In *Pseudomonas solanacearum* the loss of virulence is associated with a decreased synthesis of extracellular polysaccharides. However, this phenomenon is probably accompanied by a loss of several enzyme characters. For instance, our results point to a loss of functional catalase.

Since, *in vitro*, avirulent mutants are able to recover catalase activity when provided with the preformed prosthetic haem group, it seems that avirulent mutants are able to form the apoenzyme of catalase but not the haem coenzyme.

The results indicate that only the bacterial cells with a high level of catalase activity can establish themselves in the xylem of tomato plant and multiply to such an extend that they may damage the vascular system.

The significance of the bacterial catalase activity could be that it plays a role in the critical first stage of the establishment of the pathogen by competing with peroxydase of the host for peroxyde substrates (Rudolph & Stahman, 1964).

It is suggested that bacterial catalase indeed plays a role in the virulence of *Pseudomonas solanacearum*, but that other factors, still unknown, may also be involved.

References

Buddenhagen, I. & A. Kelman, 1964. Biological and physiological aspects of bacterial wilt caused by *Pseudomonas solanacearum*. Ann. Rev. Phytopathol. 2: 203–230.

Husain, A. & A. Kelman, 1958. Relation of slime production to mechanism of wilting and pathogenicity of *Pseudomonas solanacearum*. Phytopathology 48: 155–165.

Husain, A. & A. Kelman, 1968. The role of pectic and cellulolytic enzymes in pathogenesis by *Pseudomonas solanacearum*. Phytopathology 48: 377–386.

Kelman, A., 1954. The relationship of pathogenicity in *Pseudomonas solanacearum* to colony appearance on a tetrazolium medium. Phytopathology 44: 693–695.

Kelman, A. & E. B. Cowling, 1965. Cellulase of *Pseudomonas solanacearum* in relation to pathogenesis. Phytopathology 55: 148–155.

Maine, E. C., 1960. Physiological responses in the tobacco plant to pathogenesis by *Pseudomonas solanacearum*, causal agent of bacterial wilt. Doctoral thesis, N.C. State of the University of North Carolina, Raleigh.

Rudolph, K. & M. A. Stahmann, 1964. Interaction of peroxidases and catalases between *Phaseolus vulgaris* and *Pseudomonas phaseolicola* (halo blight of bean). Nature, Lond. 20: 474–475.

Sequeira, L., 1964. Inhibition of indolacetic acid oxidase in tobacco plants infected by *Pseudomonas solanacearum*. Phytopathology 54: 1078–1083.

Winstead, N. N. & A. Kelman, 1952. Inoculation techniques for evaluating resistance to *Pseudomonas solanacearum*. Phytopathology 42: 628–634.

Development of pathogenic and saprophytic bacterial populations in plant tissue

J. M. Young and A. M. Paton

Bacteriology Division, School of Agriculture, University of Aberdeen, Aberdeen, U.K.

Abstract

When *Pseudomonas phaseolicola* and saprophytes are inoculated together into bean plants, the saprophyte population is stimulated by the pathogen and the pathogen does not appear to be affected by the presence of the saprophyte. Saprophyte populations decline very slowly when inoculated alone into host plants. These results suggest that the pathogen is able to induce the release of nutrients and saprophytes are not. This may account for the failure of saprophyte development in plant tissue.

Introduction

Of all bacterial species, only plant pathogens are capable of rapid multiplication in living plant tissue. These pathogenic species, whether in the host plant species to which they are normally pathogenic (homologous), or in some other plant species (heterologous), are able to multiply significantly, though to different levels (Ercolani & Crosse, 1966), whereas saprophytes do not develop at all (Klement *et al.* 1964).

Several mechanisms involving the induction of inhibitors have been suggested to account for the failure of the development of saprophytes in the apparently congenial environment of the host tissue. Cruickshank (1962) suggested that entry by any organism into living plant tissue induced the production of a phytoalexin – a broad-spectrum antibiotic characteristic of the host species, which is effective against most organisms other than the homologous pathogens of the host. Moustafa & Whittenbury (1970), studying *Pseudomonas* species, suggested that oxidase positive saprophytes induce the conversion of plant phenols to bactericidal quinones and thus inhibit their own development, while the pathogenic species, lacking this character, are able to multiply.

The behaviour of pathogenic and saprophytic bacterial populations from pure and mixed inocula in plants were studied in an effort to determine the factors regulating bacterial multiplication.

Method

Three isolates of *Pseudomonas phaseolicola* were each paired with an isolate of *P. fluorescens*, *P. putida* (both oxidase positive) and *Erwinia herbicola* (oxidase negative).

Inoculum was prepared in 1:2,000 peptone solution at a standard optical density equivalent to 2×10^8 cells/ml and diluted in 1:20,000 Tween 80 immediately prior to inoculation into plants.

Susceptible bean plants *(Phaseolus vulgaris* cv. Masterpiece) were inoculated into their trifoliate leaves by the vacuum infiltration method of Stolp (1961). The pathogen was inoculated into plants at 10^6 cells/ml both in pure suspension and with the saprophyte present at concentrations of 10^6 and 10^7 cells/ml. Saprophytes were also inoculated in pure suspension. Plants were thereafter maintained at 22°–25°C.

Leaf discs were excised after 0, 2, 4, 6 and 8 days, macerated in buffer and the suspension decimally diluted. Drop plates were prepared following the method of Miles & Misra (1938).

The different bacterial species were easily discriminated in mixed culture on the basis of colony morphology, colour and rate of development.

Results

The development of a representative pathogen-saprophyte pair is shown in the figure. All experimental results were similar in general trend.

The growth of the pathogen was apparently unaffected by the presence of the oxidase negative or oxidase positive saprophytes.

The saprophyte populations fluctuated during the course of the experiment, in some cases showing a slight decline but in others remaining relatively constant.

When the saprophyte was present together with the pathogen, a stimulation regularly amounting to greater than one order of magnitude was observed.

There was no obvious difference in the reaction or effect of oxidase positive and oxidase negative saprophytes.

Discussion

The most striking result of these experiments is the consistent stimulation of the saprophyte when co-inoculated with the pathogen. The fact that cell permeability increases following inoculation with pathogenic bacteria (Williams & Keen, 1967; Cook & Stall, 1968) probably accounts for the release of nutrients from host cells into the common environment of the two species in the intercellular spaces. This characteristic of the pathogen appears to be important in distinguishing it from the saprophyte. If the saprophyte is unable to obtain nutrients then its gradual decline in plant tissue would be expected.

It has been suggested that the co-inoculation of a pathogen with saprophytes leads to the induction of the plant defence mechanism and hence the suppression of the pathogen (Crosse, 1965; Goodman, 1967). Such a conclusion is not supported by this work. Similarly these results also seem to negate the theory proposed by Moustafa & Whittenbury (1970) that differences in bacterial oxidase activity are directly associated with the induction of plant defence mechanisms.

The results reported here militate against mechanisms involving bactericidal inhibitors in the limitation of saprophytic bacteria in plants. Although all other possible

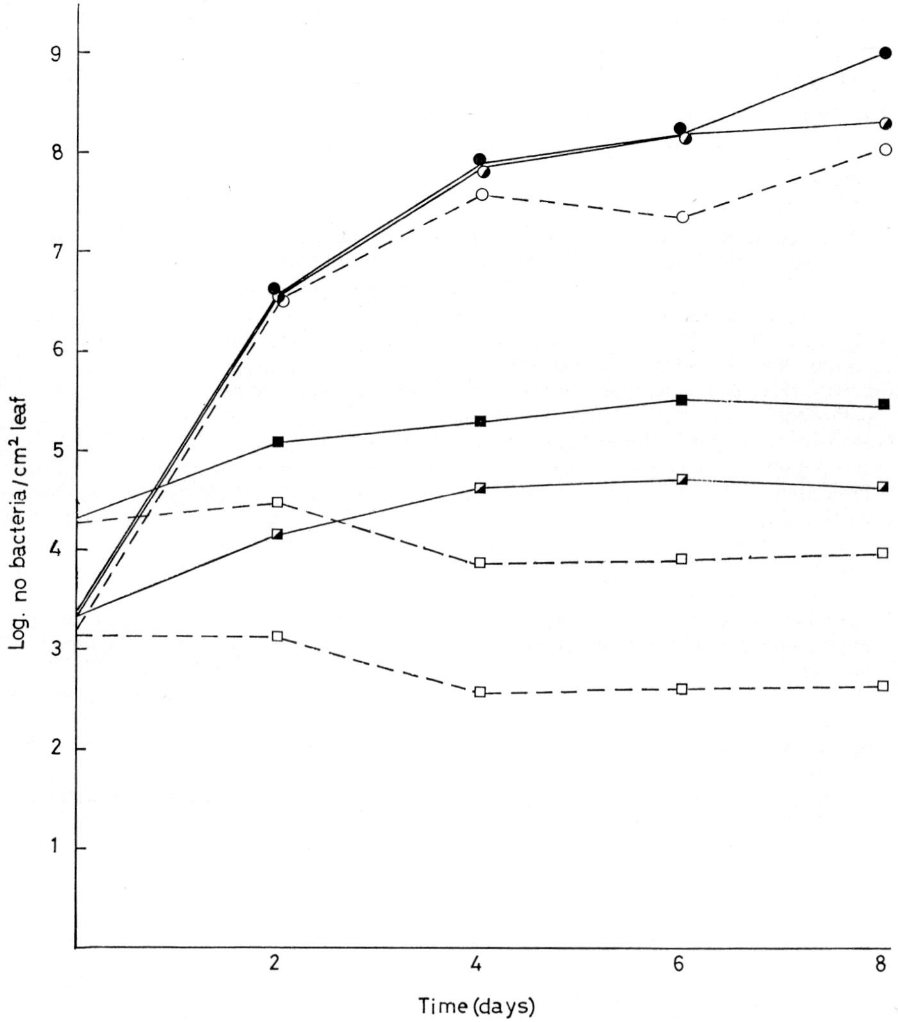

The growth of a pathogen (*Pseudomonas phaseolicola* NCPPB 604) and a saprophyte (*P. fluorescens* NCIB 9046) from pure and mixed inocula in plant tissue. Pathogen inoculated at 10^6 cells/ml, saprophyte inoculated at 10^6 and 10^7 cells/ml.

○– – – – – – ○ pathogen alone
◐——————◐ pathogen with saprophyte at 10^6 cells/ml
●——————● pathogen with saprophyte at 10^7 cells/ml
□– – – – – – □ saprophyte alone
◪——————◪ saprophyte (10^6 cells/ml) with pathogen
■——————■ saprophyte (10^7 cells/ml) with pathogen

mechanisms are not excluded by these results, the absence of any reaction by the plant to the presence of saprophytic bacteria and the absence of any reaction by the saprophyte as expressed by changes in viable cell numbers suggests that their failure to develop is most easily explained by their inability to obtain nutrients.

References

Cook, A. A. & R. E. Stall, 1968. Effect of *Xanthomonas vesicatoria* on loss of electrolytes from leaves of *Capsicum annuum*. Phytopathology 58: 617–619.

Crosse, J. E., 1965. Bacterial canker of stone-fruits. VI. Inhibition of leaf-scar infection of cherry by a saprophytic bacterium from the leaf surfaces. Ann. appl. Biol. 56: 149–160.

Cruickshank, I. A. M., 1962. Studies on phytoalexins. IV. The antimicrobial spectrum of pisatin. Aust. J. biol. Sci. 15: 147–159.

Ercolani, G. L. & J. E. Crosse, 1966. Growth of *Pseudomonas phaseolicola* and related plant pathogens *in vivo*. J. gen. Microbiol. 45: 429–439.

Goodman, R. N., 1967. The infection process — infection by bacteria. In: The biochemistry and physiology of infectious plant disease. Ed. R. H. Goodman, Z. Király & M. Zaitlin. Princeton.

Klement, Z., G. L. Farkas & L. Lovrekovitch, 1964. Hypersensitive reaction induced by phytopathogenic bacteria in the tobacco leaf. Phytopathology 54: 474–477.

Miles, A. A. & S. S. Misra, 1938. The estimation of the bactericidal power of the blood. J. Hyg., Camb. 38: 732–749.

Moustafa (Darweish), F. A. & R. Whittenbury, 1970. Properties which appear to allow phytopathogenic pseudomonads to counteract plant defence mechanisms. Phytopath. Z. 67: 214–224.

Stolp, H., 1961. Neue Erkenntnisse über phytopathogene Bakterien und die von ihnen verursachten Krankheiten. I. Verwandtschaftsbeziehungen zwischen phytopathogenen *Pseudomonas*-'Arten' und saprophytischen Fluoreszenten auf der Grundlage von Phagenreaktionen. Phytopath. Z. 42: 197–262.

Williams, P. H. & N. T. Keen, 1967. Relation of cell permeability alterations to water congestion in cucumber angular leaf spot. Phytopathology 57: 1378–1385.

The effect of ionizing irradiation on bacteriophages of Pseudomonas glycinea and Xanthomonas phaseoli var. sojense

R. M. Numić

Laboratory of Plant Pathology, Faculty of Agriculture, Sarajevo University, Sarajevo, Yugoslavia

Abstract

The experiments were carried out on a series of bacteriophages, four Pg10-1, PgII-2, PgIII-4 and PgI-9 of *Pseudomonas glycinea* and six Xpg30-3, Xpg5-5, XpgI-6, Xpg10, Xpg11 and Xpg8-12 of *Xanthomonas phaseoli* var. *sojense*. The test was designed to study the rates of inactivation of bacteriophage infectivity by gamma rays in T_1 and nutrient broth (10%) media. Samples were irradiated in glass tubes with a maximal irradiation dose of 5×10^5 rad.

The slope of the inactivation curve is very different for the various bacteriophages; so does the inactivation dose D_{37}. The inactivation dose was considerably higher when the bacteriophages were irradiated in nutrient broth.

The XpgI-6 bacteriophage had the highest inactivation dose D_{37} (74.730 rad in nutrient broth and 47.043 rad in T_1 media). The bacteriophage PgI-9 had the lowest inactivation dose (15.045 rad and 14.455 rad, respectively).

It can be concluded that gamma rays produce a direct effect on irradiated bacteriophages in nutrient broth, and direct and indirect effects if the bacteriophages are irradiated in T_1 media.

Introduction

The effect of ionizing irradiation on the inactivation of bacteriophages has been studied by many authors.

Wollman & Lacassagne (1940) were the first who investigated the inactivation of bacteriophages, suspended in broth, by X-rays; they found their sensitivity to be directly proportional to the size of the particles. This work, and the later researches carried out by Wollman *et al.* (1940), Luria & Exner (1941), Lea & Salaman (1946), gave an explanation of the process of bacteriophage inactivation in the light of the 'target theory'. The same authors calculated the 'sensitive volume', i.e. the size of bacteriophage particles. On this basis they found a reasonable concordance between the size of the bacteriophage particles obtained by the irradiation of bacteriophages with X-rays suspended in nutrient broth, and by ultracentrifugation, ultrafiltration and electrone microscopy.

Watson (1950, 1952) studied the properties of the T group bacteriophage of *Escherichia coli* after irradiation by X-rays and found them to be quite different when the X-rays produced the direct or the indirect effect.

The aim of the work reported on here was to study the effect of gamma rays on the inactivation of bacteriophages when exposed to irradiation in two different media: nutrient broth 10% (Difco) and T_1 medium (Matsushiro *et al.*, 1964).

Materials and methods

The experiments included four bacteriophages (Pg10-1, PgII-2, PgIII-4 and PgI-9) of *Pseudomonas glycinea* and six bacteriophages (Xpg30-3, Xpg5-5, XpgI-6, Xpg10, Xpg11 and Xpg8-12) of *Xanthomonas phaseoli* var. *sojense*. Corresponding strains of *Pseudomonas glycinea* Pg1, Pg4, Pg2 and Pg30, and of *Xanthomonas phaseoli* var. *sojense* X2, X18, X24 and X55 were used for preparing the lizat of bacteriophages and plaque essays.

Table 1. Bacteriophage titers before and after a maximal irradiation dose of gamma rays (5×10^5 rad).

Bacteriophage	Before irradiation		After irradiation	
	nutrient broth	T_1 medium	nutrient broth	T_1 medium
Pg10-1	6.8×10^9	1.4×10^{10}	1.7×10^5	8.0×10^4
PgII-2	4.3×10^9	1.0×10^{10}	4.0×10^4	3.9×10^4
PgIII-4	6.4×10^9	1.2×10^{10}	3.9×10^3	1.0×10^4
PgI-9	4.0×10^8	9.0×10^9	0	0
Xpg30-3	2.5×10^8	5.7×10^8	1.0×10^1	0
Xpg5-5	1.8×10^9	1.2×10^9	1.6×10^5	2.6×10^4
XpgI-6	1.0×10^8	2.0×10^8	1.5×10^5	7.0×10^3
Xpg10	1.5×10^9	3.4×10^8	5.8×10^4	2.0×10^3
Xpg11	1.7×10^8	1.6×10^8	1.2×10^3	1.0×10^3
Xpg8-12	3.6×10^7	7.8×10^7	1.0×10^3	5.0×10^2

Table 2. Inactivation dose of bacteriophages irradiated with gamma rays in nutrient broth and T_1 medium.

Bacteriophage	Dose in relation $N/N_0 = 1/e$ in rad	
	nutrient broth (10%-Difco)	T_1 medium
Pg10-1	41.312	34.984
PgII-2	41.448	27.787
PgIII-4	39.637	30.915
PgI-9	15.115	14.455
Xpg30-3	53.957	15.739
Xpg5-5	54.745	46.836
XpgI-6	74.730	47.043
Xpg10	51.535	40.750
Xpg11	43.435	41.850
Xpg8–12	52.143	41.667

The nutrient broth served to study the direct effect of gamma rays on the bacteriophages. The T_1 medium, containing 0.15 g $MgSO_4 \cdot 7H_2O$, 5×10^{-4}M $CaCl_2$, and 6×10^{-3}M Tris buffer (pH 7.3) per liter of water without gelatine, was used to study the indirect effect of gamma rays on bacteriophages.

The phages were purified by differential centrifugation (three times with 40.000 rpm, ultracentrifuge 'Hitachi'-55 PA for 30 minutes, and three times with 4000 rpm for 15 minutes) with addition of 1 μg DN-ase, RN-ase and trypsin within one hour.

Prior to irradiation, 0.5 ml of each bacteriophage was suspended in five small test tubes in nutrient broth or in five tubes of T_1 medium. The tubes were closed with a cotton plugh.

The source of gamma rays was ^{60}Co with an emission of 1.240 rad per minute and supplied through filters to remove other kinds of radiation.

Every 80 minutes one series of irradiated bacteriophage was used, the last ones after 6 hours and 40 minutes; the maximum dose of irradiation was 5×10^5 rad.

The bacteriophage titer was determined by the plaque count method (Adams, 1959) with five repetitions; the average number of plaques in one Petri dish was 250. The titer before and after a maximal dose of irradiation (5×10^5 rad) is shown in Table 1. The data were calculated with the aid of the 'Gamma 30' Bull computer, of the ERC 'Energoinvest' at Sarajevo, Yugoslavia.

The trends were calculated according to the exponential equation $y = AB^x$, the inactivation dose D_{37} as proposed by Lea (1962); they both are presented on semi-logarithmic paper in Figures 1 and 2 (see also Tables 1 and 2).

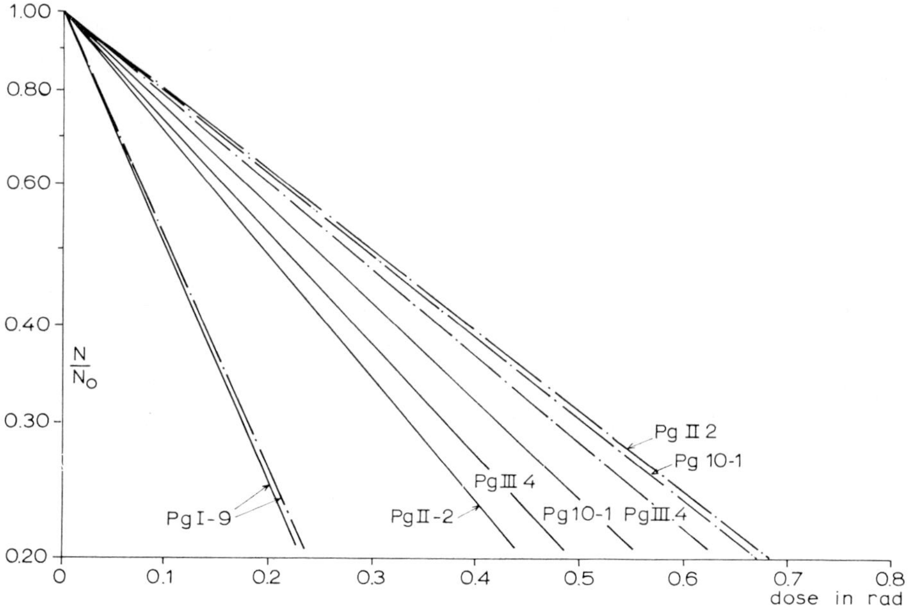

Fig. 1. The effect of γ-rays on bacteriophages of *Pseudomonas glycinea*.
—— = T1 media; —·—· = nutrient broth (10%) Difco.

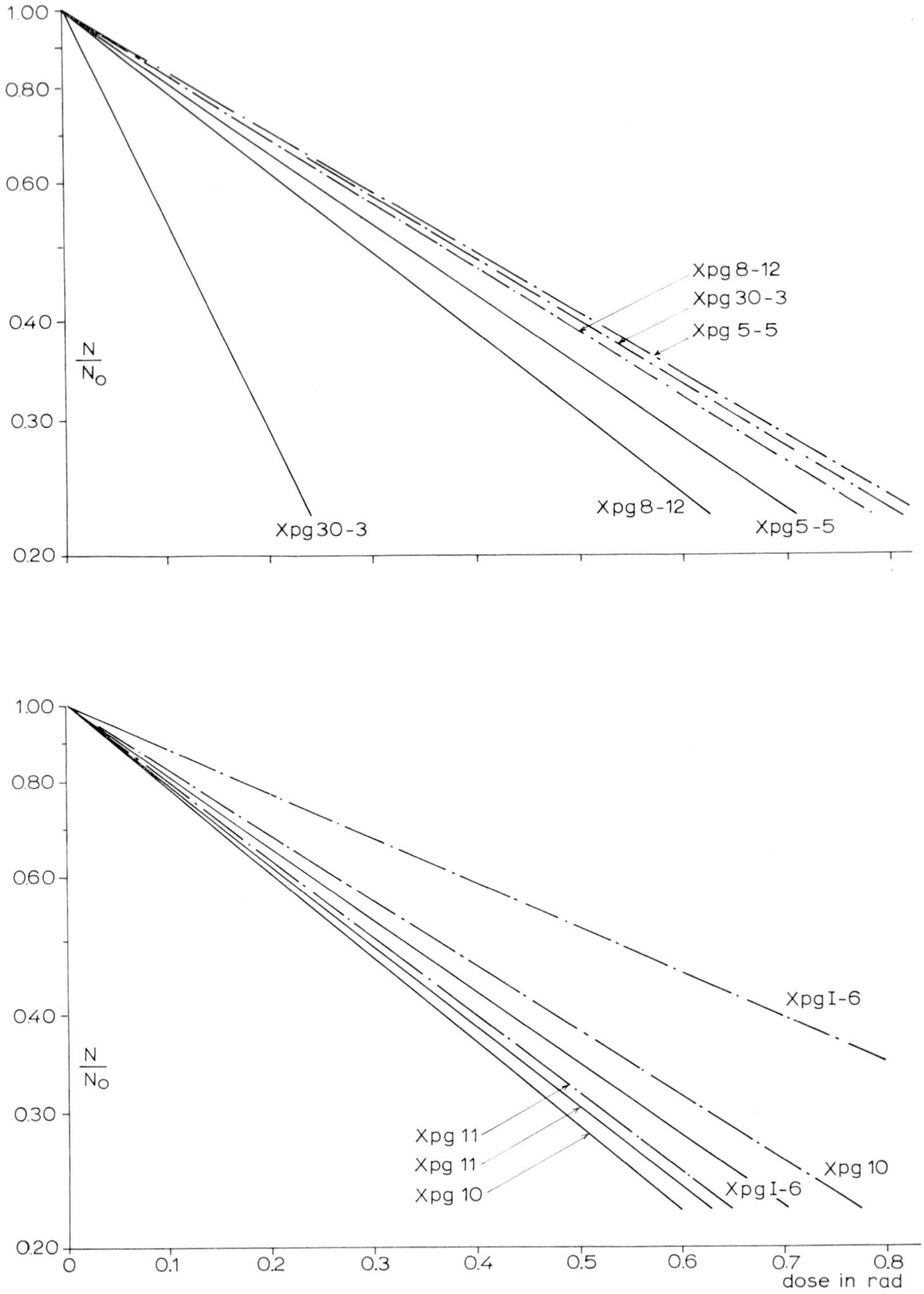

Fig. 2. The effect of γ-rays on bacteriophages of *Xanthomonas phaseoli* var. *sojense*. —— = T_1 media; —·—· = nutrient broth (10%) Difco.

Results

The bacteriophage PgI-9 of *Pseudomonas glycinea* was completely inactivated when irradiated, both in nutrient broth and T_1 medium, with the maximum dose of irradiation of 5×10^5 rad, meaning that the presence of bacteriophage particles able to produce plaques was not detectable. The bacteriophage Xpg30-3 of *Xanthomonas phaseoli* var. *sojense*, when irradiated with a maximal dose of 5×10^5 rad was completely inactivated only if suspended in the T_1 medium.

The inactivation curves of the bacteriophages showed different slopes, no matter in which media they were suspended. The greatest difference in the slope of the inactivation curve was found in the bacteriophage Xpg30-3 of *Xanthomonas phaseoli* var. *sojense*, where it was very steep when irradiated in T_1 medium, as could be expected from the size of the particles (150 $\times$ 17 um).

The bacteriophage XpgI-6 of *Xanthomonas phaseoli* var. *sojense* showed the highest inactivation dose, both in nutrient broth and T_1 medium. The bacteriophage PgI-9 of *Pseudomonas glycinea* had the lowest inactivation dose also in both media.

The inactivation dose is considerably higher if bacteriophages are irradiated in nutrient broth, while the bacteriophages PgI-9 of *Pseudomonas glycinea* and Xpg11 of *Xanthomonas phaseoli* var. *sojense* make an exception, their inactivation dose being only slightly higher when irradiated in nutrient broth.

Discussion

The obtained results show that the effect of gamma rays on the inactivation of infectivity of the bacteriophages of *Pseudomonas glycinea* and *Xanthomonas phaseoli* var. *sojense* is twofold. If bacteriophages are exposed to the irradiation of gamma rays in nutrient broth, the gamma rays produce only the direct effect; when exposed in the T_1 medium they produce both the direct and the indirect effect. This is in agreement with earlier results from the investigations on bacteriophages of *Escherichia coli* in different media and with X-ray irradiation (Luria & Exner, 1941; Wollman & Lacassagne, 1940; Lea & Salaman, 1946; Watson, 1950, 1952).

The bacteriophages PgI-9 and Xpg11, when exposed to the irradiation of gamma rays both in nutrient broth and T_1 medium, show a small difference in inactivation dose. This can be a consequence of the reaction of their DNA, but this needs further study.

The mutational changes of the irradiated bacteriophages will be considered in a future paper.

References

Lea, D. E., 1962. Action of radiation on living cells. Cambridge Univ. Press, Cambridge, England.

Lea, D. E. & M. H. Salaman, 1946. Experiments on the inactivation of bacteriophages by radiations, and their bearing on the nature of bacteriophages. Proc. Roy. Soc. London 133B: 434.

Luria, S. E. & F. M. Exner, 1941. The inactivation of bacteriophages by X-rays – Influence of the media. Proc. natl. Acad. Sci. USA 27: 370.

Matsushiro, A., S. Koki & Kida Setsuko, 1964. Characteristics of the transducing elements of bacteriophage 80. Virology 23(3): 299.

Wollman, E. & A. Lacassagne, 1940. Recherches sur le phénomène de Twort-d'Herelle. VI. Evalution des dimensions des bactériophages au moyen des rayons X. Ann. Ins. Past. Paris 64: 5.

Wollman, E., P. Holweek & S. Luria, 1940. Effect of radiations on bacteriophage C_{16}, Nature, Lond. 145: 935.

Watson, J. D., 1950. The properties of X-rays inactivated bacteriophages. I. Inactivation by direct effect. J. Bact. 60: 697.

Watson, J. D., 1952. The properties of X-rays inactivated bacteriophages. II. Inactivation by indirect effects. J. Bact. 63: 473.

Variation in inoculum level of Pseudomonas mors-prunorum persicae on the leaf surface of peach trees

L. Gardan, J. Luisetti and J.-P. Prunier

Institut National de la Recherche Agronomique, Station de Phytobactériologie, 49 – Beaucouze, France

Abstract

Since 1967, severe damage to peach trees has been caused in France by a pseudomonad non-fluorescent on King B medium. Though its biochemical, serological and phage typing characters place it near *Pseudomonas syringae* and *P. mors-prunorum*, it was provisionally named *P. mors-prunorum persicae*.

The study of its occurrence on leaf surfaces has proved that the bacterium is solely present in diseased orchards. In spring as many as 10^5 to 10^7 specimens may be present per leaf, in summer it is below 10^4, in autumn it rises again (10^4–10^5). This corresponds with the following life cycle: during the growing season the bacterium is on the leaf surface, later on it enters the plant through the leaf scars, in winter it multiplies in the tissues of the host.

Introduction

In 1967, our attention was drawn to a new bacterial disease on peach trees in the valley of the river Rhône. Its main centres were located during the winter 1968–1969 (Gardan *et al.*, 1969). Its symptomatology (Fig. 1) can be described as follows (Vigouroux, 1970; Prunier *et al.*, 1970).

In winter, the dormant buds appear as if soaked with water. In January and February the shoots show easily recognizable blackish spots, during blossom time changing into reddish brown cankers which cause the death of the shoots and, eventually, the whole tree. In addition, in spring necrotic spots appear on the leaves, later on changing into holes causing early fall of the leaves. In spring and summer the fruits show superficial damages with abundant gummosis.

The culprit appeared to be a pseudomonad, non-fluorescent on King B medium which, though according to its biochemical, phage typing and serological characteristics closely allied to *Pseudomonas syringae* and *P. mors-prunorum*, was provisionally named *Pseudomonas mors-prunorum persicae*.

As the points of attack were supposed to be in the dormant buds, the biology of this canker was compared with that of the bacterial canker of the cherry tree. Artificial inoculations indeed proved that the leaf scar was the place of entrance (in autumn). In 1969 and 1970 the occurrence and behaviour of the pseudomonad on leaf surfaces was studied; the results will be discussed here.

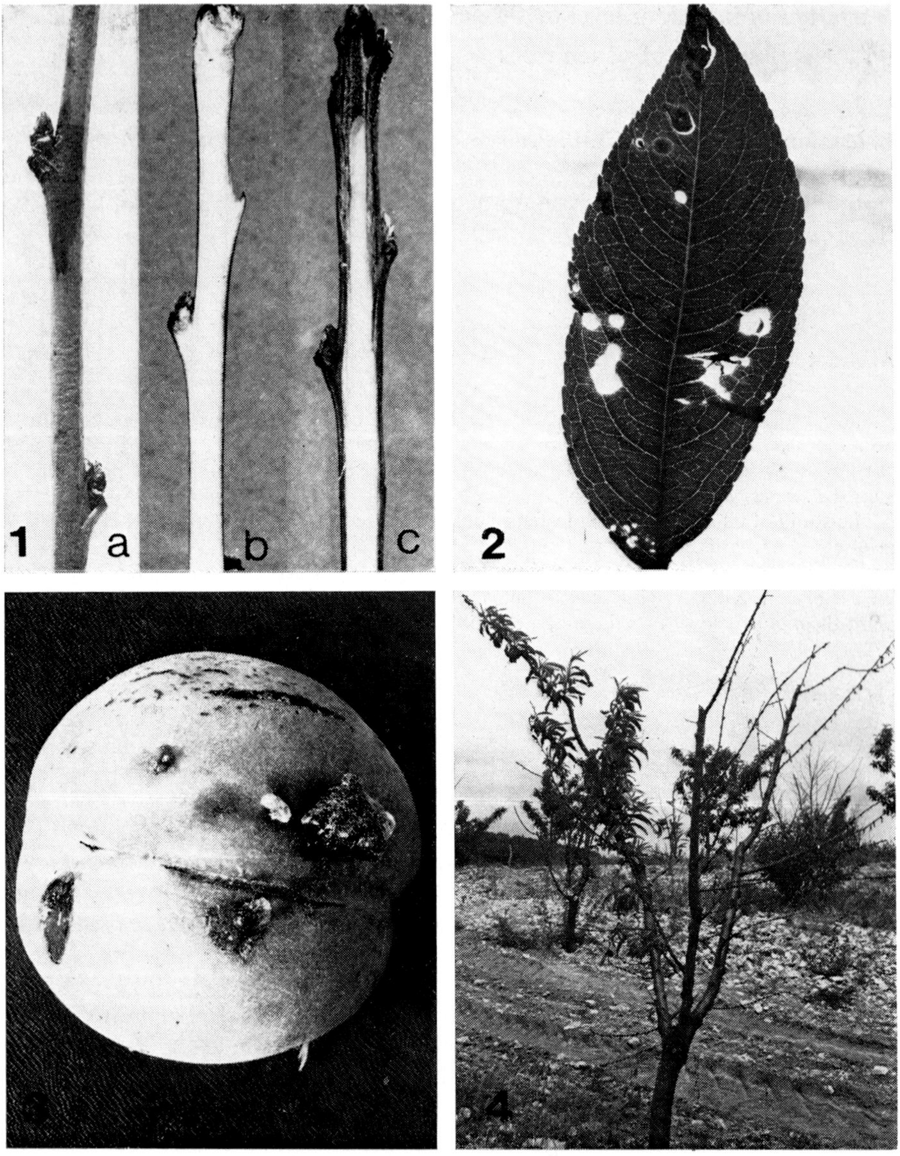

Fig. 1. Symptoms of the bacterial disease:
1. (a) Necrosis around a dormant bud, (c) on the tip of young shoot, (b) healthy shoot.
2. Leaf spot and leaf hole on leaf.
3. Superficial necrosis and guminosis on fruit.
4. A severely attacked tree.

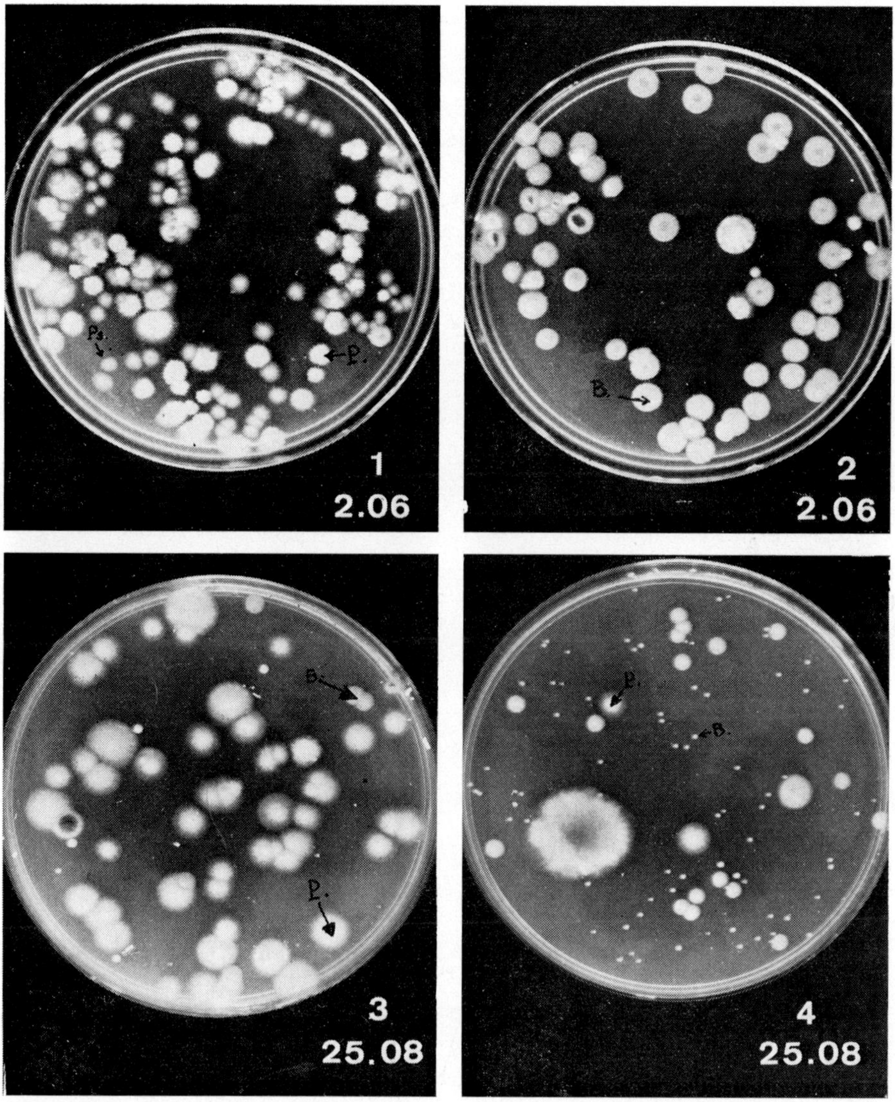

Fig. 2. Results of plate cultivation with material from
diseased orchard: analysis on June 2 (1) and August 25 (3)
healthy orchard: analysis on June 2 (2) and August 25 (4)
Ps = *Pseudomonas mors-prunorum persicae*, P = *Pullularia pullulans*, b = bacterium, F = fungus.

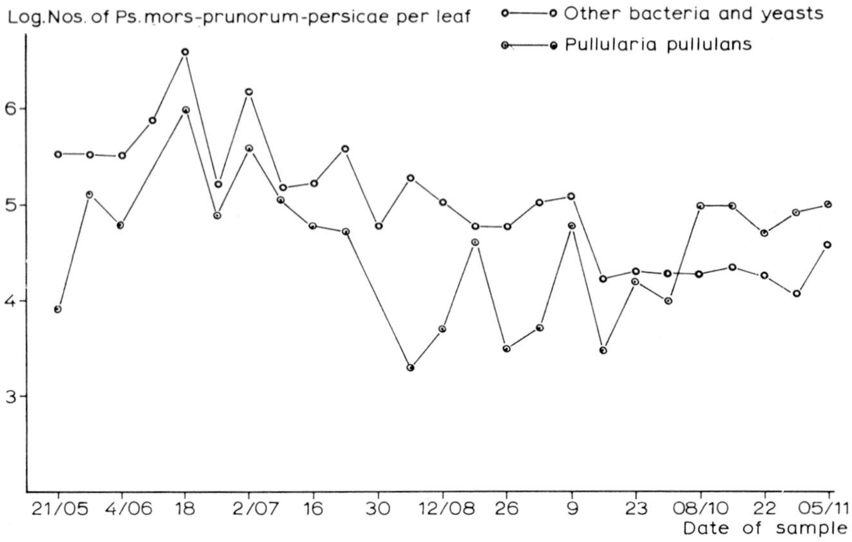

Fig. 3. Log of numbers of *P. mors-prunorum persicae* isolated from leaf surface of J. H. Hale (healthy variety). 1969.

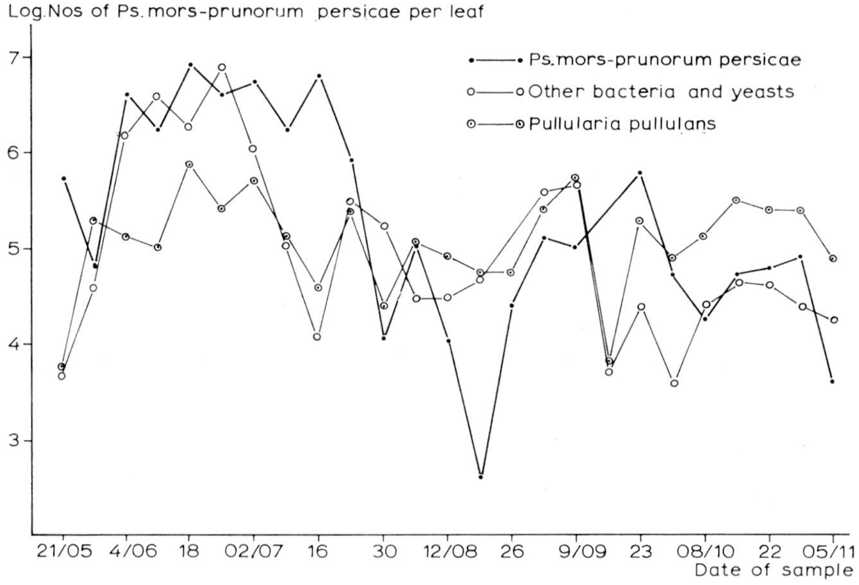

Fig. 4. Log of numbers of *P. mors-prunorum persicae* isolated from leaf surface of J. H. Hale (diseased variety). 1969.

Material and methods

Peach trees belonging to the cultivar JH HALE were studied in two orchards: (1) four years old trees severely attacked during the winter 1968–1969, (2) three years old trees, 2 km from the first, which had never shown symptoms of bacterial disease.

Every week a sample of 50 leaves (in 1969) or 64 leaves (1970) was taken from four trees 1.50 m above the soil, according to the directions north, south, east and west.

To obtain a bacterial suspension, the technique applied by Crosse (1959) as modified by Luisetti & Paulin (1971) for the study of the epiphytic microflora of pear trees was used. The leaves were shaken in a one litre Erlenmeyer with peptoned water 1g/l. After 4 hours, 1/20 ml was sprayed over a petri dish with a medium containing 10g dextrose, 5g Difco proteose pectone No. 3 and 5g Difco yeast extract per liter.

After 65 h incubation at 25 °C the colonies showed to be different from those of other pseudomonads: their diameter was about 2 mm, they were creamy white and had an irregular outline (Fig. 2). For an analysis of these colonies one at a time was taken out and submitted to the biochemical tests described by Lelliot *et al.* (1966). Three groups of germs were counted: *Pseudomonas mors-prunorum persicae*, other bacteria and yeast, and *Pullularia pullulans*; the results were expressed as numbers of organisms per leaf.

Results

During the growing season of the year 1969 (see Fig. 3), no pathogenic bacteria could be detected on the leaves from the healthy orchard. The number of other bacteria and yeast cells varied between 7.7×10^4 (May 21) and 4.7×10^6 (September 9); from September 16 on it remained at 2×10^4. *Pullularia* varied more; its number was lower than that of the bacteria, except in autumn.

In the infected orchard (see Fig. 4), the pathogen was always present on the leaves, in high numbers between June 4 and July 23 (over 10^6), with an important decrease in August (10^3). In September it rose again to 10^5–10^6 and it stabilized during fall at 10^4–10^5. The number of other bacteria and yeasts was nearly always lower than or equal to that of the pathogen, except in August. The number of Pullularia's varied around 10^5, except that it was distinctly lower in August.

During the growing season of the year 1970, the healthy orchard (see Fig. 5) showed a low number of bacteria and yeasts up till June 9; than it gradually rose to steady again between 10^5 and 10^6. The number of *Pullularia* showed a reduction between July 7 and August 25.

In the infected orchard (Fig. 6) the pathogen was present up till July 14; afterwards the applied method did not allow the demonstration of its presence. It returned at the end of September. The other bacteria and the yeasts dominated up till the end of July, then their number varied between 10^4 and 10^5. From August 11 on their number was again greater than that of the pseudomonads.

These data indicate that the pathogen greatly decreases in number or totally disappears during some summer months, but that it returns in autumn. At fall it is

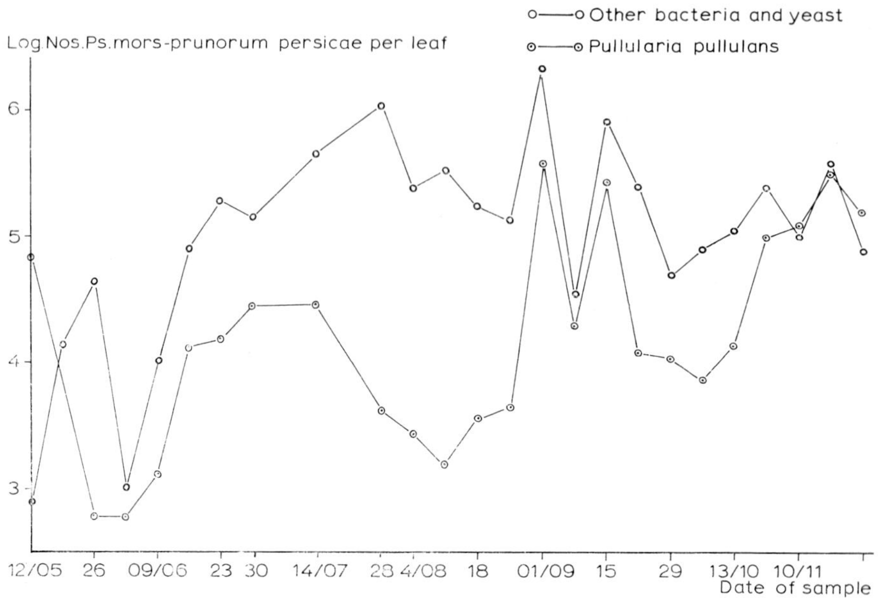

Fig. 5. Log of numbers of *P. mors-prunorum persicae* isolated from leaf surface of J. H. Hale (healthy variety). 1970.

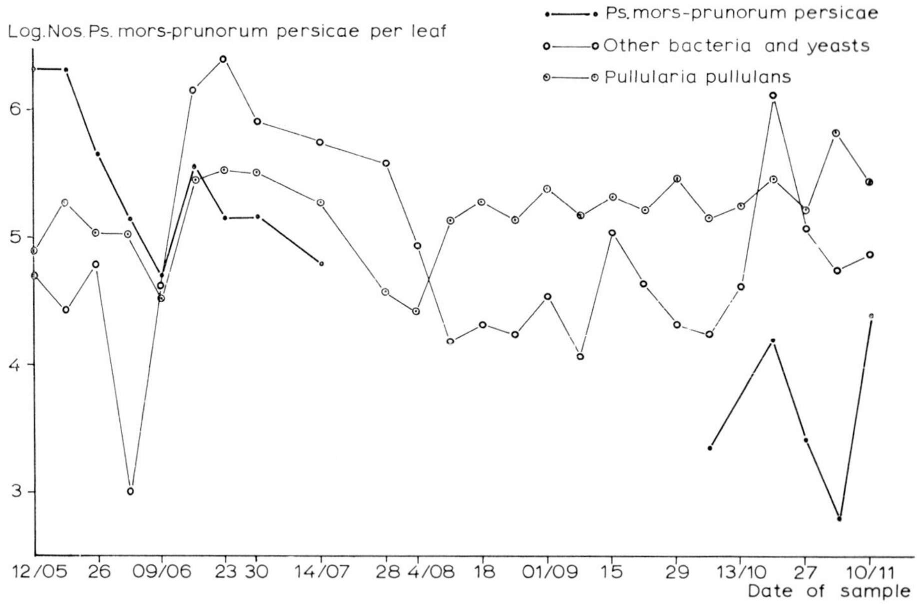

Fig. 6. Log of numbers of *P. mors-prunorum persicae* isolated from leaf surface of J. H. Hale (diseased variety). 1970.

always present to infect leaf scars. On the other hand, *Pullularia pullulans* is a normal component of the peach tree's epiphytic flora. As are the other bacteria and yeasts, their numbers are always high, but their composition changes during the season.

Further data on the microflora

Several other pseudomonads were observed, among them *Pseudomonas viridiflava* (on flowers and leaves in spring and autumn) and two forms of *P. syringae* (group Ia in Lelliot's classification of 1966) of which one resembling *P. mors-prunorum persicae* in serological and phage typing characters.

In summer, a chromogenic flora, mainly consisting of yellow bacteria dominates.

According to our observations, and in agreement with those mentioned by Panagopoulus (1966), abrupt changes in the numbers of pathogens and other micro-organisms may occur within a short period (seven days). This is due to differences in the atmospheric conditions, especially in the temperature, the air humidity and the radiation: the year 1969 had a cold and rainy spring and a wet summer, 1970 had a dry summer. Studies under controlled conditions still have to confirm this conclusion.

The bacterial and yeast flora of tree leaves has never been systematically studied, except by Jenssen (1970) for beech. On peach trees its composition is rather simple, though the organisms may play an important role by interacting with the pathogen. This may be of fundamental importance, especially in view of the effects of pesticides, also in connection with wheather conditions.

Discussion

Crosse (1956) found, in autumn, comparable numbers of *P. mors-prunorum* on cherry tree leaves. English & Davis (1960) proved the presence of *P. syringae* on leaves and fruits of peach trees attacked by bacterial decay, but, though stating that the number of bacteria was higher in winter and in spring than in autumn, they did not supply quantitative data. Luisetti *et al.* (1971) found 10^6 to 10^7 cells of *P. syringae* on flowers and leaves of pear trees in spring, 10^2 to 10^3 per leaf in summer. Panagopoulos (1966) counted 10^3 to 10^5 specimens of *P. syringae* per cm^2 leaf surface of lemon and orange trees in winter and spring.

These observations differ from ours in two aspects:

1. we never found *P. mors-prunorum persicae* on healthy trees in healthy orchards,
2. our numbers of pathogens in affected orchards were high both in spring and in autumn.

These inconsistencies may be explained as follows.

The fast multiplication of the inoculum in winter in the exudate of the injuries results in contamination of flowers and leaves at blossom time. The increase of the pathogen in the leaf tissues in spring keeps up the high population. Though varying in numbers, the bacteria remain present there till autumn, than penetrate the leaf scars: 50 specimens are sufficient to cause infection (Luisetti *et al.*, 1971), a number guaranteed by the presence of 10^4–10^5 bacteria per leaf.

References

Crosse, J. E., 1956. Leaf scar infection of cherry. J. hort. Sci. 51: 212–224.

Crosse, J. E., 1959. Investigation of a method for measuring the inoculum potential of cherry trees. Ann. appl. Biol. 47: 306–317.

Crosse, J. E., 1962. A comparison of leaf surface populations of *Pseudomonas mors-prunorum* in autumn of two cherry varieties. Ann. appl. Biol. 52: 97–104.

English, H. & J. R. Davis, 1960. The source of inoculum for bacterial canker and blast of stone fruit trees. Phytopathology 50: 634 (abstr.).

Gardan, L., J.-P. Prunier & J. Luisetti, 1969. Premiers résultats concernant l'extension de la bactériose du pêcher. Arboric. fruit. 190: 25–30.

Gardan, L., J.-P. Prunier & J. Luisetti, 1971. Recherche de *Ps. mors-prunorum persicae* en surface des organes aériens du pêcher et étude de ses variations quantitatives. Ann. Phytopath. (in print).

Jensen, V., 1970. The bacterial flora of beech leaves. Symp. on Ecology of Leaf Surface 7–11 Sept. 1970 (in print).

Lelliot, R. A., E. Billing & A. C. Hayward, 1966. A determinative scheme for the fluorescent plant pathogenic pseudomonads. J. appl. Bact. 29: 470–489.

Luisetti, J. & J.-P. Paulin, 1971. Recherche de *Pseudomonas syringae* en surface des organes aériens du poirier et étude de ses variations quantitatives. Ann. Phytopath. (in print).

Luisetti, J., L. Gardan & J.-P. Prunier, 1971. Étude expérimentale du pouvoir pathogène de *Pseudomonas mors-prunorum persicae*, agent du dépérissement bactérien du pêcher. Ann. Phytopath. (in print).

Panagopoulos, C. G., 1966. Studies on the source of inoculum for blast and black pit of citrus. Proc. 1st Congr. Medit. Phytopath. Union: 340–345.

Prunier, J. P., J. Luisetti & L. Gardan, 1970. Caractérisation d'un *Pseudomonas* non fluorescent agent d'une bactériose nouvelle du pêcher. Ann. Phytopath. 2: 168–197.

Vigouroux, A., 1970. Une nouvelle bactériose du pêcher: description, étiologie, développement du parasite. Ann. Phytopath. 2: 155–167.

The ultrastructure of potato tubers infected by the soft rot organism Erwinia carotovora var. atroseptica[1]

R.T.V. Fox,

ICI Plant Protection Ltd, Jealott's Hill Research Station, Bracknell, Berkshire, England

Abstract

Bacterial entry into potato tubers is affected by the environment, particularly the relative humidity. Light and electron microscopy, and physiological studies, show that resistance to invasion is due to the production of a barrier of suberin in the intercellular spaces and to two bands in the cell wall. If complete, the suberin barrier prevents invasion, but if no so, pectolysis spreads through the gaps and then the bacteria bypass the suberin barrier. The periderm is resistant to bacterial invasion due to the lack of both pectin and intercellular spaces. The bacteria are completely intercellular. Infection of xylem and phloem is restricted by the suberised vascular parenchyma, although localised pectolysis and melanin formation occurs. Tissue degradation involves destruction of the cytoplasm as well as cell wall maceration. Cell membranes are ruptured, together with an increase in size of the microbodies. The effect of the environment on the physical barriers is complex. Observations on the formation of suberin and melanin at the infection interface indicate that the suberin may hinder bacterial spread; the role of the melanin is obscure.

Introduction

It is known that *Erwinia carotovora* var. *atroseptica* can enter potato tubers via fresh wounds and lenticels (Smith, 1905; Smith & Ramsey, 1947), but that the periderm is resistant (Harper *et al.*, 1963). The physical factors regulating bacterial entry through these channels are not fully understood. Electron microscopy of tubers during resistant and susceptible phases of wound healing or lenticel development provides a means of elucidating the fundamental effects of these factors. The actual mode of entry into and the invasion of the potato tissues has not before been studied at the ultrastructural level.

Maceration of potato cells by *Erwinia* spp. has been well-defined biochemically and reviewed by Wood (1967). Many workers have described separate symptoms of potato soft rot using light microscopy, maceration (Smith, 1905), cytoplasmic damage

1. The work on this subject was carried out at the Department of Botany, University of Southampton, on a Potato Marketing Board Post Graduate Course. I want to thank Dr J. G. Manners and Dr. A. Myers for the help and encouragement given to me, and Mr J. Baile for technical assistance with the electron microscopy. I should also like to thank Plant Protection Limited for the help and encouragement given in presenting this demonstration.

(Artschwager, 1927), osmotic damage (Tribe, 1955) and changes in nuclear size (Butler & Jones, 1943). Both suberisation (Rudd Jones & Dowson, 1950) and polyphenol oxidation (Lovrekovich *et al.*, 1967) have been suggested as host defence mechanisms.

The investigations on which this paper is based (Fox, 1969) attempted to bridge the gap between the biochemical and light microscopy aspects of soft rot research, and to clarify the nature of the host defence mechanisms.

Materials and methods

Potato tubers (cv. King Edward), previously stored at 5 °C and 78 % relative humidity, were surface sterilised with 0.1 % $HgCl_2$, and rinsed in sterile distilled water. Physiological studies (Fox *et al.*, 1971) were carried out to determine the effect of pre-inoculation variation in relative humidity on lenticel entry, length of exposure time between wounding and subsequent inoculation together with suberisation and periderm formation processes, and other effects of the environment on the entry of bacteria into the potato (Fox, 1969). An 18 h culture of strain NCPPB 138 was used to inoculate tubers as a suspension containing either approximately 10^9 or 10^{11} cells/ml for various periods under controlled sterile conditions. At the same time samples of infected tissue were harvested with a razor blade and portions were (a) treated with 1 % 2,3,8-tripheny tetrazolium chloride (TTC) to determine the approximate extent of infection (Lovrekovich *et al.*, 1967), (b) fixed in Randolph's CRAF and embedded in ester wax (Steedman, 1960), (c) prepared for electron microscopy by fixing overnight in 1 % osmic acid in Kellenberger buffer with 5 % sucrose and embedding in 'Araldite' with the 1,2-epoxypropane/'Araldite' phase being prolonged to 7 days at 20 °C, (d) in certain cases samples were fixed and stained for pectinate using Albersheim's method (Albersheim *et al.*, 1960).

Results

Bacterial entry through lenticels Post-treatments at different relative humidities did not appear to have lasting effects, but different pretreatments followed by inoculation for 10 min resulted in different degrees of infection. At relative humidities below 80 % the spread was very limited; at higher relative humidities extensive invasion occurred. Under the electron microscope the intercellular spaces of filling tissue of lenticels stored at low relative humidities and the periderm were more or less completely blocked by suberin (Figs 1, 2 and 3). This was not true of the phelloderm which appeared more lightly suberised as judged by the ammoniacal gentian violet test of Artschwager (1927) and electron microscopy (Figs 4 and 5). Plastid clusters (Fig. 6) were typically present in such suberising cells. Under 100 % relative humidity the lenticel may proliferate phelloderm cells (Fig. 7), which are highly susceptible to bacterial entry (Bétancourt & Prunier, 1965), these cells are highly vacuolate and contain melanin; suberin was not present in the large intercellular spaces in any quantity (Fig. 8). Lenticels are individually more variable as channels for bacterial entry than fresh wounds (Fox *et al.*, 1971), due perhaps to varying microclimates during tuber development and storage.

Bacterial entry through wounds The susceptibility of fresh wounds is limited by the development of suberin rather than by that of wound periderm. The suberisation and wound periderm formation occur more readily at high temperatures or high relative humidities than at lower values (Fox *et al.*, 1971). As in lenticels, the suberin is present as two bands in the cell wall (Fig. 9) and as a block in the intercellular region. Clusters of plastids similar to those in the suberising cells of the lenticel occur (Fig. 10).

The periderm as a resistant tissue The periderm is resistant to soft rotting (Wood, 1967). By the use of Albersheim's method (Albersheim *et al.*, 1960) to demonstrate pectinic substances in the cell wall, it can be seen that the periderm has only a thin layer of such substances and probably other pectic substances (Fig. 11), and few intercellular spaces compared to storage parenchyma. It is also highly suberised, and the cells in the periderm and below contain polyphenols. The distribution of suberin may be causally related to that of polyphenolics. As polyphenol-rich cells are present in the highly susceptible proliferated lenticel, it appears unlikely that these substances act as a direct barrier to *Erwinia* (Fox *et al.*, 1971).

The location of the bacteria Throughout the investigation of all stages of colonisation of the tuber, infection is almost entirely intercellular. At early stages of infection within the intercellular space, a central zone can be seen occupied by the bacteria, an intermediate zone, and an outer zone comprising the remains of the cell wall (Figs 12, 13 and 14). This zonation is lost in older infections (Figs 15, 16). When both intercellular space and cell wall suberisation are incomplete, suberin blockages may be by-passed without the aid of a specific enzyme system by pectolysis of the surrounding cell wall, thus permitting further invasion of the host (Figs 17, 18). Limited intracellular colonisation may occur in the vascular elements and calcium oxalate cells of the phelloderm.

Cell wall and cytoplasmic damage The middle lamella region of the cell wall is destroyed by pectolysis but starch grains are apparently not utilised (Fig. 19). Other cell contents, apart from microbodies, are degraded (Fig. 20) and the series of changes is complex (Fox, 1969). The intercellular spaces are enlarged and pectinates removed as demonstrated by the use of Albersheim's method (Figs 21, 22).

Host reaction Under aerobic conditions both suberin and melanin are produced at the infection interface (Figs 23, 24); under anaerobic conditions only suberin is laid down. Under both environmental conditions a reduction in the rate of spread occurs after 40 hours, regardless of whether oxidised polyphenols are present or not (Fox, 1969). Electron microscopy demonstrates the presence of melanin granules in the tonoplast (Fig. 25), suberin present as two bands in the cell wall and blocking the intercellular spaces (Figs 26, 27), and clusters of plastids (Fig. 28) in cells at the infection interface. These cells appear to have abundant relatively undamaged cytoplasmic contents, including the 'myelin' bodies seen in potato bud tissue by Marinos (1967).

References

Albersheim, P., K. Mühletaler & A. Frey-Wyssling, 1960. Stained pectin as seen in the electron microscope. J. biophys. biochem. Cytol. 8: 501–516.

Artschwager, E. F., 1927. Wound periderm formation in the potato as affected by temperature and humidity. J. agric. Res. 35: 995–1000.

Bétencourt, A. & J. P. Prunier, 1965. À propos de la parriture sèche lenticellaire des tubercules de pommes de terre provoquée par *Erwinia carotovora* (Jones) Holland. Eur. Potato J. 8: 230–242.

Butler, E. J. & J. G. Jones, 1949. Plant Pathology. Macmillan, London.

Fox, R. T. V., 1969. Ph.D. thesis, University of Southampton.

Fox, R. T. V., J. G. Manners & A. Myers, 1971. Ultrastructure of entry and spread of *Erwinia carotovora* var. *atroseptica* into potato tubers. Potato Res. 141: 61–73.

Harper, P. C., A. E. W. Boyd & D. C. Graham, 1963. Growth cracking and bacterial soft rot in potato tubers. Pl. Pathol. 12: 139–142.

Jones, D. R. & W. J. Dowson, 1950. On the bacteria responsible for soft rot in stored potatoes and the reaction of the tuber to invasion by *Bacterium carotovorum* (Jones) Lehmann et Neumann. Ann. appl. Biol. 37: 563–569.

Lovrekovich, L., H. Lovrekovich & M. A. Stahmann, 1967. Inhibition of phenol oxidation by *Erwinia carotovora* in potato tissue and its significance in disease resistance. Phytopathology 57: 737–742.

Marinos, M. G., 1967. Multijunctioned plastids in the meristematic region of potato tuber buds. J. ultrastruct. Res. 17: 91–113.

Smith, E. F., 1950. Bacteria in relation to plant diseases. Carnegie Inst. Washington, DC.

Smith, M. A. & G. B. Ramsey, 1947. Bacterial lenticel infection of early potatoes. Phytopathology 37: 225–242.

Tribe, H. T., 1955. Studies in the physiology of parasitism. XIX. On the killing of plant cells by enzymes from *Botrytis cinerea* and *Bacterium aroideae*. Ann. Bot. Lond., N.S. 19: 351–368.

Wood, R. K. S., 1967. Physiological plant pathology. Botanical Monographs 6, Blackwell, Oxford.

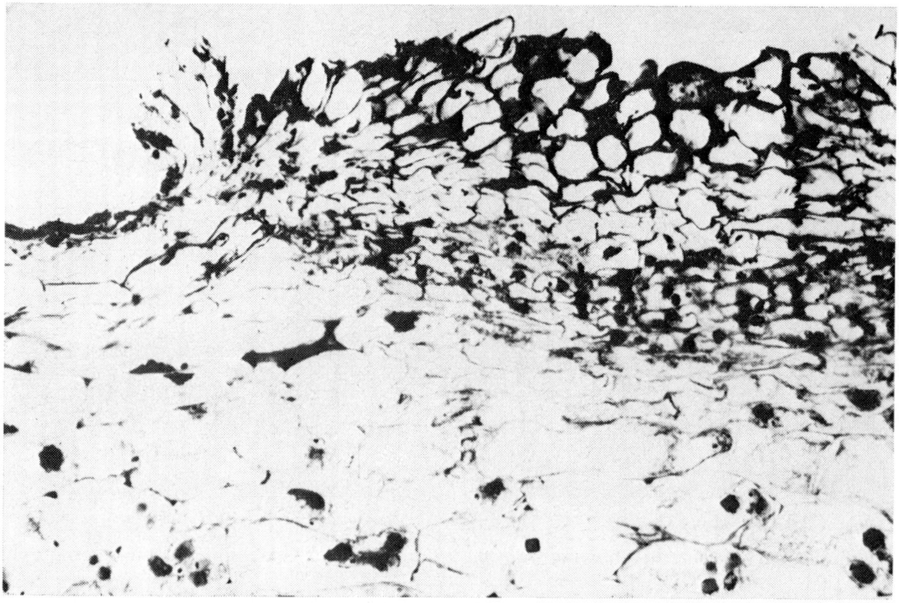

Fig. 1. Lenticel from potato tuber stored at low relative humidity (78%). Light micrograph, × 145.

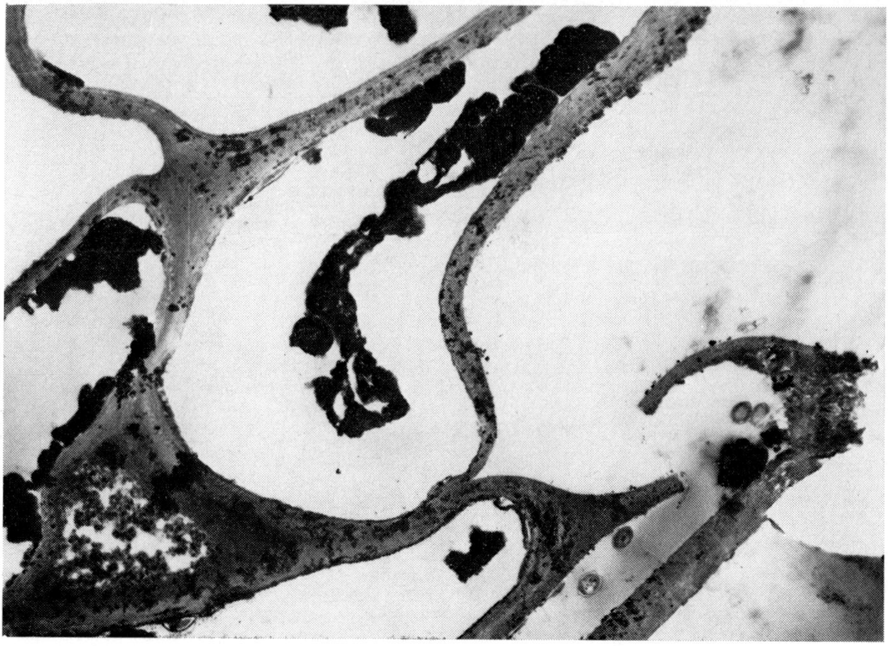

Fig. 2. Filling tissue cells from lenticel stored at low relative humidity (78%) showing suberin in intercellular spaces. Electron micrograph, × 7.500.

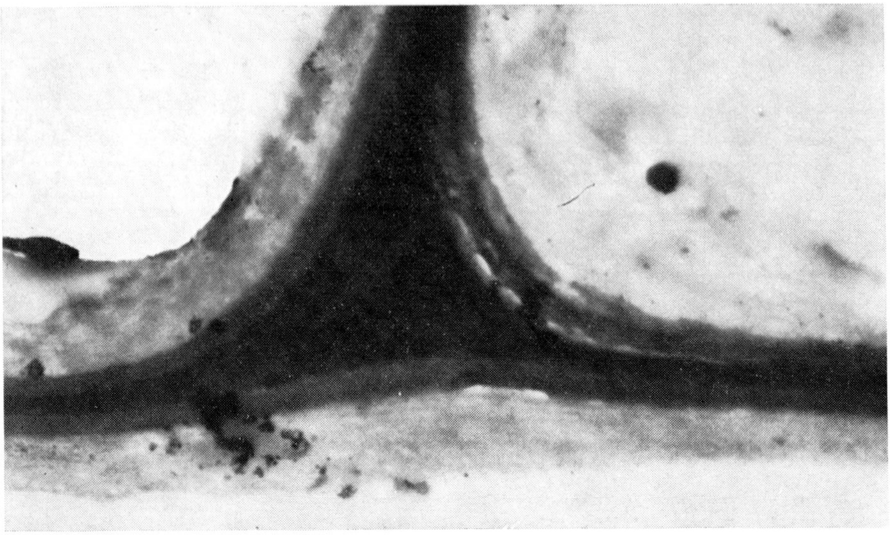

Fig. 3. Primary periderm showing suberin in intercellular spaces. Electron micrograph, × 11,500.

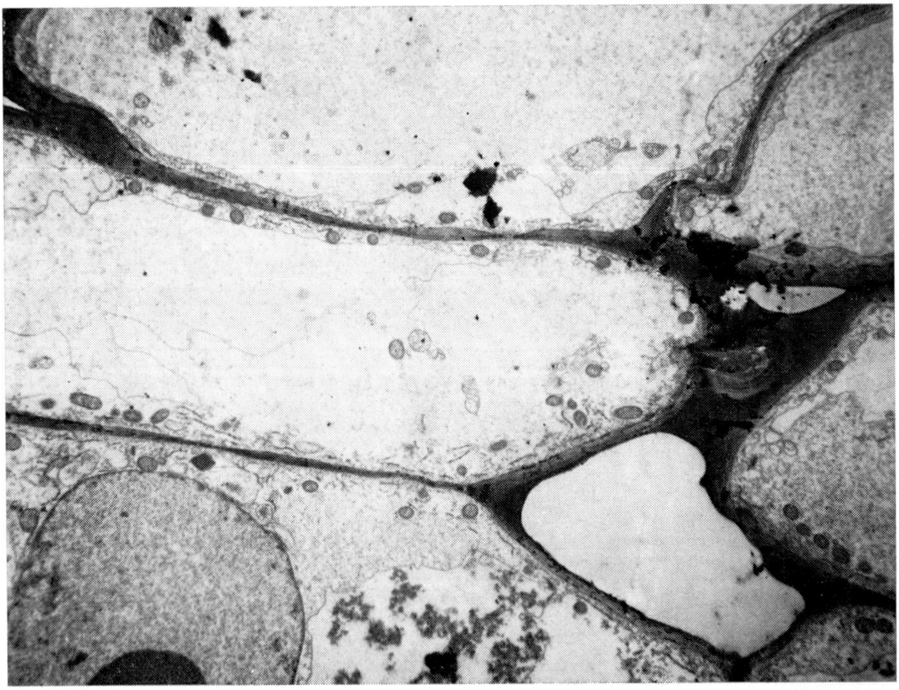

Fig. 4. Phelloderm from lenticel stored at low relative humidity, showing incomplete suberisation of intercellular spaces. Electron micrograph, × 2,500.

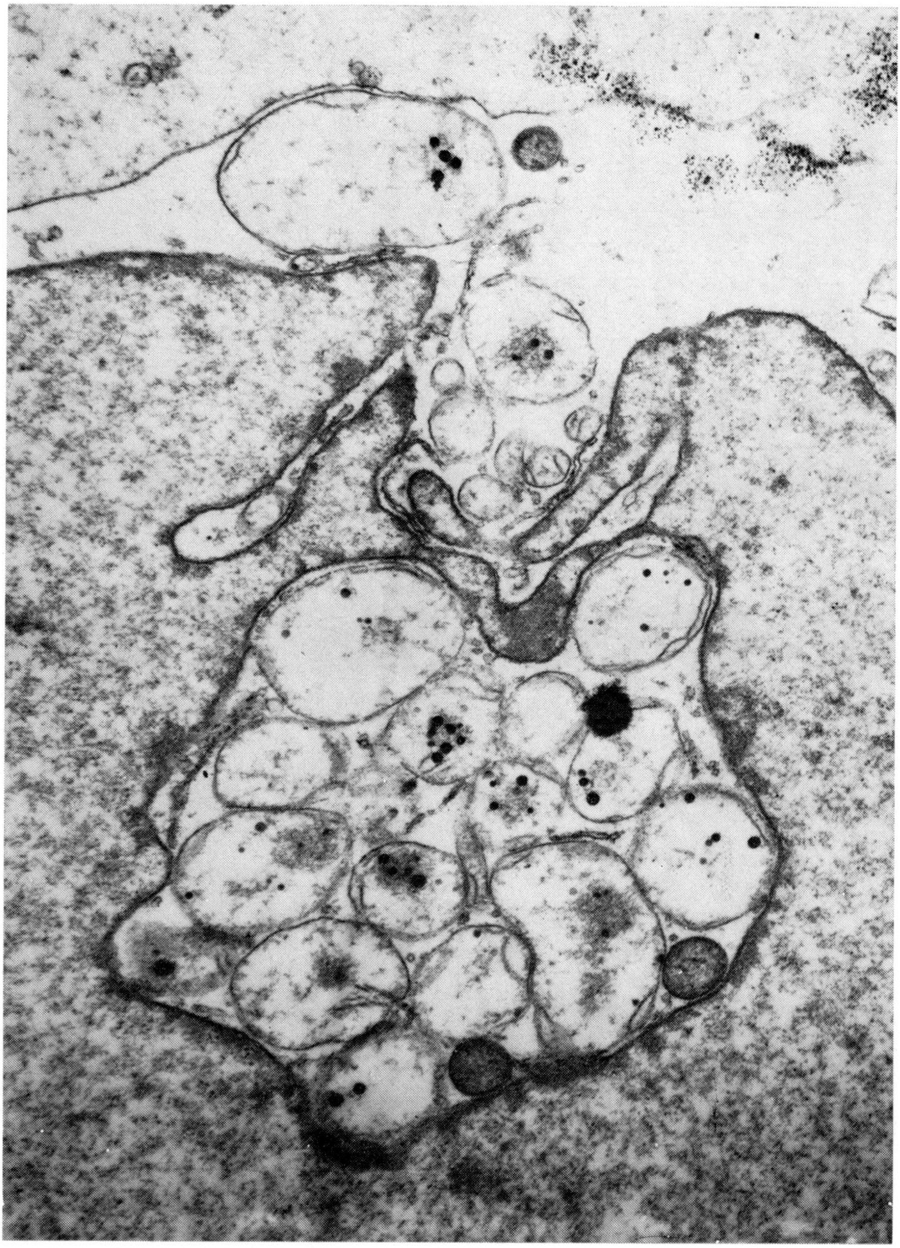

Fig. 5. Plastid cluster in suberising lenticel phelloderm cell. Electron micrograph, × 11,000.

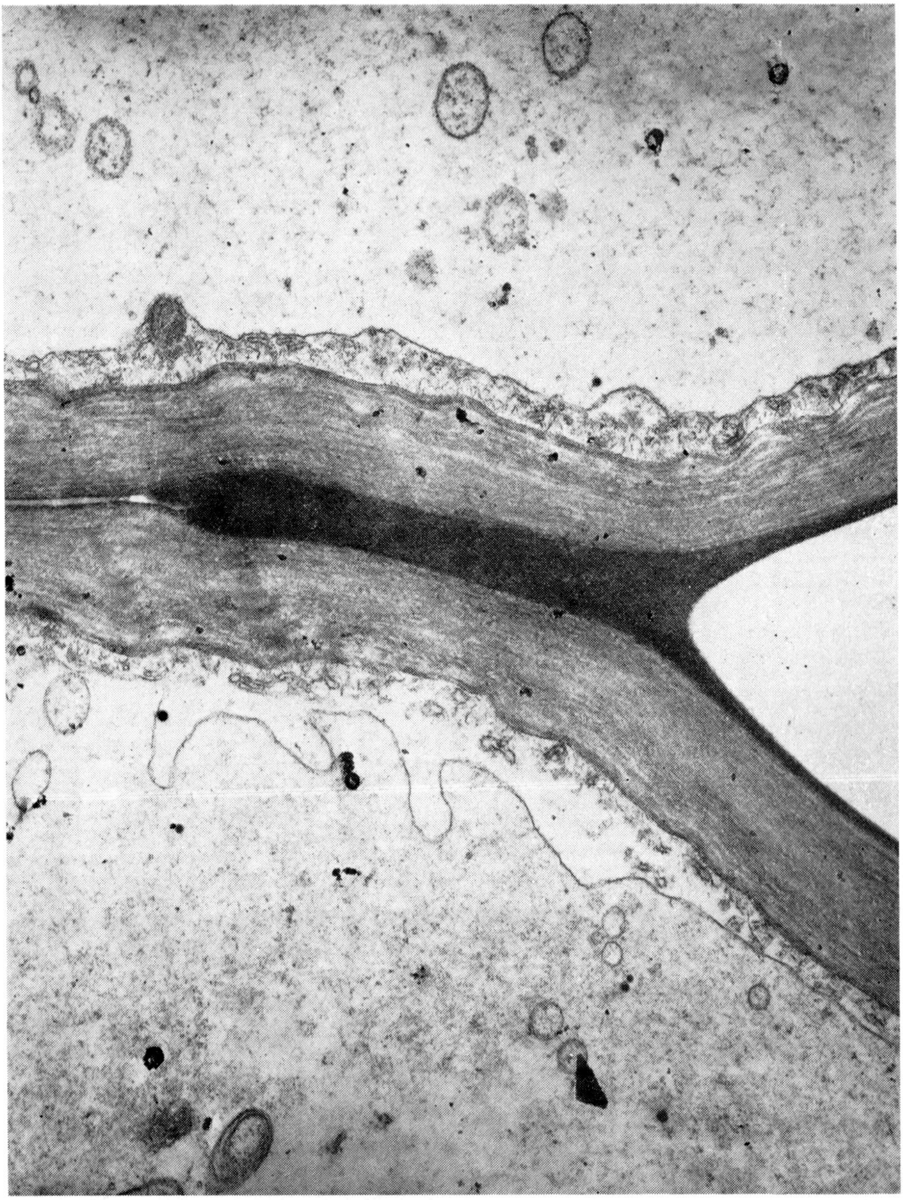

Fig. 6. Early stage in suberisation of intercellular space of lenticel phelloderm cell. Electron micrograph, × 30,000.

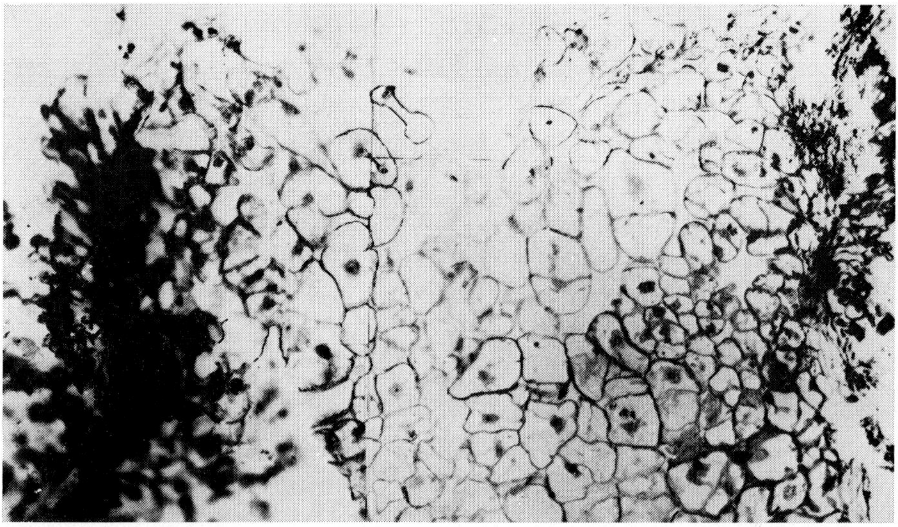

Fig. 7. Proliferated lenticel from tuber stored at 100% relative humidity. Light micrograph, × 140.

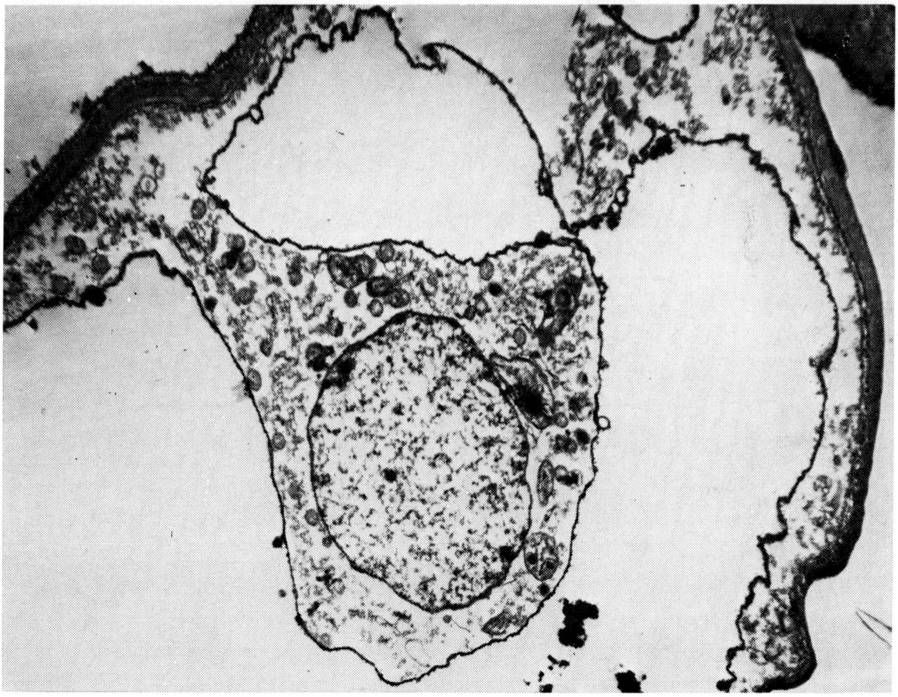

Fig. 8. Cell from proliferated lenticel. Electron micrograph, × 2,700.

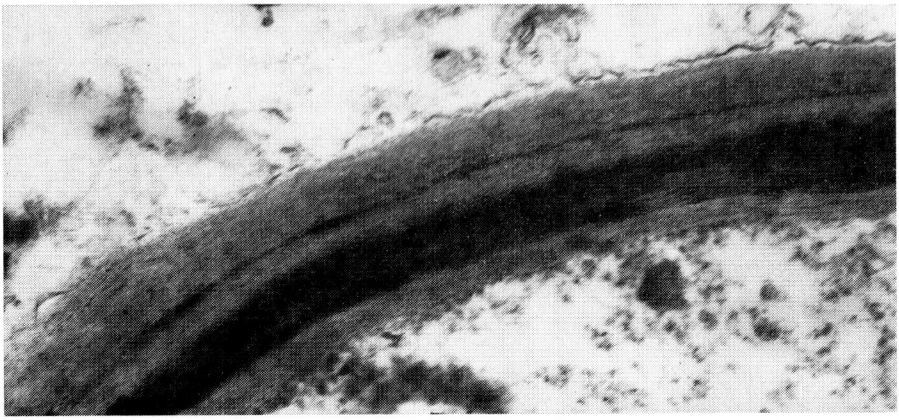

Fig. 9. Suberised cell wall from healed wound surface. Electron micrograph, × 11,500.

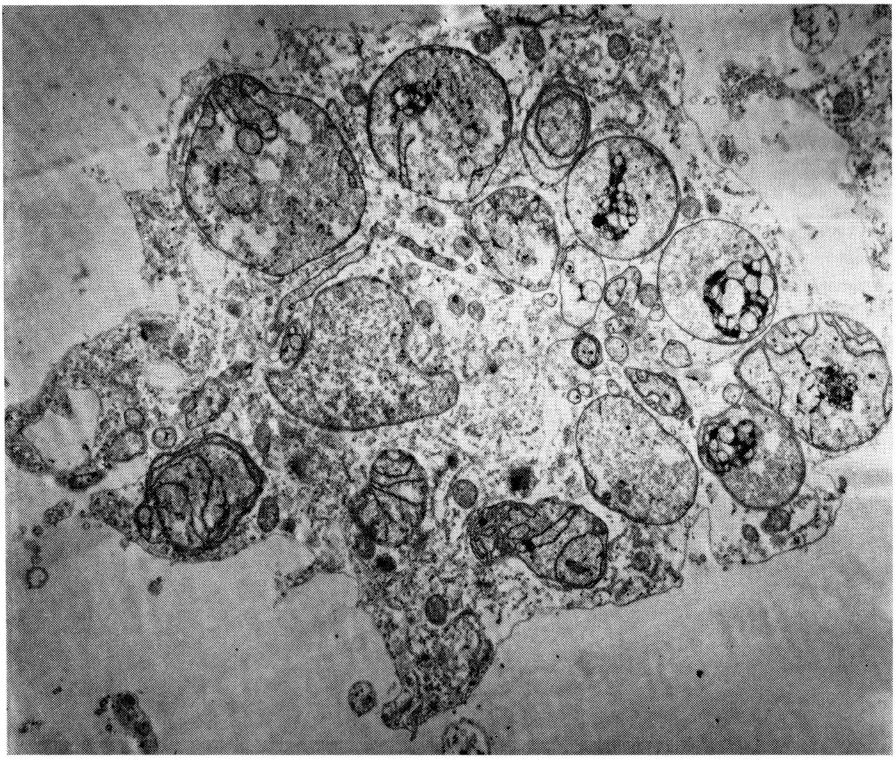

Fig. 10. Plastid cluster from suberising cell of healed wound. Electron micrograph, × 11,500.

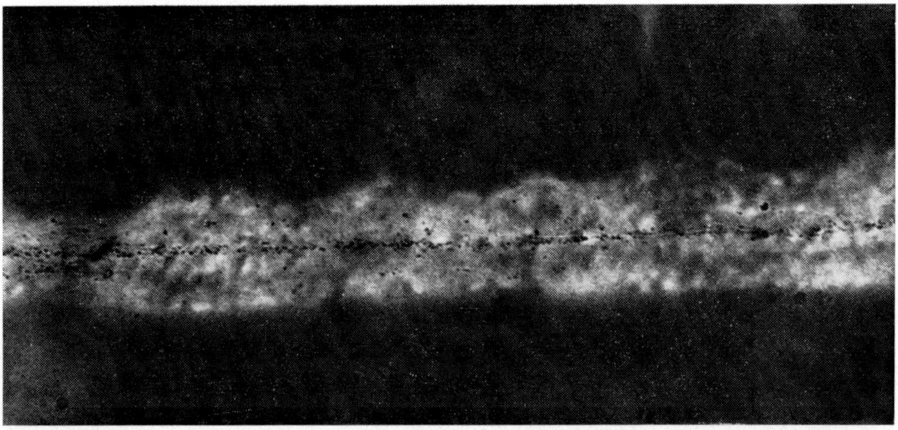

Fig. 11. Periderm cell wall stained by Albersheim's method. Electron micrograph, × 36,500.

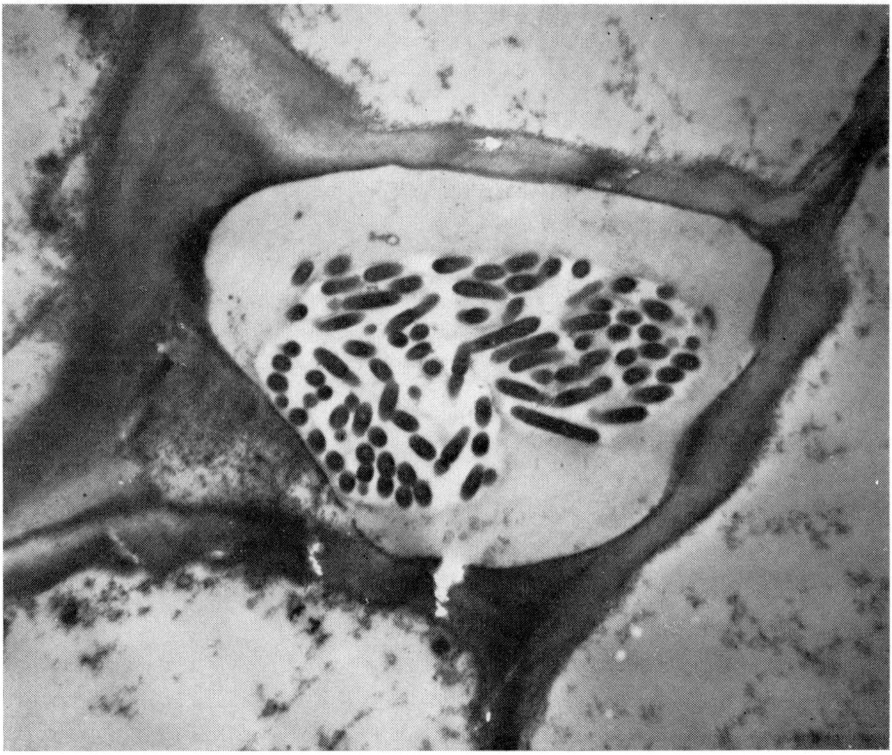

Fig. 12. Early stage of colonisation of potato tissue. Zoned colony of *Erwinia* cells in potato intercellular space. Electron micrograph, × 3,200.

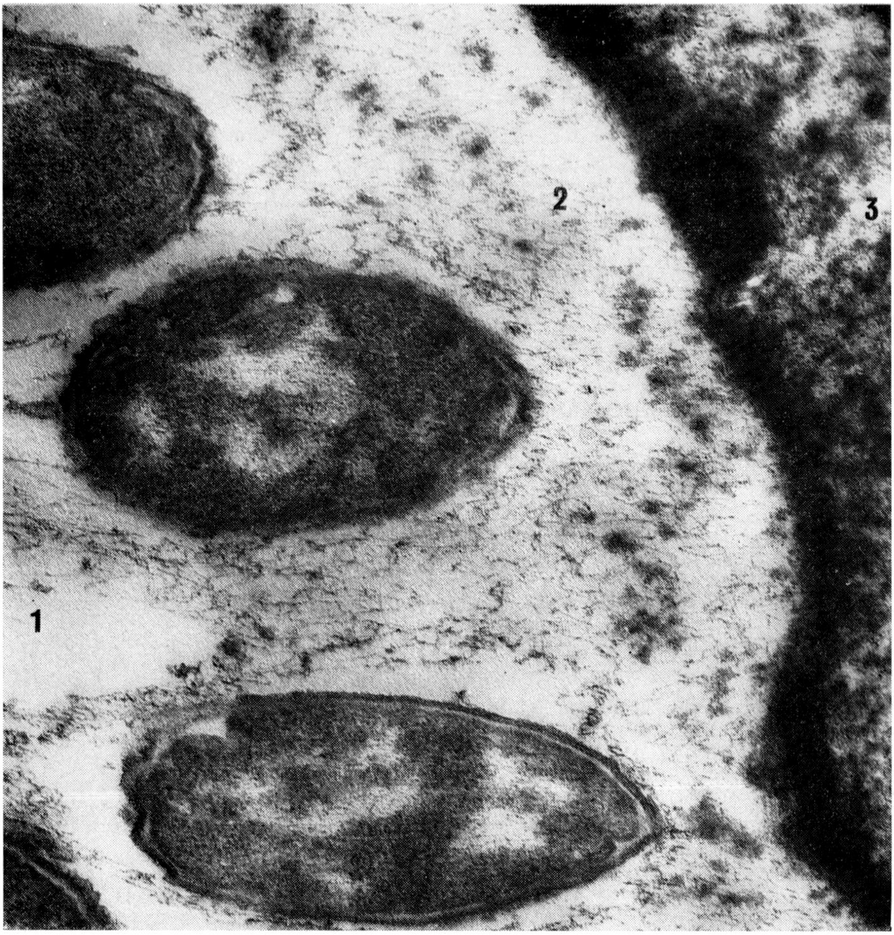

Fig. 13. Colony of *Erwinia* cells showing clear zone 1, fibrillar zone 2, and remains of cell wall 3. Electron micrograph, × 77,000.

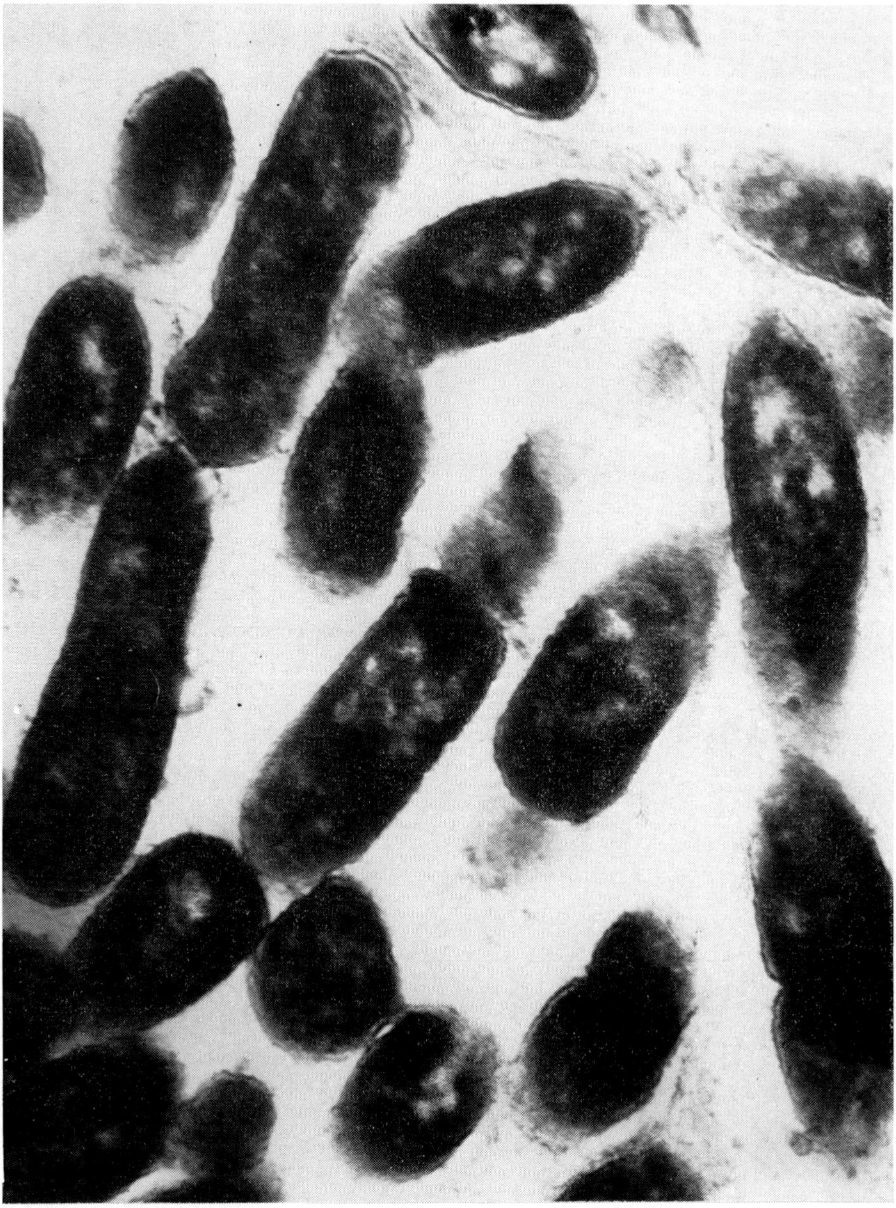

Fig. 14. *Erwinia* cells in clear zone. Electron micrograph, × 30,000.

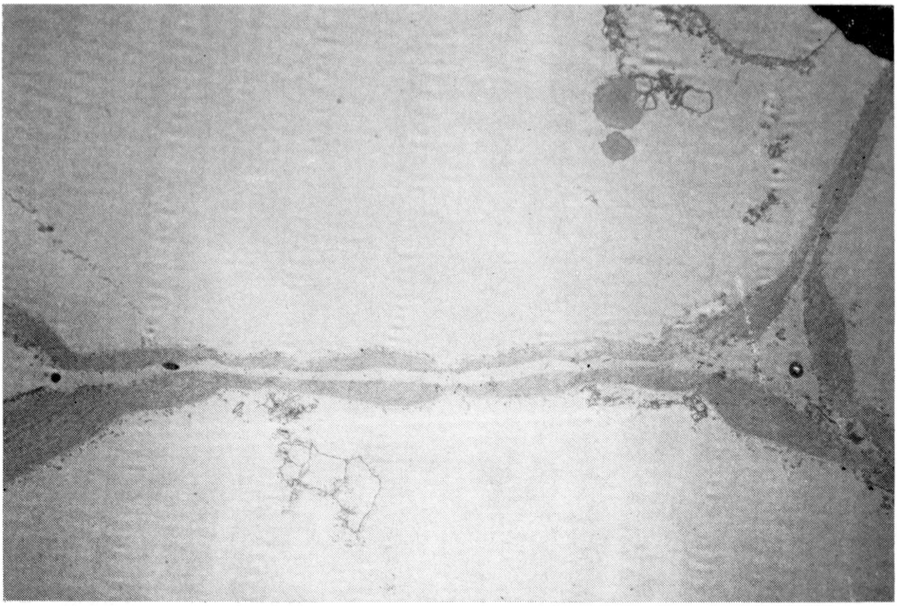

Fig. 15. Late stage of colonisation of potato tissue. Complete maceration of potato cells. *Erwinia* entirely intercellular. Electron micrograph, × 4,500.

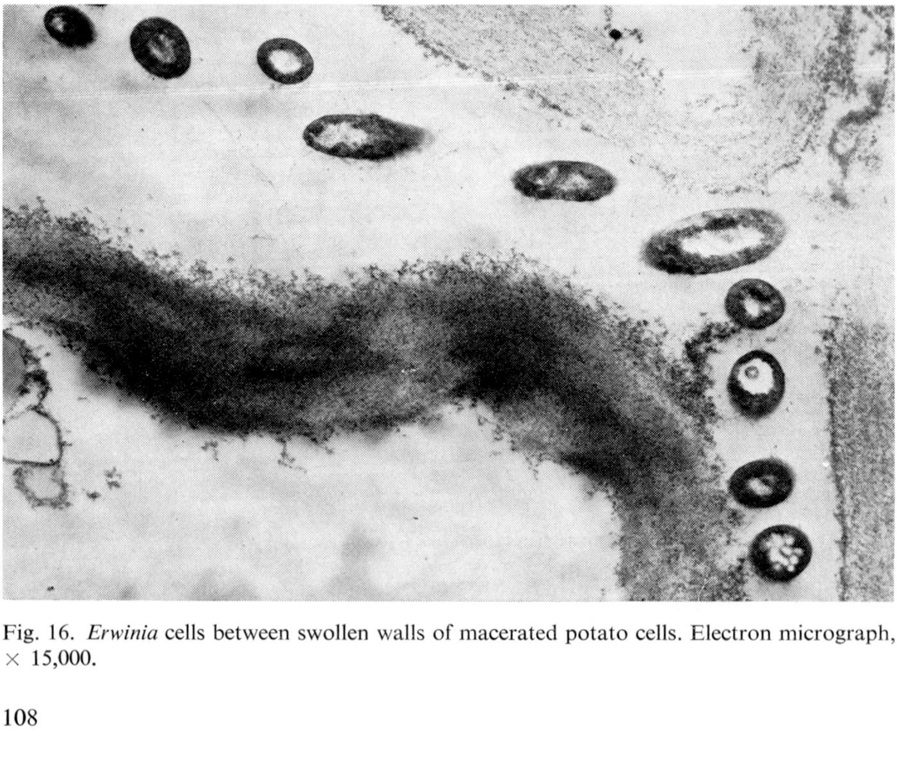

Fig. 16. *Erwinia* cells between swollen walls of macerated potato cells. Electron micrograph, × 15,000.

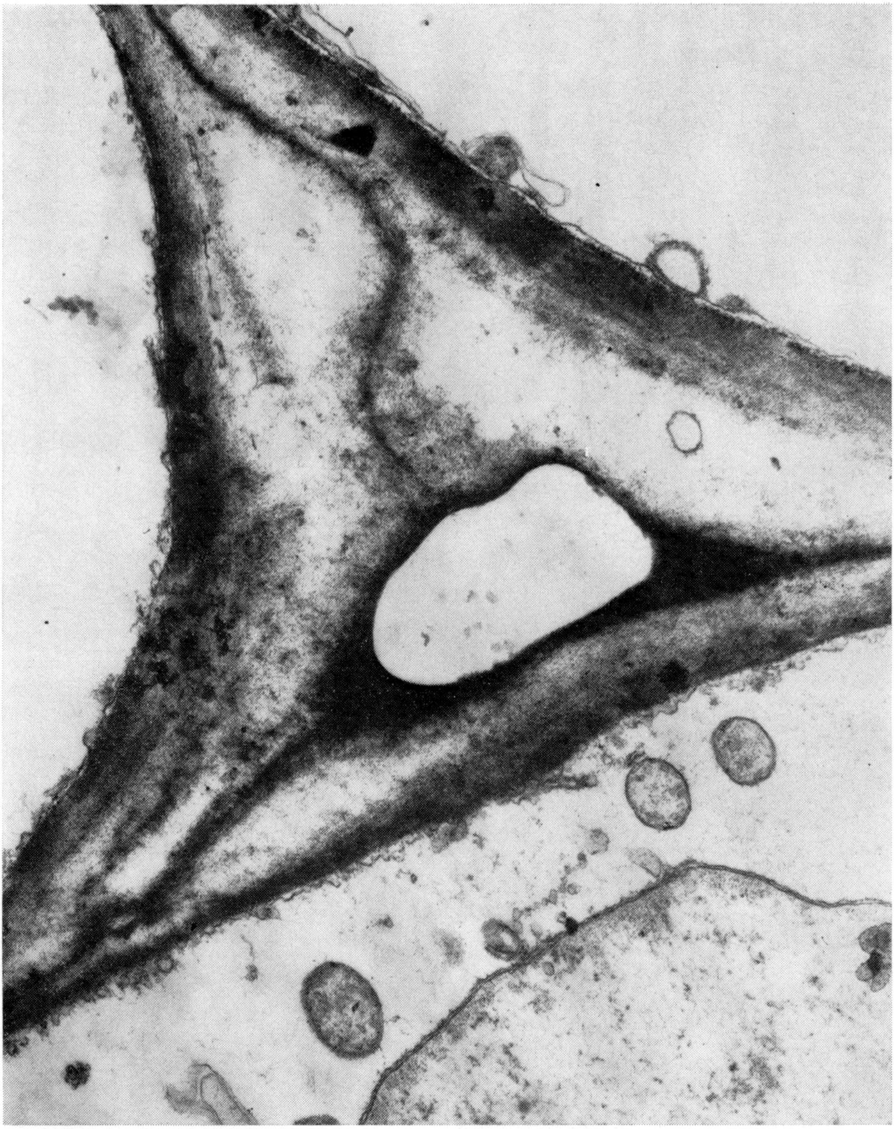

Fig. 17. Suberised intercellular plug separated from disintegrated cell wall. Electron micrograph, × 11,000.

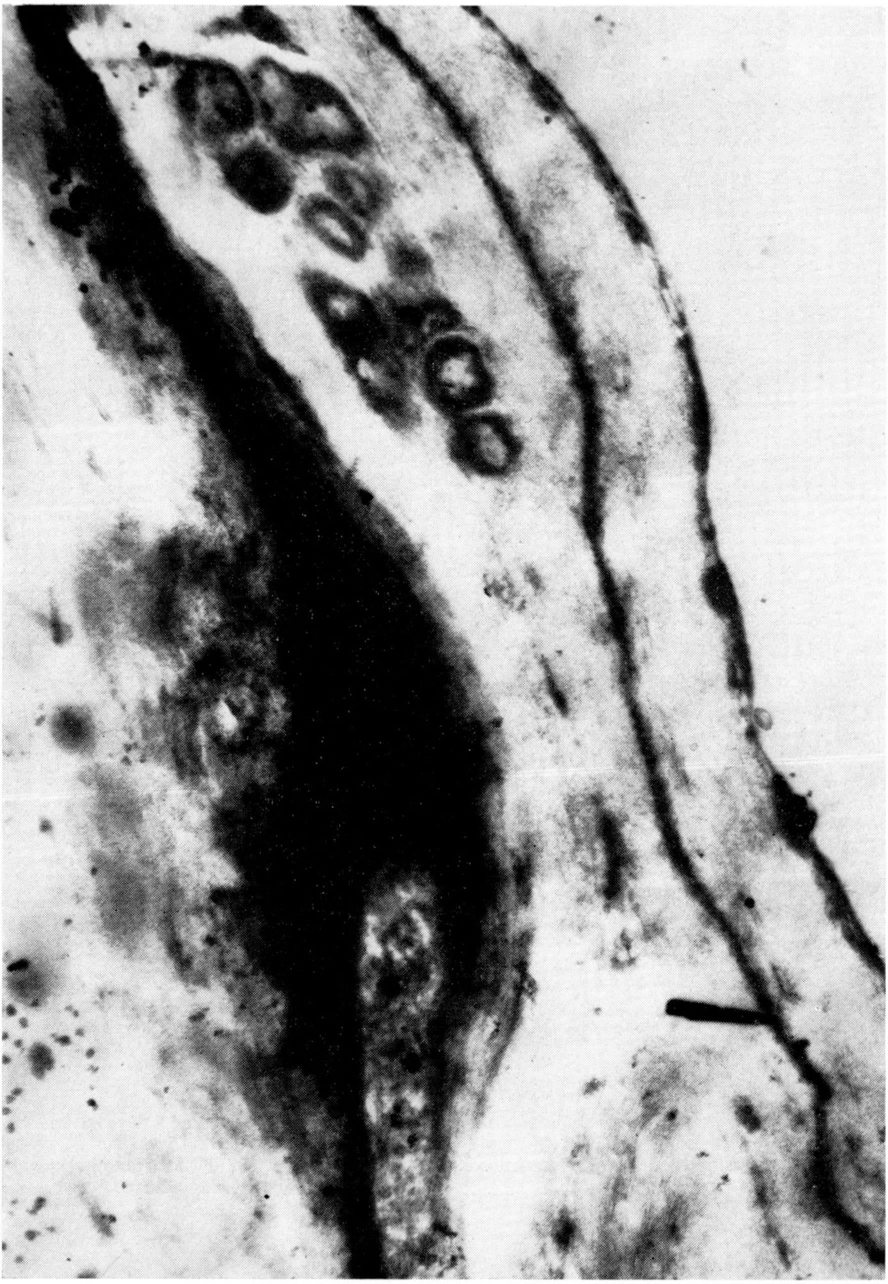

Fig. 18. Suberised intercellular plug surrounded by bacteria. Electron micrograph, × 18,000.

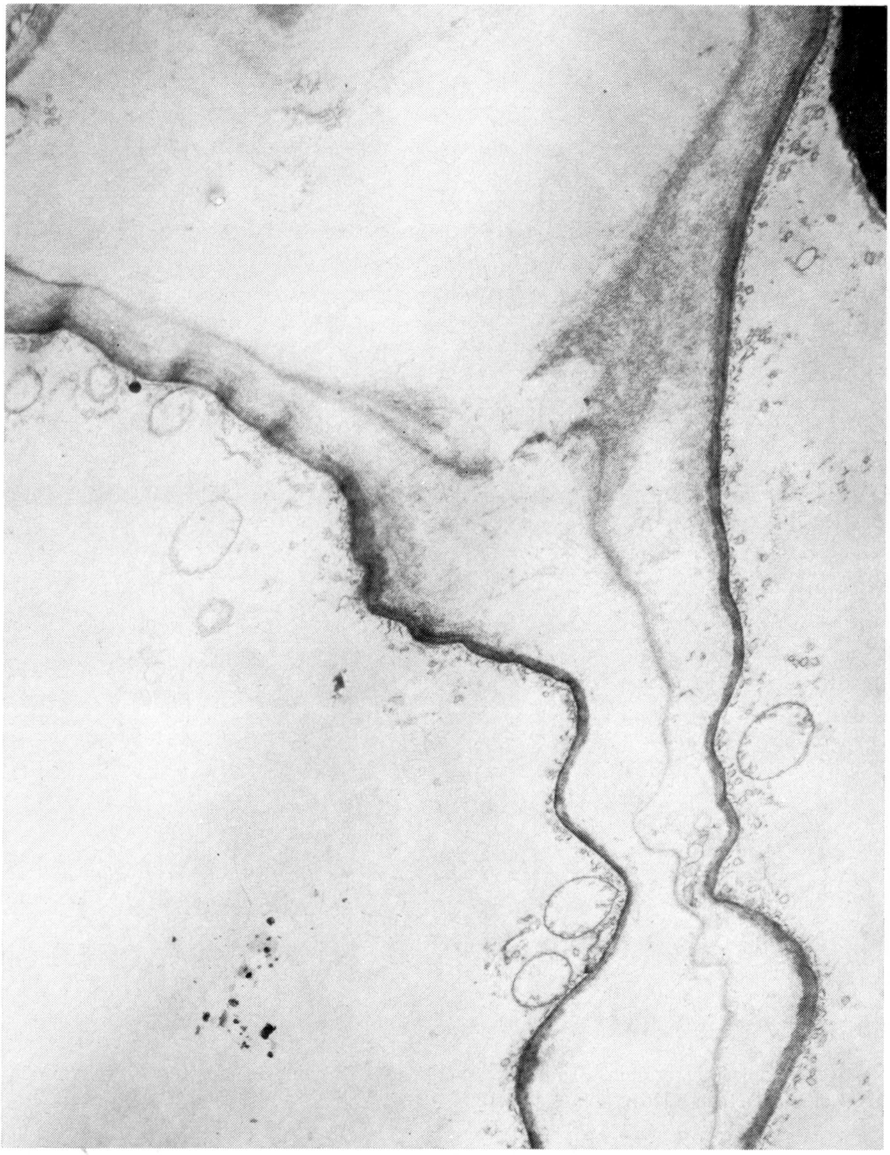

Fig. 19. Macerated cell wall showing zones of disintegrated wall tissue. Electron micrograph, × 7,500.

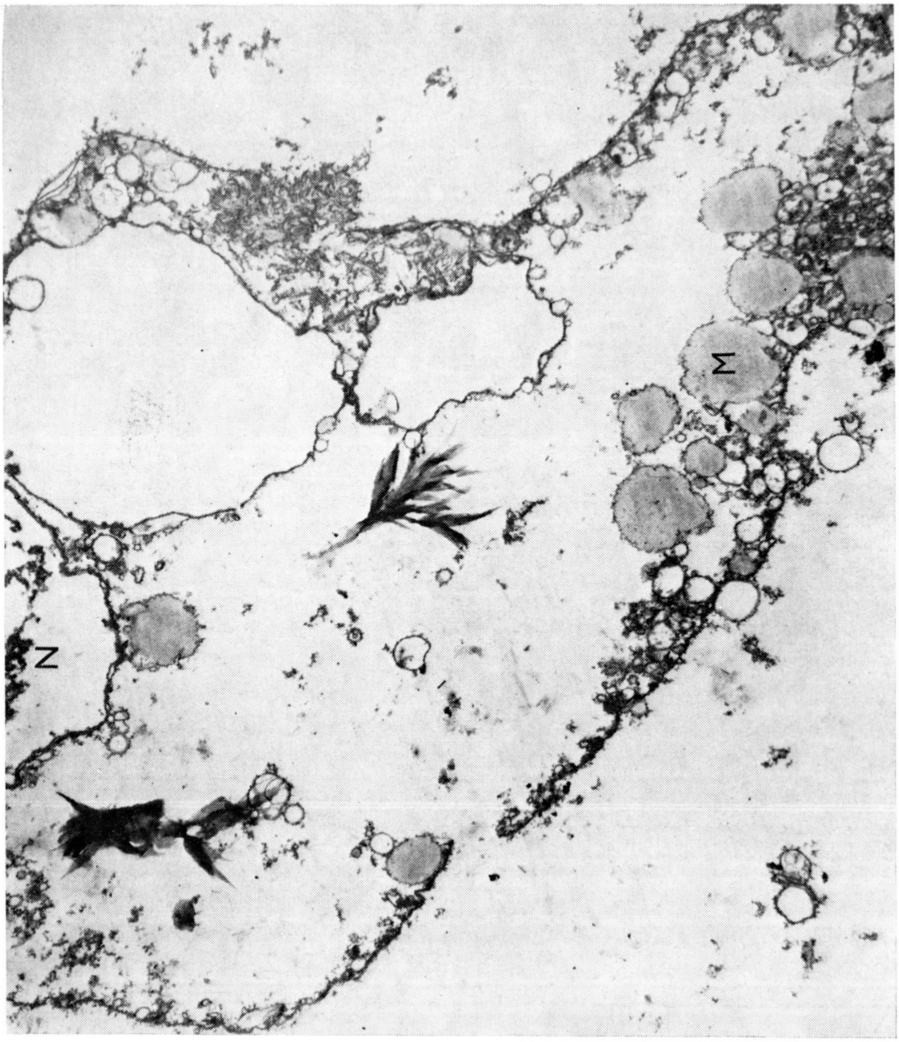

Fig. 20. Cytoplasmic remains from macerated cell. Remains of nucleus Z, microbodies arrowed Σ. Electron micrograph, × 6,400.

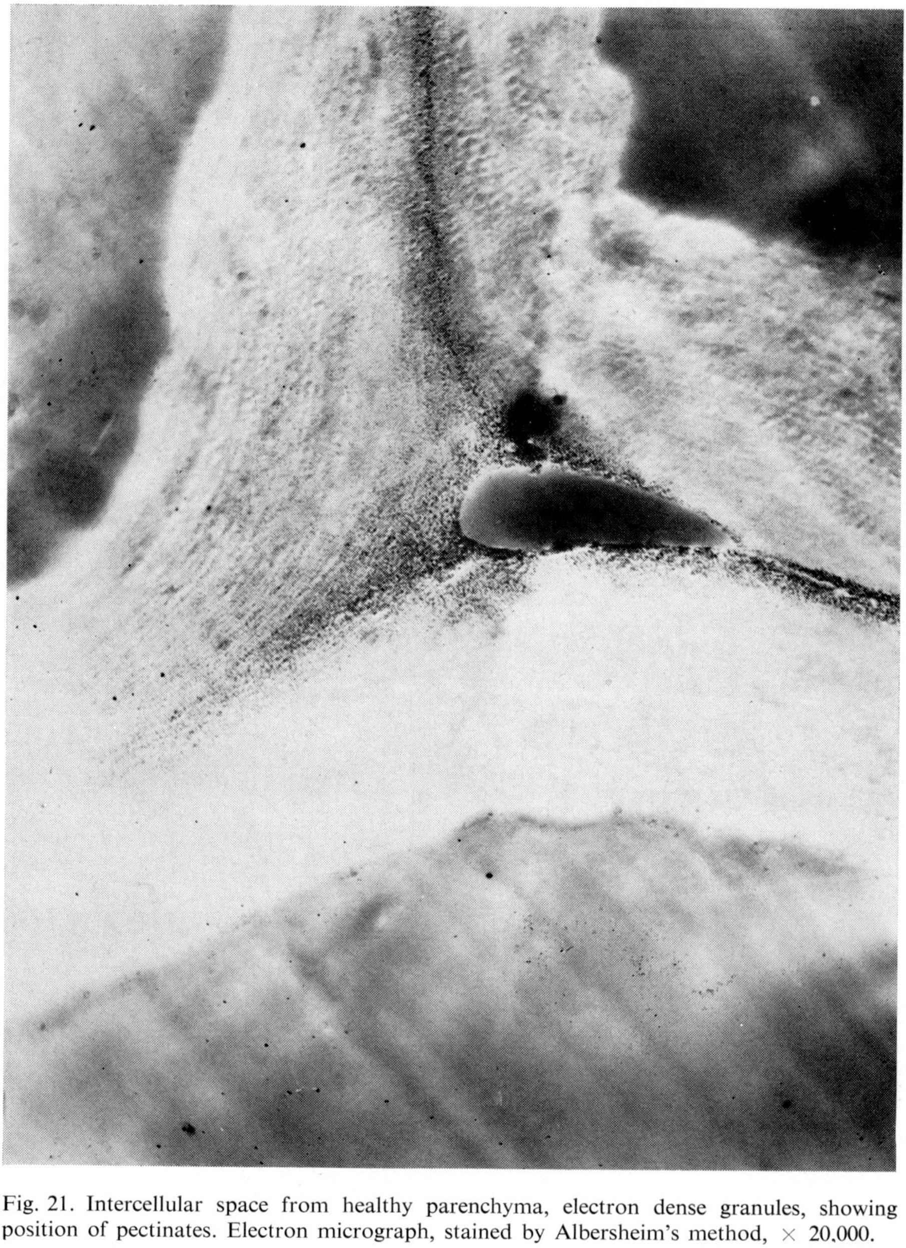

Fig. 21. Intercellular space from healthy parenchyma, electron dense granules, showing position of pectinates. Electron micrograph, stained by Albersheim's method, × 20,000.

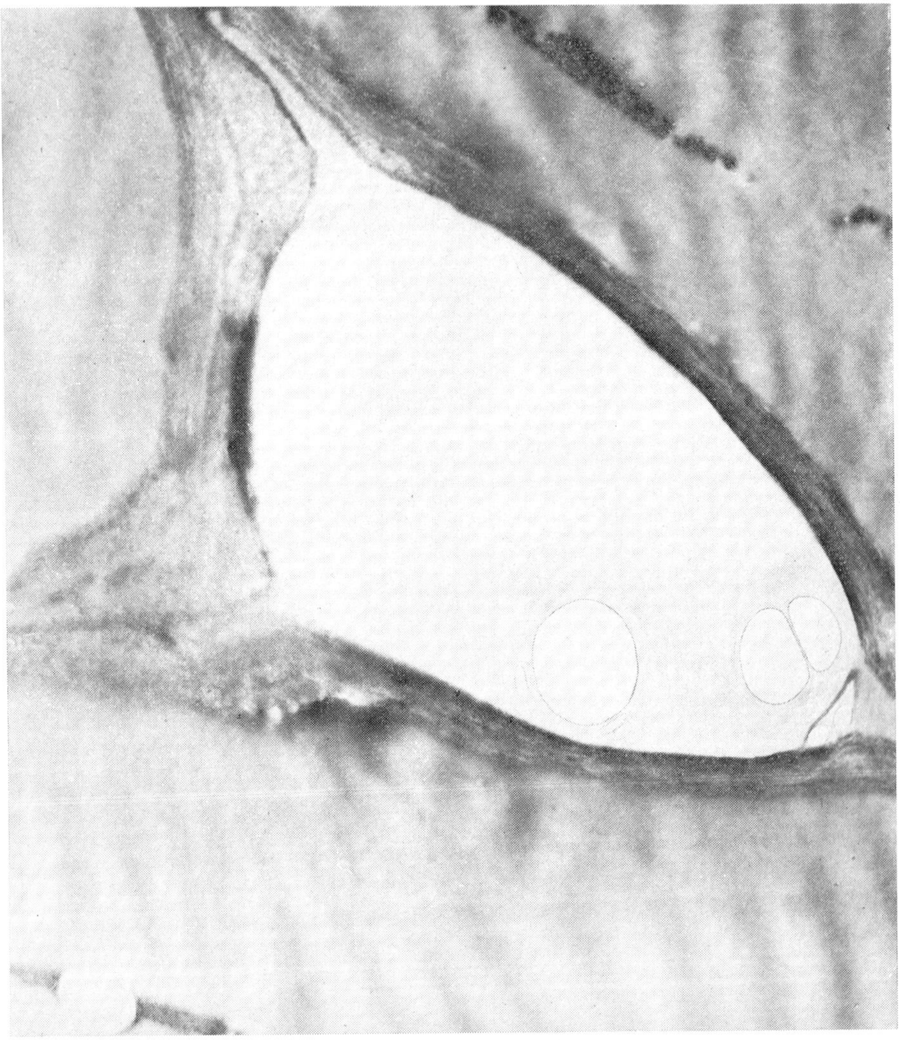

Fig. 22. Intercellular space from diseased tissue. Electron micrograph, stained by Albersheim's method, × 8,900.

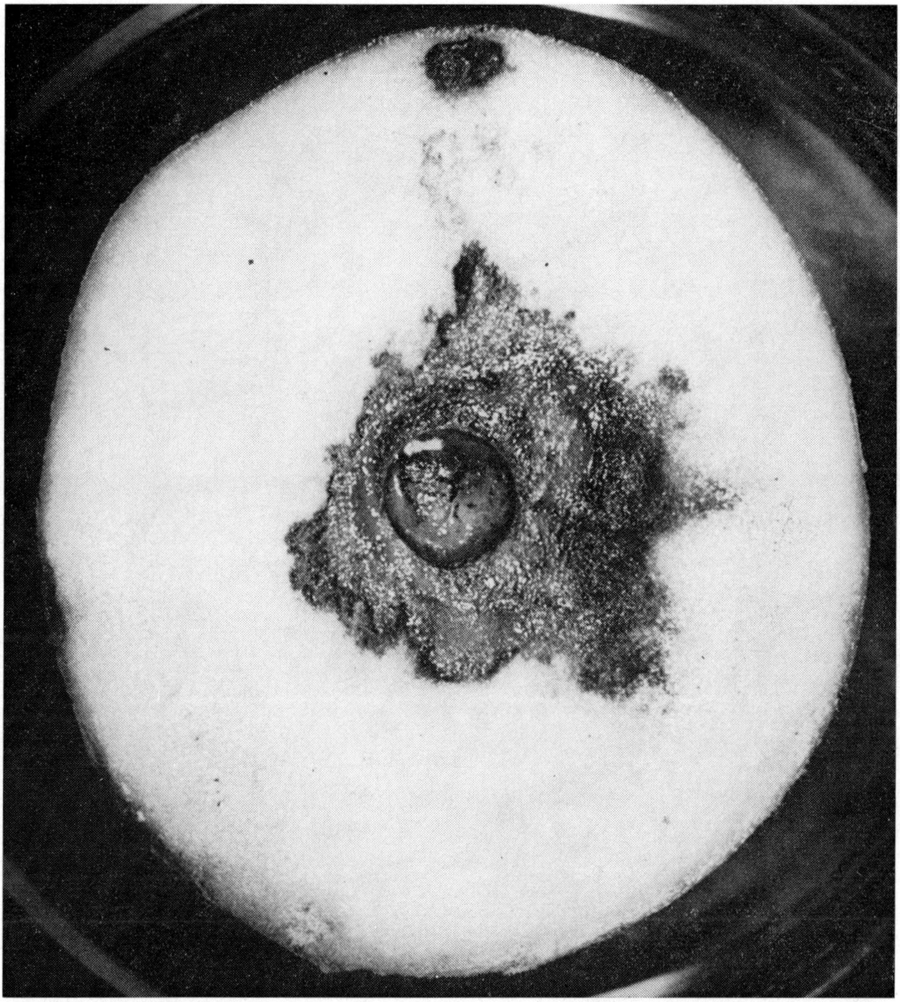

Fig. 23. Artificially inoculated tuber stored at 25 °C under aerobic conditions, × 4.2.

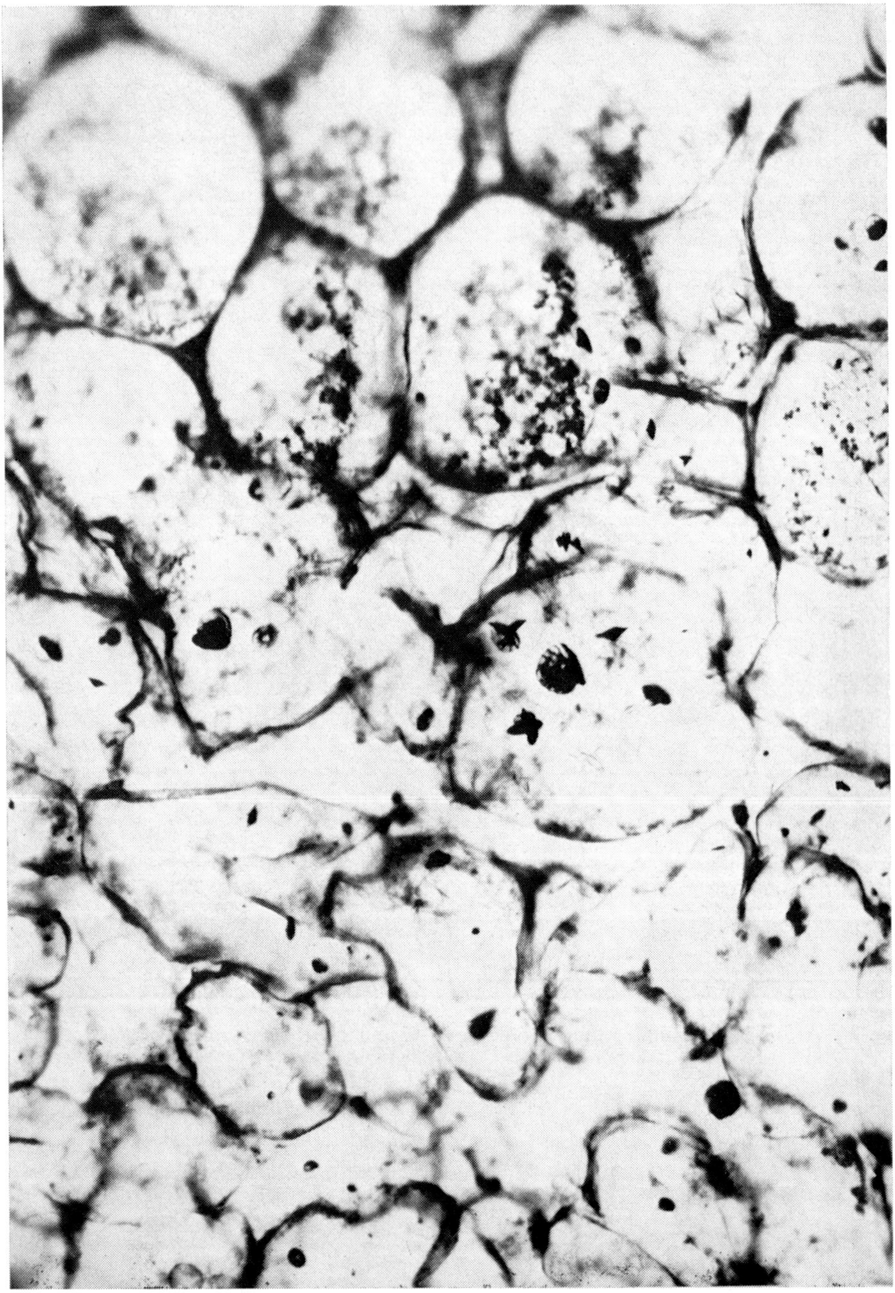

Fig. 24. Transverse section through infection interface cells. Light micrograph, × 180.

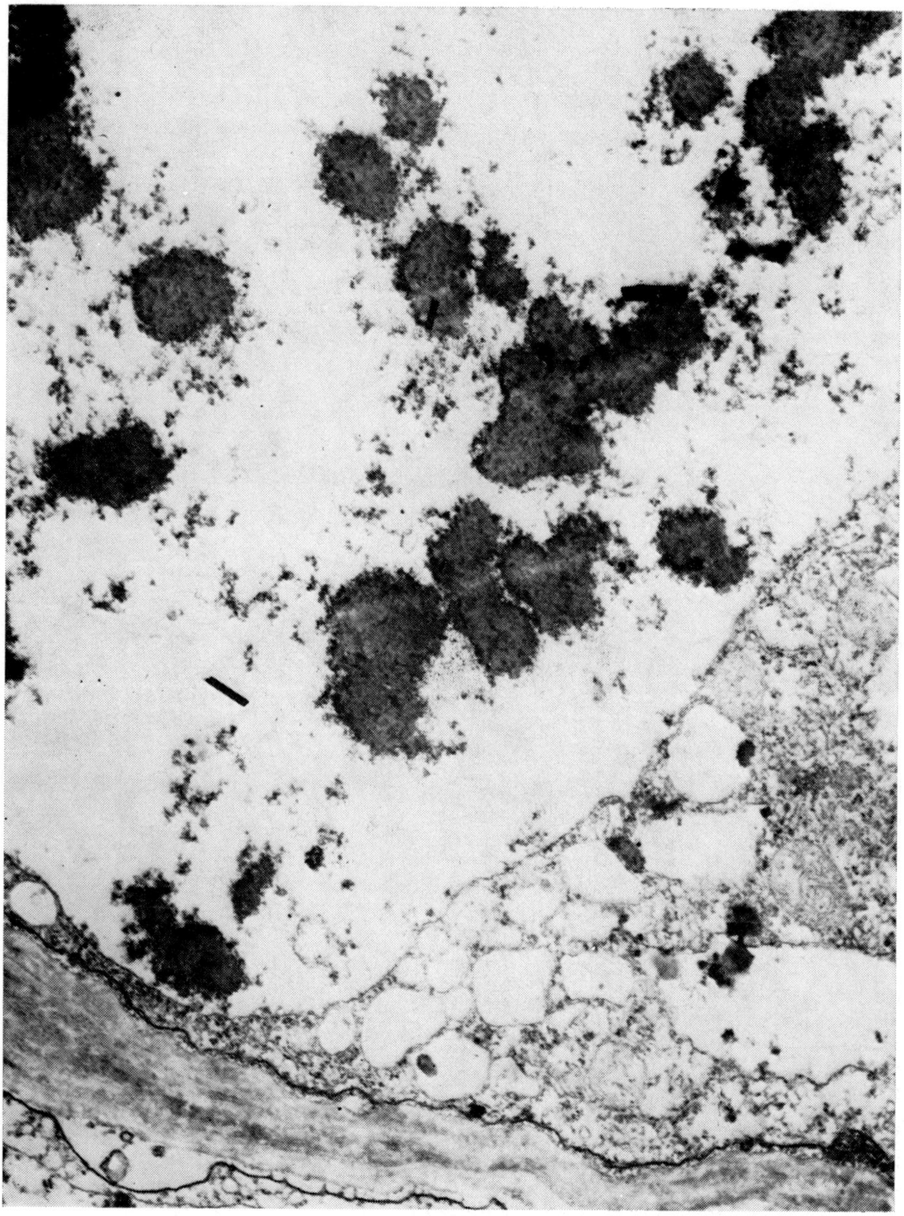

Fig. 25. Granules of oxidised polyphenols in tonoplast of infection interface cell. Electron micrograph, × 15,400.

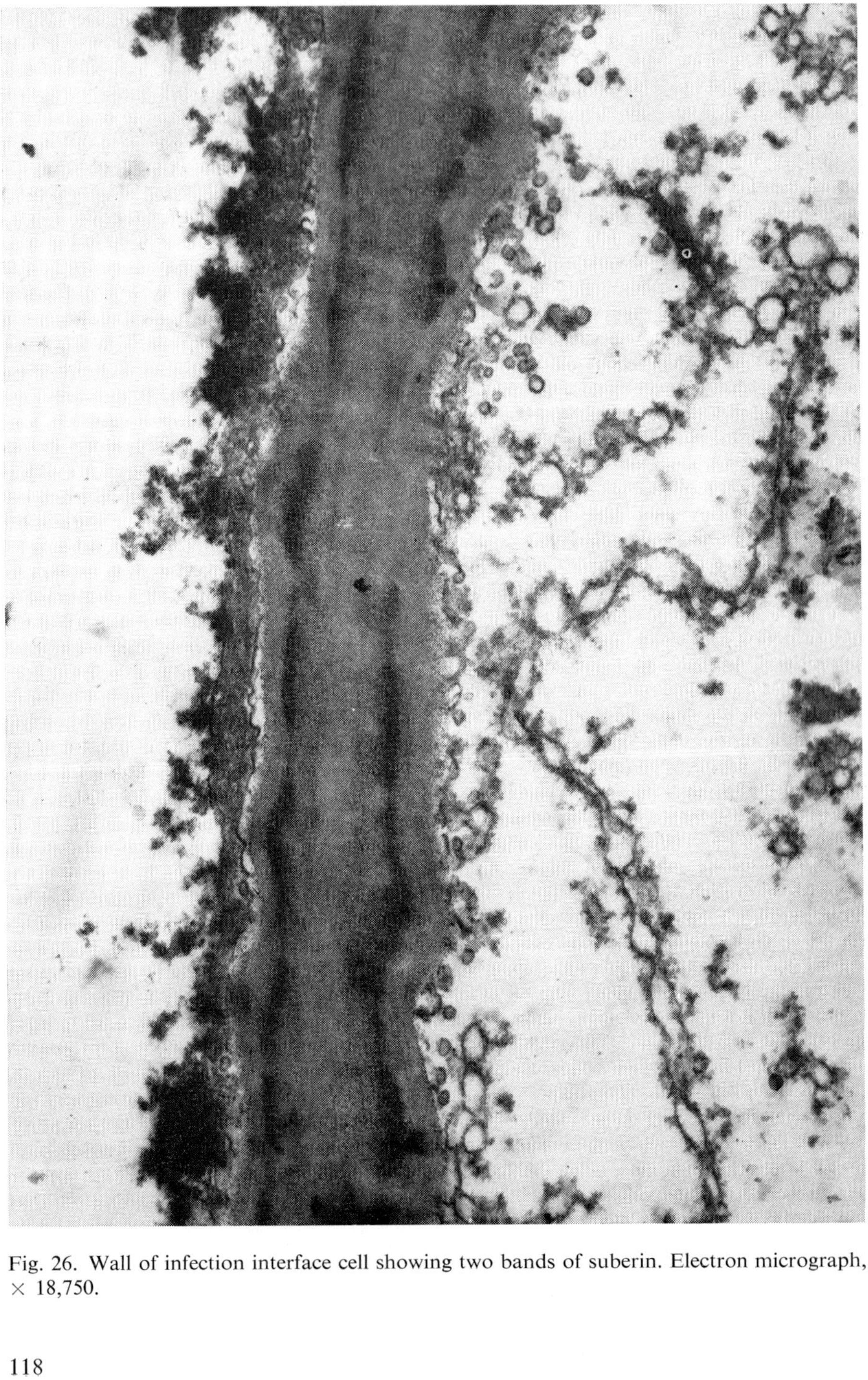

Fig. 26. Wall of infection interface cell showing two bands of suberin. Electron micrograph, × 18,750.

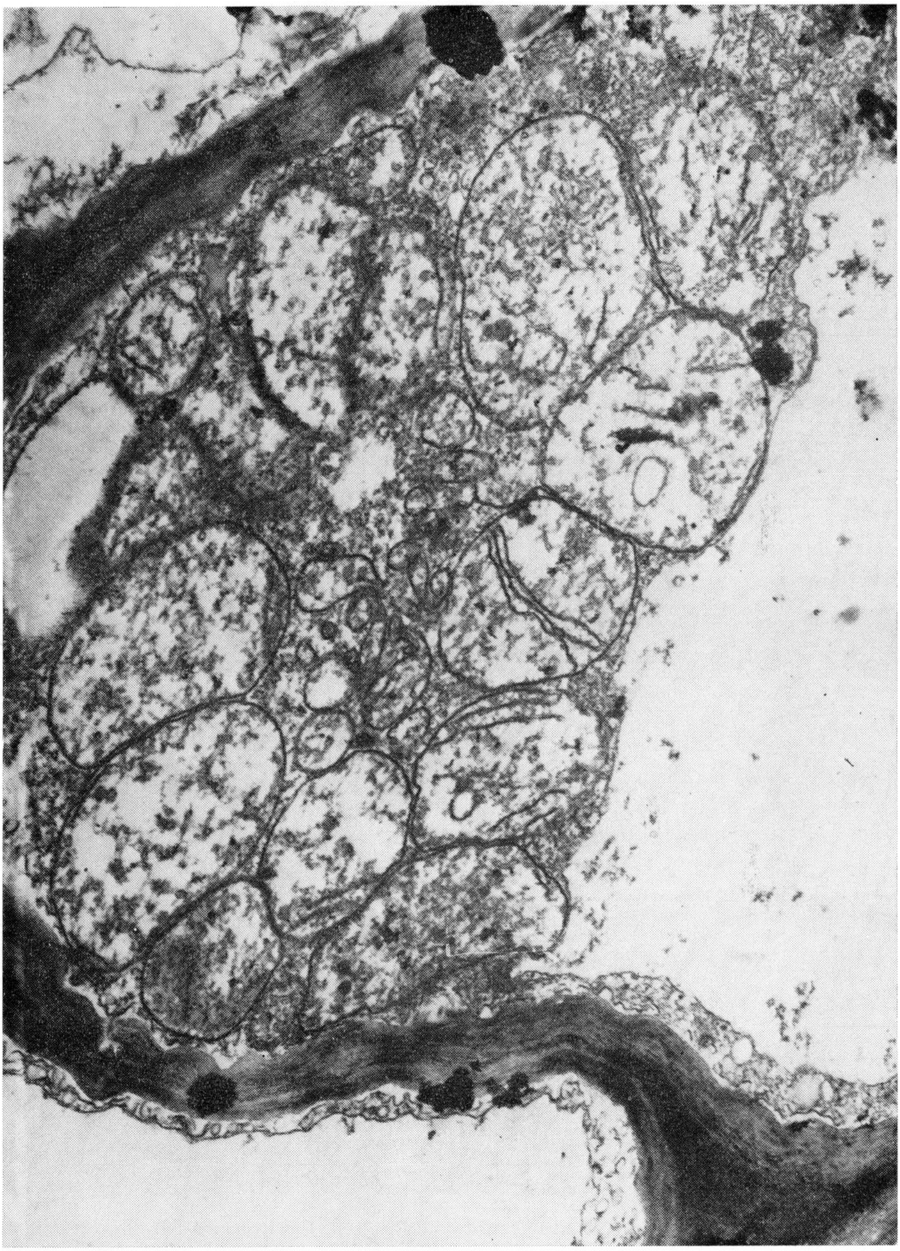

Fig. 27. Plastid cluster in suberised infection interface cell. Intercellular space plugged with suberin. Electron micrograph, × 7,000.

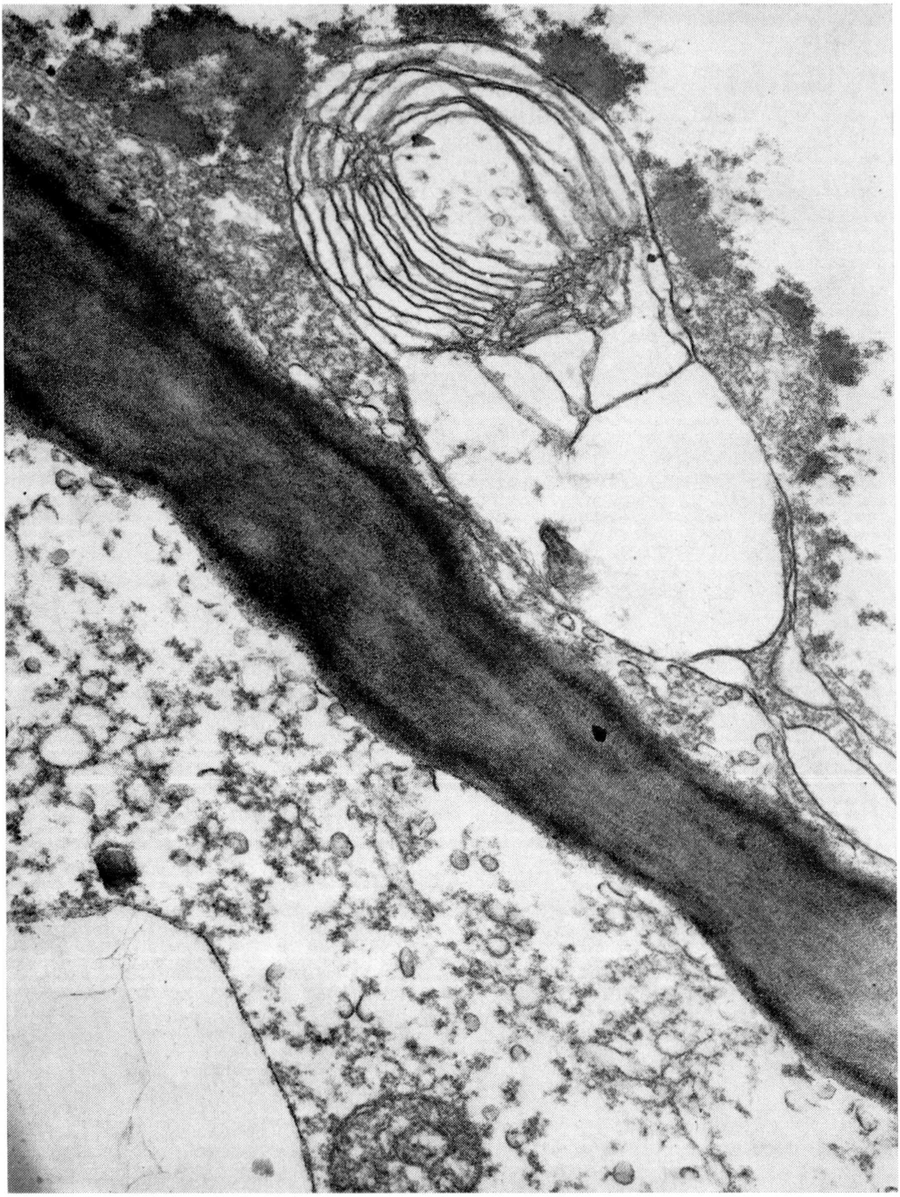

Fig. 28. 'Myelin' body in suberised infection interface cell. Electron micrograph, × 20,000.

Gel electrophoresis as a tool in diagnosing Pseudomonas isolates

H. R. Cameron and B. C. Palmer

Department of Botany and Plant Pathology, Oregon State University, Corvallis, Oregon 9733, U.S.A.

Abstract

Similarity of protein band patterns obtained by gel electrophoresis suggested that the oxidase-negative plant pathogenic isolates form a group distinct from the non-pathogenic isolates. Within the pathogenic group similar band patterns indicate possible synonymy. Pattern differences could not be correlated with host specificity.

Introduction

Recently there has been an increased use of gel electrophoresis for taxonomic purposes. Several studies have been made on higher organisms but the method has been particularly applicable to micro-organisms. Genera of bacteria studied include *Xanthomonas* (Gill & Khare, 1968; El-Sharkawy & Huisingh, 1968), *Agrobacterium* (Huisingh & Durbin, 1967), *Corynebacterium* (Robinson, 1966), *Erwinia* (Smith & Powell, 1968), *Streptococcus* (Lund, 1965), and *Streptomyces* (Gottlieb & Hepden, 1966). Hoitink *et al.* (1968) also used electrophoresis to compare two species of *Pseudomonas*.

We have been using gel-electrophoresis since 1969 to compare isolates of *P. syringae* and other pathogenic and saprophytic isolates of fluorescent pseudomonads to provide a more rapid identification of pathogenic isolates.

Methods

Each isolate was purified by streaking on a plate of King's medium B (King *et al.*, 1954) and the growth from a single colony was used to prepare a stock culture according to the method of DeVay & Schnathorst (1963). Cultures for electrophoresis were grown for three days at 25 °C in one-liter flasks containing 400 ml of nutrient glucose broth (3 g beef extract, 10 g peptone, 10 g glucose per liter). At the time of harvesting, the cultures were checked for purity by streaking a sample from each flask on a plate of medium B. The harvested cells were centrifuged for 10 minutes at 5800 *g*, washed by resuspending them in 50 volumes of 0.05 M TRIS-HCl buffer at pH 7.2 and then centrifuged again. The best buffer and pH were determined from a series of preliminary experiments comparing TRIS, HEPES, and phosphate buffers over a range of pH.

Soluble proteins were extracted by suspending cells in equal volumes of buffer and sonicating with a MSE Ultrasonic Disintegrator for 3.5 minutes. To prevent heating during sonication, the suspension was surrounded by dry ice and sonication was done in 20-second intervals. The broken cell suspension was then centrifuged at 4 °C for 60 minutes at 100,300 *g* on a Spinco model L ultracentrifuge. The supernatant was pipetted off, divided into 0.5 ml aliquots, and frozen at −20 °C until needed.

Acrylamide gels were prepared according to the method of Davis (1964) using a spacer gel and a 7.5% running gel. The sample was prepared by mixing approximately 0.1 ml of sample with 0.9 ml of large pore solution, giving a protein concentration of 500–900 μg of protein per tube as determined by the method of Lowry *et al.* (1951). Bovine serum was used as the standard. The stock TRIS-glycine reservoir buffer was diluted 1:1 for use. The tubes used were 0.5 × 8.0 cm internal diameter.

Electrophoresis was carried out at 25 mamp for five minutes, then amperage was increased to 55 mamp (4.6 mamp/tube) and continued until the front of the twelve tubes in the container had moved through the running an average distance of four centimeters. This took about 50 minutes. Position of the front was indicated by bromphenol blue marker dye.

The gels were removed from the tubes by rimming them with a jet of water forced through a hypodermic needle. The fixation and staining procedure was that of Chrambach *et al.* (1967). The gels were fixed for a minimum of 40 minutes in 12.5% trichloroacetic acid and stained for 20 minutes with 0.05% Coomassie brilliant blue R 250. The gels were destained for two days in 10% TCA, then transferred to tubes containing 7% acetic acid for permanent storage. Densitometer tracings were made of each gel with a Schoeffel Model SD-3,000 spectrodensitometer set at a wavelength of 650 mm.

Results

Approximately 25 bands were visible in each gel. Duplicate gels were identical in band pattern although there were small uncontrollable differences in migration velocity between some tubes.

A comparison between samples of the same isolate grown under different conditions (48 hours at 30 °C *v.* 72 hours at 25 °C) revealed very similar patterns. The variation was no greater than that when the same extract was run after a period of several days or months.

Uncontrollable variation in migration velocities between tubes presented a problem when attempting to determine which bands constitute homologs. This was especially critical when comparing tubes from different runs. E_f values for a number of gels were calculated, but the results were completely unsatisfactory. Based on E_f values, there was greater variation among isolates of *P. syringae* than between *P. syringae* and other species.

The evaluation of results by the method of Johnson *et al.* (1967) was found to give more accurate comparisons. Band patterns of 1:1 protein mixtures of paired isolates were compared with those of each separate isolate. Band position and staining intensity were used to establish homologous bands. Once two or more widely separated pairs of homologous bands per pair of gels had been established, the densitometer tracings were adjusted photographically to show the homologous bands at the same point. A band-by-band comparison could then be made.

Our results are in agreement with others in that a clear separation of nomenspecies can be made using electrophoresis; however, no differences were detected that could be correlated with differences in pathotype, serotype or host specificity. The method appears to be of value as a relatively rapid and simple means of identification species of *Pseudomonas*.

References

Chrambach. A., R. A. Reisfeld, M. Wyckoff & J. Zaccari, 1967. A procedure for rapid and sensitive staining of protein fractionated by polyacrylamide gel electrophoresis. Analyt. Biochem. 20: 150–154.

Davis, B. J., 1964. Disc electrophoresis. II. Method and application to human serum proteins. Ann. N.Y. Acad. Sci. 121: 404–427.

DeVay, J. E. & W. C. Schnathorst, 1963. Single-cell isolation and preservation of bacterial cultures. Nature, Lond. 199: 775–777.

El-Sharkawy, T. A. & D. Huisingh, 1968. Identification of *Xanthomonas* species by acrylamide gel disc electrophoresis of soluble bacterial proteins. Phytopathology 58: 1049 (Abstr.).

Gill, H. S. & M. N. Khare, 1968. Disc electrophoresis of sonic extracts of five phytogenic bacteria. Phytopathology 58: 1051 (Abstr.).

Gottlieb, D. & Pamela M. Hepden, 1966. The electrophoretic movement of proteins from various *Streptomyces* species as a taxonomic criterion. J. gen. Microbiol. 44: 95–104.

Hoitink, H. A. J., D. J. Hagdorn & Elizabeth McCoy, 1968. Survival, transmission, and taxonomy of *Pseudomonas syringae* van Hall, the causal organism of bacterial brown spot of bean *(Phaseolus vulgaris* L.*)*. Can. J. Microbiol. 14: 437–441.

Huisingh, D. & R. D. Durbin, 1967. Physical and physiological methods for differentiating among *Agrobacterium rhizogenes*, *A. tumefaciens*, and *A. radiobacter*. Phytopathology 57: 922–923.

Johnson, B. L., D. Barnhart & Hall Ove, 1967. Analysis of genome and species relationships in the polyploid wheats by protein electrophoresis. Am. J. Bot. 54: 1089–1098.

King, Elizabeth O., Martha K. Ward & D. E. Raney, 1954. Two simple media for the demonstration of pyocyanin and fluorescin. J. Lab. clin. Med. 44: 301–307.

Lowry, O. H., Nira J. Rosebrough, A. L. Farr & Rose J. Randall, 1951. Protein measurement with the Folin phenol reagent. J. biol. Chem. 193: 265–275.

Lund, Barbara M., 1965. A comparison by the use of gel electrophoresis of soluble protein components and esterase enzymes of some group D streptococci. J. gen. Microbiol. 40: 413–420.

Robinson, K., 1966. An examination of *Corynebacterium* spp. by gel electrophoresis. J. appl. Bact. 29: 179–184.

Smith, J. H. & D. Powell, 1968. A disc electrophoretic comparison of protein patterns of *Erwinia amylovora* with other bacteria, including associated yellow forms. Phytopathology 58: 972–975.

Studies on the preparation of anti-Erwinia sera in rabbits

I. Lazăr

Tr. Săvulescu Institute of Biology, Spl. Independentia 296, Bucharest, Romania

Abstract

Various methods of immunizing rabbits were tested to obtain agglutinant or precipitant anti-Erwinia sera. The most efficient methods were:
for agglutinant sera: one intramuscular inoculation with incomplete Freund antigen, followed after 20–24 days by a series of 3 or 4 intravenous inoculations with bacterial suspension in increasing doses starting with 0.2–0.5 ml;
for precipitant sera: three subcutaneous inoculations with incomplete Freund antigen at 21 day intervals, followed by a series of 4 or 5 intravenous inoculations with antigenic extracts in increasing doses starting with 0.2–0.5 mg.

Details are given of the antigens used, route of administration, number of inoculations and the interval between them, and the dosage for less toxic and very toxic strains.

Introduction

In the last twelve years, besides research on plant pathogenic bacteria, the program of Tr. Săvulescu Institute of Biology has included studies on serological relationships between various bacteria.

In contrast to such genera as *Agrobacterium*, *Corynebacterium*, *Pseudomonas* and *Xanthomonas*, the species of the genus *Erwinia* presented serious difficulties, as already recorded by other investigators (Graham, 1963; Goto & Arsenijević, pers. comm.), in the preparation of antisera, due to the high death-rates of inoculated rabbits. Therefore, besides researches on the toxic action of these bacteria (Popovici & Lazăr, 1964; Săvulescu & Lazăr, 1964, 1967), attempts were made to establish some methods of immunizing that would give antisera with corresponding titres in antibodies, and reduce the death-rate of immunized rabbits.

Here the results of the various immunization methods will be discussed.

Materials and methods

The following species were used: *E. atroseptica* (10 strains), *E. carotovora* (9), *E. carotovora* f. sp. *zeae* (1), *E. aroideae* (9), *E. chrysanthemi* (9), *E. cytolytica* (1), *E. carnegieana* (1), *E. amylovora* (3), *E. quercina* (2), *E. herbicola* (3), *E. lathyri* (3), and 6 strains of the soft rot type isolated by Dr. M. Arsenijević (Faculty of Agriculture,

Table 1. Immunization schemes used in the preparation of agglutinant sera in rabbits.

Antigen[1]	Way of antigen administration	Number of inoculations	Intervals between inoculations (days)	Administered antigen doses (ml)	Death rate of rabbits (%)	Quality of antiserum
1 Unwashed bacterial suspension	intravenous	5–6	4– 5	0.5/1/1.5/2/2.5	80–85	satisfactory
2 Bacterial suspension, 3 washings by centrifugation	intravenous	5–6	2– 3	1/1.5/3/3.5/4.5	75	good
			4– 5	0.5/1/1.5/2/2.5	45	good
3 Bacterial suspension, 3 washings by centrifugation 1 hour (60 °C)	intravenous	5–6	4– 5	0.5/1/1.5/2/2.5	10	unsatisfactory
4 Bacterial suspension, 3 washings	subcutaneous	1	12–14	1		satisfactory
by centrifugation	intravenous	4	2– 3	1/1.5/2/2.5	25–30	
5 a. Incomplete Freund antigen	intramuscular	1	12–14	2–4	10–15	good or
b. Bacterial suspension, 3 washings by centrifugation	intravenous	3–4	4	1/2/2.5/3	70	very good
	intramuscular	1	12–14	2–3	10–15	good or
	intravenous	3–4	4	0.3/0.5/1/2.5		very good
	intramuscular	1	20–24	2–4	15	very good[2]
	intravenous	3–4	3– 4	1/1.5/2/2.5	15–20	
	intramuscular	1	20–24	2–4	15	very good[2]
	intravenous	3–4	3– 4	0.3/1/1.5/2	very sporadic	
	subcutaneous	1	21	4	0	good
	intravenous	3–4	3	0.5/1/2/3	very sporadic	

1. Bacterial suspension concentration about 1.5×10^9 cells/ml. Freund antigen concentration about $4–5 \times 10^9$ cells/ml.
2. Antiserums can likewise be used with satisfactory results for precipitation reactions in tube or even in agar gel.

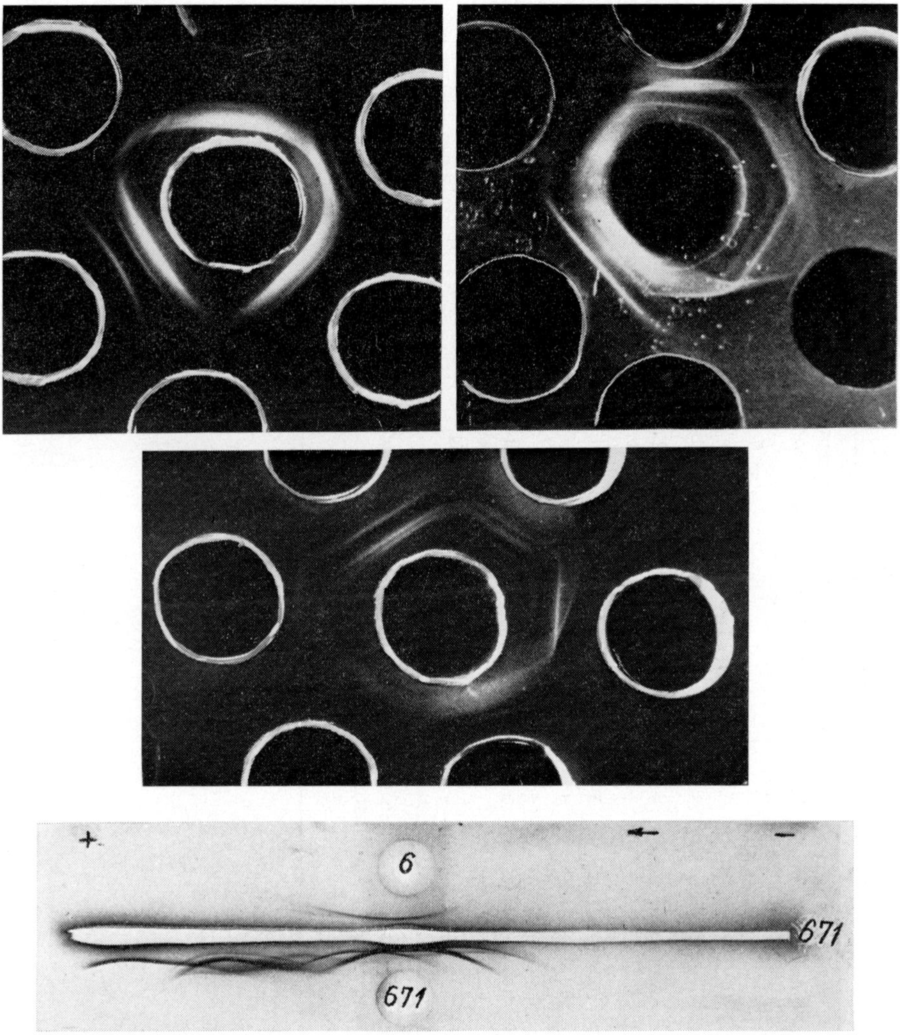

Serological relationships resulting from administering antisera obtained by hyperimmunization of rabbits with antigenic extracts.

Upper figures: double gel diffusion, with the antisera in the centre, surrounded by the antigenic extracts.

Lower figure: immunoelectrophoresis, with the antigenic extracts in the middle and the antiserum along the ditches.

Table 2. Immunization schemes used in the preparation of precipitant sera in rabbits.

Antigen[1]	Way of antigen administration	Number of inoculations	Intervals between inoculations (days)	Administered antigen doses	Death rate of rabbits (%)	Quality of antiserum
1 Incomplete Freund antigen	subcutaneous[3]	3	12–14	4 ml	usually 0	
2 Antigenic extracts of nucleoproteinic nature	intravenous	4–5	3– 5	0.5/1/1.5/2.5/3.4 mg or 0.2/0.4/1/1.5/2–2.5 mg	about 50 10–15	good
	subcutaneous[3] intravenous	3 4–5	21 3– 5	4 ml 1/2/3/4/5 mg	usually 0 55–65[2]	very good
	subcutaneous[3] intravenous	3 4–5	21 3– 5	4 ml 0.5/1/2/3/4 mg or 0.2/0.4/0.8–1/1.5/ 2/2.5–3 mg	usually 0 15 usually 0	very good

1. Concentration of Freund antigen about 4–5 $\times$ 10^9 cells/ml. Dry substance in antigenic extract 0.002–0.007 g/ml, depending on strain.
2. Rabbits usually died within 12 hours after first inoculation.
3. Nape zone.

Novi-Sad, Yugoslavia) from sunflower, pepper, tomato and cactus. Most strains used, nearly 40, were obtained from Dr D. C. Graham (Agricultural Scientific Services, East Craigs, Edinburgh, Scotland), Dr R. A. Lelliott (Plant Pathology Laboratory, Harpenden, Herts., England), Dr D. C. Hildebrand (University of California, Berkeley, U.S.A.) and Dr M. Kocur (J. E. Purkyne University, Brno, Czechoslovakia).

The types of antigen and the main immunization methods are given in Tables 1 and 2. Bacterial suspensions were prepared from 24 hour cultures on solid mediums, while the antigenic extracts employed were obtained by the technique previously described by Lazăr (1968).

The weights of the immunized rabbits were between 2.5 and 2.8 kg. They belonged to the breeds 'Large Belgian' (brown), grey Chinchilla, and white New Zealand. During the last four years 5 to 10 rabbits were inoculated with each strain.

Results and conclusions

The death rate percentages among the inoculated rabbits and the quality of the antisera in the rabbits that resulted in lasting immunization under the various methods are given in Tables 1 and 2 and in the figure.

The analysis of these data, supported by the experience acquired during the numerous trials of the last four years, lead to the following conclusions.

1. Anti-Erwinia agglutinant sera with high titres in antibodies that reduced death rate among inoculated rabbits are obtained when methods are used based on the administration of preferably intramuscular Freund antigen, followed by intravenous inoculations with bacterial suspension. It is essential that at least three weeks elapse between the inoculation of the Freund antigen and the series of three to four inoculations with bacterial suspension, and that the doses administered for less toxic strains should begin with 0.5 ml. For the highly toxic strains a start should be made with 0.2 to 0.3 ml, except for *E. atroseptica*, which usually is not toxic for rabbits, contrary to the strains of the other species which as a rule are very toxic.

2. To obtain high titre precipitant sera immunization methods should be based on rabbit hyparimmunization techniques with antigenic extracts, corresponding to the use of immunodiffusion in agar gel.

Two factors seem to be essential here.

The first is the interval between the three inoculations with Freund antigen, as well as the interval of about 21 days between the last of these three and the series of intravenous inoculations with antigenic extracts of nucleoproteinic-nature.

The second is that the doses of antigenic extracts for strains with little or no toxic action should be of 0.5, 1, 2, 3 and 4 mg, while for those with strong toxic action they should be 0.2, 0.4, 0.8, 1.5–2 and 2.5–3 mg.

3. As to the behaviour of the different races of rabbits to the toxic action of bacterial cells, or to cellular extracts of bacteria of the genus *Erwinia*, it was found from repeated experiments that rabbits of the Large Belgian and Chinchilla breeds are much more resistant than those of the New Zealand breed. The relevant experiments included about 35 attempts on some 10 rabbits for each strain.

References

Graham, D. C., 1963. Serological diagnosis of potato black-leg and tuber soft rot. Pl. Path. 12: 142–144.

Lazăr, I., 1968. Serological relationships of Corynebacteria. J. gen. Microbiol. 52: 77–88.

Popovici, I. & I. Lazăr, 1964. Research work on the zoopathogenic action of several species of bacteria from the *Erwinia* genus. Revtă rom. Biol. (Zool.) 6: 379–383.

Săvulescu, A. & I. Lazăr, 1964. Studies on the relation between the pathogenicity of certain species of *Erwinia* parasitic on plants and their toxic action on animals. Symp. hung. Acad. Sci. 18–22 Oct. 1964, Host parasitic relations in plant pathology: 151–157.

Săvulescu, A. & I. Lazăr, 1957. Investigations on bacterium *Erwinia cytolytica* Chester. Revtă rom. Biol. (Bot.) 5: 379–383.

Serological relationships between the 'amylovora', 'carotovora' and 'herbicola' groups of the genus Erwinia

I. Lazăr

Tr. Săvulescu Institute of Biology, Bucureşti, Romania

Abstract

The relationships between some 60 strains of ten *Erwinia* species is studied by immunological techniques. Involved are *E. atroseptica*, *E. carotovora*, *E. aroideae*, *E. chrysanthemi*, *E. cytolytica*, *E. carnegieana*, *E. amylovora*, *E. quercina*, *E. herbicola* and *E. lathyri*. These relationships are established both by cross-agglutinations and by precipitation with the double gel-diffusion technique.

The results indicate a close serological relationship of the species that are at present included in the genus. They also agree with results obtained in biochemistry (including nucleotide base-composition of DNA) by others.

The extension of such a serological study to other enterobacteria might lead to a better understating of the taxonomic relationships between phytopathogenic and other enterobacteria.

Introduction

Over the years much data have accumulated on the genus *Erwinia* which can be used in establishing its taxonomic relationships with other enterobacteria. Evidence includes data on nutrition (Starr & Mandel, 1950), end-products of glucose dissimilation (Suton & Starr, 1959), uronic acid catabolism (Kilgore & Starr, 1959), pectolytic enzymes (Kraght & Starr, 1953; Starr & Moran, 1962; Nasuno & Starr, 1966; Moran *et al.*, 1968a, b), pigment formation (Starr *et al.*, 1966), bacteriophages and bacteriocins (Goto, 1967), nucleotide composition of the DNA (Rosypal & Rosypalova, 1966; Mandel, 1967; De Ley, 1968; Starr & Mandel, 1969), physico-chemical analysis of cytoplasmatic components (Grou & Lazár, 1968; Grou, 1971), characterization or description of new organisms (Billing & Baker, 1963; Muraschi *et al.*, 1965; Whitcomb *et al.*, 1966; Graevenitz & Strause, 1966; Graham & Hodgkiss, 1967), and comparative taxonomy (Waldoe, 1945; Martinec & Kocur, 1963, 1968; Graham, 1964; Lockhart & Koenig, 1965; Krieg & Lockhart, 1966; Dye, 1968a, 1969a, b, c).

Our knowledge on the serology of this group is still incomplete, particularly if compared with what is known about the enterobacteria of medical interest. Nevertheless the studies of Lacey (1926), Matsumoto (1929, 1930), Elrod (1941), Stapp (1928), Mushin *et al.* (1959), Okabe & Goto (1955, 1956, 1957, 1958), Novakova (1957), and Ha Li Yoan (1961) should be mentioned.

To gather some additional data for a better understanding of the taxonomic

Erwinia species and material.

Species	Culture number	Obtained from
E. atroseptica	SR2/2, SR4/1, G110, HTi, QB4, E24, C399, MG146/42, 75V	Dr D. C. Graham
	531, 277	own isolates
E. carotovora	G123, G147, 1172, G117, ENA114, T2, E32, MG117/38	Dr D. C. Graham
	438	Dr R. A. Lelliott
E. carotovora	1008, 1011	Dr M. Kocur
f.sp. *zeae*	1065	Dr R. A. Lelliott
E. aroideae	119V, 911V, G173, 140V, J2, H2, A8IV, G115	Dr D. C. Graham
	550, 1231	Dr R. A. Lelliott
E. chrysanthemi	EP3, 910, 516, 377, ENA49, 517, 1516, 393, 402	Dr D. C. Graham
	M.380	H. P. Maas Geesteranus
E. cytolytica	6	own isolate
E. carnegieana	671	Dr R. A. Lelliott
E. amylovora	FB1, FB-2, FB-9	Dr D. C. Hildebrand
	1114, 1133	Dr M. Kocur
E. quercina	Ac1, Ac2, Ac4, AcC	Dr D. C. Hildebrand
E. herbicola	G138, G141	Dr D. C. Graham
	2405, 2406, 2407	Dr M. Kocur
E. lathyri	G144, G146, G150, G151, G153, G155, G157	Dr D. C. Graham

relationships between organisms included in the genus Erwinia, we started (some five years ago) a comparative serological study on soft rot coliform. The first results, which are included in the present paper, have already been presented to the 1st Congress of Plant Pathology (London, 1968). Other species of the 'amylovora' and 'herbicola' groups were then studied, the latter including Erwinia-like bacteria isolated from plants, animals and man.

The present study was parallel with the complex biochemical investigations undertaken by Dr Graham (personal communication), in so far that part of his strains will be serologically characterized here.

Materials and methods

Cultures Some 60 strains were included (see table), except three supplied by Dr D. C. Graham (Agricultural Scientific Services, Edinburgh, Scotland), Dr R. A. Lelliott (Plant Pathology Laboratory, Harpenden, England), Dr D. C. Hildebrand (University of California, Berkely, USA), Dr M. Kocur (J. E. Purkyne University, Brno, Czechoslovakia) and Mr Maas Geesteranus (Institute of Plant Pathology, Wageningen, the Netherlands).

Antigens The antigens used in the preparation of agglutinant sera consisted of bacterial suspensions obtained from 24-hour cultures grown on nutrient broth-glucose or sucrose-agar medium. The bacterial cells used in the preparation of suspensions for intravenous inoculations or, in the incomplete Freund antigen test for subcutaneous inoculations, were washed at least three times with physiological saline (0.85%) by centrifuging.

For cross-agglutinations bacterial suspensions in a 10^{-8}–10^{-9} dilution were used; for precipitations in agar gel, antigenic extracts were obtained by treating acetone-dried bacterial cells with a 2% sodium desoxycholate solution according to a previously described technique (Lazar, 1968), or antigenic extracts were obtained by repeated freezing and thawing. About 2 g acetone-dried bacterial cells were mortar-ground and suspended in about 10g 0.1 M (pH = 7) phosphate buffer. They were frozen at temperatures between −10 and −15 °C. After the 6th or 7th thawing, the suspension was centrifuged for 45–50 min at 16.000 rev/min, and the supernatant was used in the precipitation reaction.

Antisera The agglutinant and precipitant sera were prepared in rabbits, according to the methods that had given the best results in previous investigations (Lazar, 1971).

Serological reactions Serological relationships were tested by cross-agglutinations and precipitations; for the latter the double gel-diffusion technique (Ouchterlossy, 1958) was used.

Results

The results are given in Figs 1, 2 and 3. Fig. 4 demonstrates how the results of the other figures were obtained. Their analysis leads to the following conclusions.

Cross-agglutinations In view of the large number of strains to be studied it was necessary to confine ourselves to dilutions up to 1:20 for 21 antisera. The results, schematically presented in Fig. 1, depict a mosaic of reactions which makes it difficult to draw any firm conclusion on the homogeneity of the antigenic action for each of the studied species. Nevertheless it is obvious that reactions between homologous antigens and antibodies, especially of the same strain, and between those of the strains of the same species (with some exceptions in strains of *E. carotovora*, *E. aroideae*, *E. chrysanthemi* and *E. lathyri*) are markedly stronger than between heterologous antigens and antibodies. This points to the existence of distinct serotypes. At the same time, however, numerous relationships were recorded (in some cases very strong ones) not only between strains of different species, but also between different groups of one species, if we consider the organisms included so far in the genus Erwinia to make up three large groups: 'carotovora', 'amylovora' and 'herbicola'.

The strains characterized as *E. atroseptica*, *E. chrysanthemi*, *E. herbicola* and *E. lathyri* seemed more homogenous from the point of view of antigenic structure. Their antigens usually reacted more weakly, and in fewer antisera of the strains of other

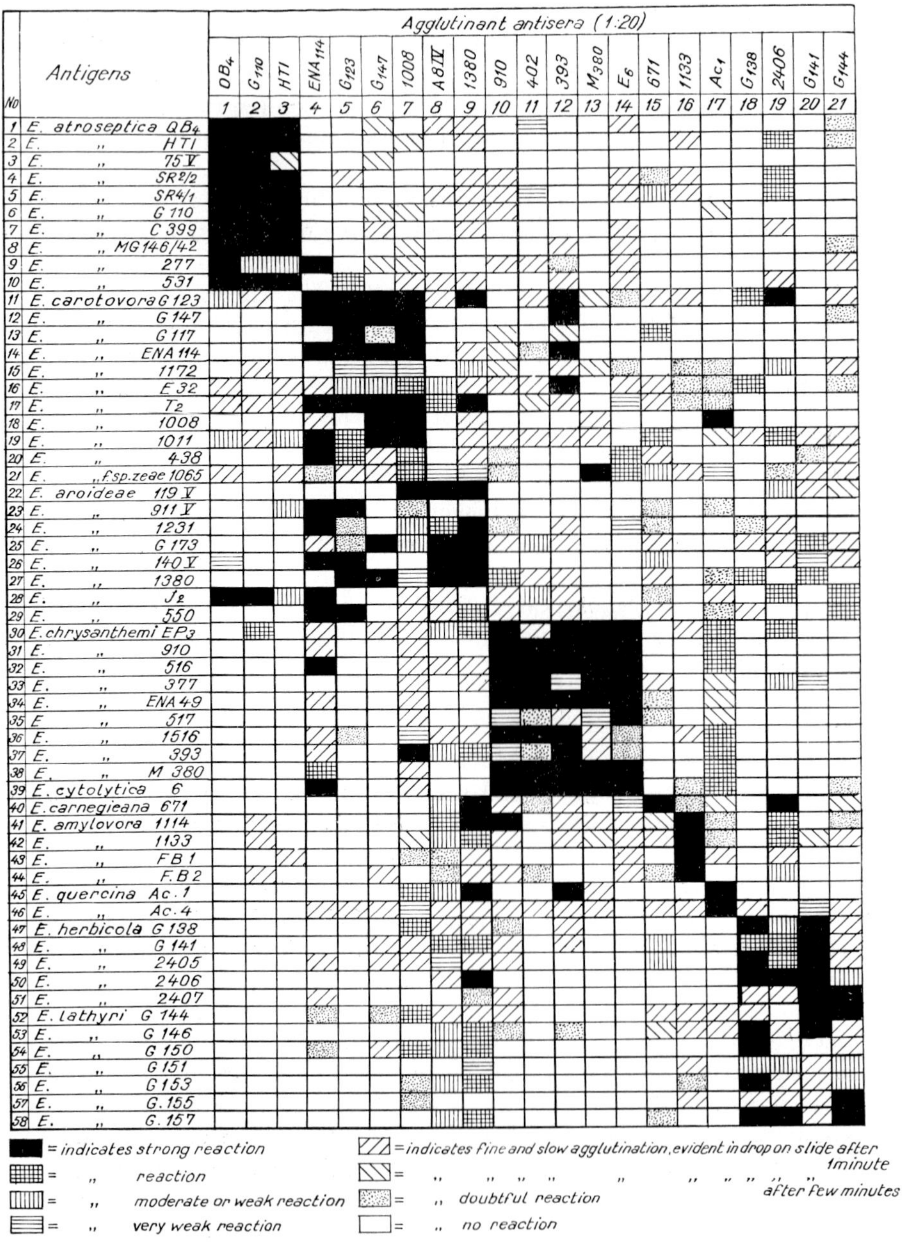

Fig. 1. Serological relationships between *Erwinia* species belonging to the 'carotovora', 'amylovora' and 'herbicola' groups, as determined by cross-agglutination reaction.

species. A great heterogeneity was recorded particularly in strains characterized as *E. carotovora* and *E. aroideae*. Likewise, from the data in Fig. 1, it could be concluded that some strains, particularly among pectobacteria, reacted more strongly in the antisera of strains of other species than in those of the species to whom they were attributed. If these strains are wrongly classified, it is premature to rely solely on serological tests; it is, however, interesting to consider also their biochemical behaviour; Dr Graham will be able to clear up this point, for he has submitted the same strains to a very wide range of biochemical tests (over 80). The species and their strains in our study did not appear as 'clean' species from the serological point of view, and if the reactions indicated in Fig. 1 with diagonal crossings are considered, it may be argued that, particularly in pectobacteria, we have to deal with strains of the same species, with numerous serotypes.

The conclusion is, that on the basis of cross-agglutinations the organisms included so far in the genus Erwinia seem to form a serologically closely related group and that, if other tests and the analyses likewise support these results, it is natural that in particular the pectobacteria have to be reduced to fewer taxons than accepted at present.

No doubt these results are far from solving the serology of the Erwinia species, but they show once again how complex and intricate the situation is. Much work will still be required to arrive at the delimitation and characterization of serotypes, as became apparent in the study of enterobacteria of medical interest.

Precipitations by the double gel-diffusion technique First a few remarks have to be made on the tested antisera and antigen extracts. The antisera, obtained by hyperimmunization of rabbits with antigenic extracts, gave good results and proved useful for detecting a large number of precipitation lines for homologous antisera and extracts, up to 8. However, the antigenic extracts obtained by autoclaving very concentrated suspensions (the cultures from the medium surface from 5 Roux dishes – 500 sq.cm, suspended after centrifuging in 10 ml double-distilled water) gave unsatisfactory results as they did not reveal more than two precipitation lines. The extracts obtained by freezing-thawing or by treatment with a 2% sodium desoxycholate solution proved very good. We recommend, nevertheless, the extracts obtained by freezing-thawing, that are not only easier to prepare, but often result in more precipitation lines being formed (Fig. 3).

The results of precipitations in agar gel, using the gel-diffusion technique, are schematically presented in Figs 2 and 3. From the analysis of the data inscribed in these figures the following two conclusions may be drawn.

1. Both in chemical and physical extracts, strong serologically relationships revealed by numerous precipitation lines were seen between strains of different species, but particularly within the same species.

2. If the reactions are compared along the diagonals in Figs 3 and 4, starting in the left upper corner, and especially those between antisera and antigens of strains of the same species, numerous precipitation lines are seen, generally more than for heterologous reactions (on both sides of the diagonal). As for cross-agglutinations, this proves the existence of distinct serotypes, as usually at least one or two additional

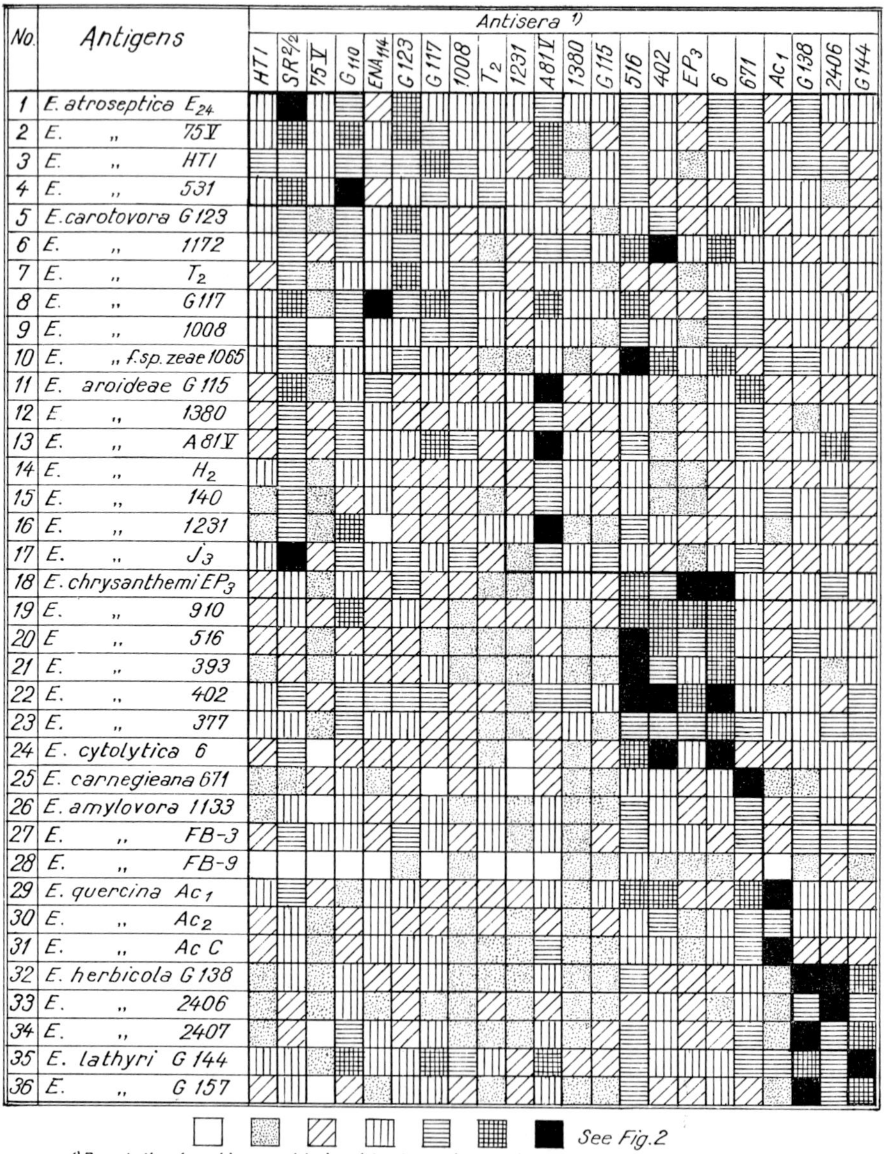

1) Precipitant antisera obtained by hyperimmunization of rabbits with antigenic extracts.

Fig. 2. Serological relationships between *Erwinia* species belonging to the 'carotovora', 'amylovora' and 'herbicola' groups, as determined by double gel diffusion, and cellular extracts from acetone-dried bacteria with 2% sodium desorychlolate solution as an antigen.

Fig. 3. Serological relationship between *Erwinia* species belonging to the 'carotovora', 'amylovora' and 'herbicola' groups as determined by double gel diffusion, using as antigen cellular extracts obtained by repeated freezing and unfreezing of acetone-dried bacteria.

No.	Antigens	SR2/2	G 110	ENA 114	G 117	T_2	A 81 V	J 3	516	402	EP_3	6	671	Ac 1	G 138	2406	G 144
		Antisera[1]															
1	*E. atroseptica* 531	6 to 8	6 to 8	4	5	5	6 to 8	4	5	5	4	5	5	2	1	1	1
2	" " G 110	4	5	2	3	3	2	2	2	2	3	3	2	1	1	0	1
3	" " SR 2/2	6 to 8	4	3	2	3	3	2	3	2	4	2	3	2	2	2	1
4	" " E 24	5	5	3	3	3	5	4	4	5	4	4	5	3	3	3	2
5	*E. carotovora* ENA 114	2	3	3	3	3	3	2	3	2	3	2	2	1	1	1	0
6	" " T_2	2	3	4	2	6 to 8	4	3	4	2	4	3	4	2	1	2	2
7	" " f. sp. zeae 1065	2	2	2	2	2	3	1	5	5	5	5	2	5	4	4	2
8	*E. aroideae* 1380	4	5	5	5	5	5	2	5	4	4	4	5	1	2	2	3
9	" " A 81 V	5	5	5	5	5	6 to 8	5	6 to 8	6 to 8	5	5	5	3	3	3	3
10	" " 140 V	3	4	4	3	3	5	4	5	4	4	4	5	3	3	2	1
11	" " H_2	2	3	3	4	3	4	3	4	4	3	4	3	1	2	3	2
12	*E. chrysanthemi* 402	1	2	1	2	2	2	1	4	4	4	3	2	2	1	2	2
13	" " 516	2	2	3	3	2	3	1	6 to 8	6 to 8	6 to 8	4	3	1	3	2	2
14	" " EP_3	2	2	3	3	2	3	1	6 to 8	6 to 8	6 to 8	4	3	3	3	4	4
15	*E. cytolytica* 6	2	1	2	2	4	2	1	3	4	2	6 to 8	2	0	4	2	2
16	*E. carnegieana* 671	2	3	3	3	2	4	1	3	4	3	2	6 to 8	2	2	3	2
17	*E. amylovora* 1133	2	1	1	2	1	3	2	2	2	2	2	2	2	1	3	1
18	*E. quercina* Ac 1	1	0	1	1	0	0	1	1	1	1	1	1	5	1	1	1
19	" " Ac 2	2	1	1	2	1	3	2	3	2	2	3	3	4	2	3	2
20	*E. herbicola* G 138	2	2	1	2	1	5	3	5	2	3	3	2	3	6 to 8	6 to 8	4
21	" " 2406	1	1	0	2	1	3	2	3	2	2	1	1	2	6 to 8	6 to 8	3
22	*E. lathyri* G 144	2	1	1	4	1	5	2	2	2	3	3	3	3	5	4	6 to 8
23	" " G 157	2	1	0	2	0	3	1	1	1	2	2	2	2	6 to 8	6 to 8	4

1) Precipitant antisera obtained by hyperimmunization of rabbits with antigenic extracts.

- (blank) = indicates no precipitation lines
- (dotted) = " 1 " line
- (diagonal hatching) = " 2 " lines
- (vertical hatching) = " 3 " "
- (horizontal hatching) = " 4 " "
- (cross-hatching) = " 5 " "
- (solid black) = " 6 to 8 " "

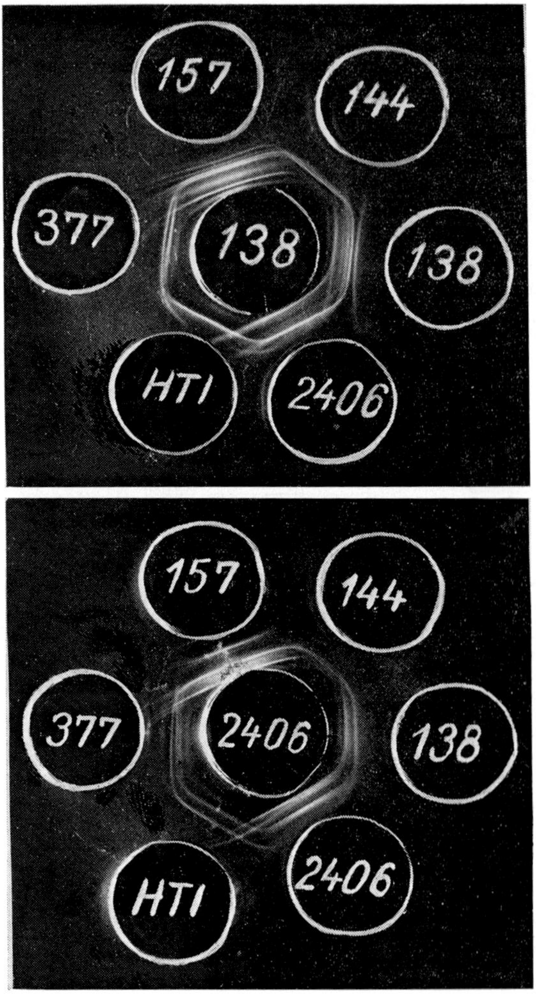

Fig. 4. Serological relationships as demonstrated by double gel diffusion technique.
In the center above: hyperimmune antiserum of *E. herbicola* strain G 138, surrounded by antigenic extracts of *E. lathyri* (G 157 and G 144), *E. herbicola* (G 138 and 2406), *E. atroseptica* (HTI) and *E. chrysanthemi* (377).
In the center below: hyperimmune antiserum of *E. herbicola* strain 2406, surrounded by the same.

precipitation lines appear that might indicate specific antigenic fractions.

At the same time, the presence in nearly all cases of common antigens between the strains of the species studied is surprising. This proves, on the one hand, the unity of the respective groups, while, on the other, it justifies the existence of the well-defined organisms distinguished over the years within the genus. No doubt additional research is still necessary, and it is rather hazardous to make any nomenclatural decisions merely on the basis of these results. Similar research, on other enterobacteria as well, is needed before any final conclusions can be drawn on the classification, or on a restriction to fewer taxons, or on other enterobacteria included at present in the genus Erwinia. Nevertheless, considering the results of this study, and taking into account the opinions brought forward by others, particularly in the field of biochemical tests (Dye, 1968a, b; 1969a, b, c; Martinec & Kocur, 1963; and others), it seems that, at least for pectobacteria, the reduction of the number of taxons is advisable. Likewise, the numerous antigenic connections recorded between pectobacteria and the species of the two other groups ('amylovora' and 'herbicola') do not seem to justify grouping the soft rot coliform ones into a separate genus, as was proposed by Waldee (1945).

Though generally the results obtained agree with those arrived at by Starr & Mandel in their recent study (1969), several remarks have to be added.

E. chrysanthemi and *E. carnegieana*, which by their GC % value (51 and 51.1, respectively) are considered by these authors to be different from the group *E. carotovora, E. atroseptica, E. aroideae, E. oleraceae* and *E. solanisapra* (GC % values around 52 %), whereas according to the present data the two species appear to differ much more from *E. carotovora, E. atroseptica* and *E. aroideae* in view of their antigenic structure than the latter three differ mutually.

The fact that the yellow-pigmented Erwinia's isolated from plants, animals and man seem related by many antigenic precipitation lines to the phytopathogenic organisms included the genus Erwinia seems hardly surprising. It may, on the contrary, assist in explaining the evolution of this so widely spread group. The resemblance of these bacteria to the phytopathogenic enterobacteria was shown by Starr & Mandel in their description (see introduction), and by the investigations on the nucleotide composition in DNA (De Ley, 1968; Starr & Mandel, 1969).

I am convinced that the results mentioned in this paper will be supported by those of Dr Graham, who investigated the biochemical properties of many more strains, including those studied by us. Doubtless his results will be a most welcome elucidation of the taxonomic relationships between phytopathogenic and other enterobacteria.

References

Billing, E. & L. A. E. Baker, 1963. Characteristics of *Erwinia*-like organisms found in plant material. J. appl. Bact. 26: 58–65.

De Ley, J., 1968. DNA base composition of yellow *Erwinia* strains. Antonie van Leeuwenhoek 34: 257–262.

Dye, D. W., 1968a. A taxonomic study of the genus *Erwinia*. 1st Int. Congr. Pl. Path. (London, July 1968), Abstracts of Papers: 51.

Dye, D. W., 1968b. A taxonomic study of the genus *Erwinia*. I. The 'amylovora' group. N.Z.Jl Sci. 11: 590–607.

Dye, D. W., 1969a. A taxonomic study of the genus *Erwinia*. II. The 'carotovora' group. N.Z.Jl Sci. 12: 81–97.

Dye, D. W., 1969b. A taxonomic study of the genus *Erwinia*. III. The 'herbicola' group. N.Z.Jl Sci. 12: 223-236.

Dye, D. W., 1969c. A taxonomic study of the genus *Erwinia*. IV. 'Atypical' Erwinias. N.Z.Jl Sci. 12: 832–839.

Elrod, R. P., 1941. Serological studies of the Erwinieae. II. Soft rot group: with some biological considerations. Bot. Gaz. 103: 266–279.

Goto, M., 1967. Relationship of *Erwinia* to Enterobacteriaceae. Summaries of The second International Conference on Plant Pathogenic Bacteria, Oeiras, Portugal.

Goto, M. & N. Okabe, 1957. Studies on the strains of *Erwinia carotovora* (Jones) Holland. IV. Antigenic variations. Bull. Fac. Agr. Schizuoka Univ. 7: 11–12.

Goto, M. & N. Okabe, 1958. Studies on the strains of *Erwinia carotovora* (Jones) Holland. V. Antigenic structures of soma, their relation to biochemical properties, and heat inactivation of agglutinins. Bull. Fac. Agr. Shizuoka Univ. 8: 1-3.

Graevenitz, A. von & A. Strouse, 1966. Isolation of *Erwinia* spp. from human sources. Antonie van Leeuwenhoek 32: 429–430.

Graham, D. C., 1964. Taxonomy of the soft-rot coliform bacteria. A. Rev. Phytopath. 2: 13–42.

Graham, D. C. & W. Hodgkiss, 1967. Identity of Gram negative, yellow pigmented fermentative bacteria isolated from plants and animals. J. appl. Bact. 30: 175–189.

Grou, E., 1971. Caracteristicile electroforetice ale proteinelor citoplasmatice de la unele specii de *Erwinia* (in press).

Grou, E. & I. Lazăr, 1968. Proprietăţile fizico-chimice şi acţiunea toxică pe iepuri a extractelor nucleoproteice de la speciile de *Erwinia* care produc putreziri moi la plante. Microbiologia 1: 183–188.

Ha Li Yoan, 1961. Nekotorîe osobennosti biologii vozbuditelja carnoj nozki kartofelja. Summary of the thesis presented to the Faculty of Biology, Univ. of Moscow.

Kilgore, W. W. & M. P. Starr, 1959. Catabolism of galacturonic and glucuronic acids by *Erwinia carotovora*. J. biol. Chem. 234: 2227–2235.

Kraght, A. J. & M. P. Starr, 1953. Pectic enzymes of *Erwinia carotovora*. Archs Biochem. Biophys. 42: 271.

Krieg, R. E. & W. R. Lockhart, 1966. Classification of enterobacteria based on overall similarity. J. Bact. 92: 1275–1280.

Lazăr, I., 1968. Serological relationships of Corynebacteria. J. gen. Microbiol. 52: 77–88.

Lazăr, I., 1971. Investigations concerning preparation of anti-*Erwinia* sera on rabbits. Proceedings Third International Conference on Plant Pathogenic Bacteria, Wageningen, The Netherlands.

Lockhart, W. R. & K. Koenig, 1965. Use of secondary data in numerical taxonomy of the genus *Erwinia*. J. Bact. 90: 1638–1644.

Mandel, M., 1967. DNA base composition and taxonomy of *Erwinia*. Summaries of The second International Conference on Plant Pathogenic Bacteria, Oeiras, Portugal.

Martinec, T. & M. Kocur, 1963. A taxonomic study of the genus *Erwinia*. Folia Publ. Fac. nat. Sci. J. E. Purkynye Univ. (Biol.) 3: 1–163.

Martinec, T. & M. Kocur, 1968. On taxonomic position of *Erwinia carotovora*. Folia Publ. Fac. nat. Sci. J. E. Purkynye Univ. (Biol.) 4: 63–67.

Matsumoto, T., 1929. On the diagnosis of certain plant infection diseases by means of serological reactions. J. Soc. trop. Agr., Taiwan 1: 14–22.

Matsumoto, T., 1930. Further studies on some putrefactive phytopathogenic bacteria by agglutinin absorbtion. J. Soc. trop. Agr., Taiwan 2: 16–25.

Moran, F., S. Nasuno & M. P. Starr, 1968a. Extracellular and intracellular polygalacturonic acid trans-eliminases of *Erwinia carotovora*. Archs Biochem. Biophys. 123: 298–306.

Moran, F., S. Nasuno & M. P. Starr, 1968b. Oligogalacturonide transeliminase of *Erwinia carotovora*. Archs Biochem. Biophys. 125: 734–741.

Muraschi, T. F., M. Friend & D. Bolles, 1965. *Erwinia*-like microorganisms isolated from animal and human hosts. Appl. Microbiol. 13: 128–131.

Nasuno, S. & M. P. Starr, 1966. Polygalacturonase of *Erwinia carotovora*. J. biol. Chem. 241: 5298–5306.

Novakova, J., 1957. A new method of isolation of blackleg pathogen from diseased plants. Phytopath. Z. 29: 72–74.

Okabe, N. & M. Goto, 1955. Bacterial plant diseases in Japan. II. Studies on soft rots due to *Erwinia aroideae* (Townsend) Holland with special reference to the antigenic structures of flagellae. Bull. Fac. Agr. Schizuoka Univ. 5: 72–86.

Ouchterlony, Ö., 1958. Diffusion in gel methods for immunological analysis. Prog. Allergy 5: 1–78.

Rosypal, S. & A. Rosypalová, 1966. Genetic, phylogenetic and taxonomic relationships among bacteria as determined by their deoxyribonucleic acid base composition. Folia Publ. Fac. nat. Sci. J. E. Purkynye Univ. (Biol.). 7: 1-131.

Stapp, C., 1928. Schwarzbeinigkeit und Knollennassfäule der Kartoffel. Mitt. biol. Reichsanst. Ld-u. Forstw. 16: 643–703.

Starr, M. P. & M. Mandel, 1950. The nutrition of phytopathogenic bacteria. IV. Minimal nutritive requirements of the genus *Erwinia*. J. Bact. 60: 669–675.

Starr, M. P. & F. Moran, 1962. Eliminative split of pectic substances by phytopathogenic soft-rot bacteria. Science, 135: 920–921.

Starr, M. M., G. Cosens & H.-J. Knackmuss, 1966. Formation of the blue pigment indigoidine by phytopathogenic *Erwinia*. Appl. Microbiol. 14: 870.

Sutton, D. D. & M. P. Starr, 1959. Anaerobic dissimilation of glucose by *Erwinia amylovora*. J. Bact. 78: 427–432.

Waldee, E. L., 1945. Comparative studies of some peritrichous phytopathogenic bacteria. Iowa St. Coll. J. Sci. 19: 435–485.

Whitcomb, R. F., M. Shapiro & J. Richardson, 1966. An *Erwinia*-like bacterium pathogenic to leafhoppers. J. Invertebr. Path. 8: 299.

A qualitative study on the antigens involved in the complement-fixation reaction among some phytopathogenic pseudomonads

A. Coleno, M. Le Normand and M. R. Barzic

Station de Pathologie Végétale I.N.R.A., 65, Rte de St Brieuc, Rennes (35), France

Abstract

The study of some phytopathogenic pseudomonads by the complement fixing technique shows the existence of several antigenic determinants some of which are common, others specific. The specificity could be used in taxonomy or seed disease detection (with the possibility of estimating contamination rates). In the case of *Pseudomonas phaseolicola, P. syringae* and *P. mors-prunorum*, the specific thermolabile antigens may be connected with precipitating or agglutinating structures.

Introduction

To characterize sera against animal or human pseudomonads mostly the somatic polysaccharidic antigens are used, involving immunodiffusion or agglutination techniques. Examples for *Pseudomonas aeruginosa* have been supplied by Müller Kleinmaier (1958), Sandvick (1960) and Köhler (1957). As a rule the number of serotypes defined in this way is large for a single species (Sandvick, 1960, mentions seven for *P. aeruginosa* of animal origin). When such antigens were used in the study of phytopathogenic pseudomonads (Lovrekovich & Klement, 1961; Lovrekovich *et al.*, 1963; Digat & Escudié, 1967; Lucas & Grogan, 1969a, b) the number of serotypes for a single species was much lower: Lucas & Grogan found two for *P. lachrymans* despite the fact that their project covered a large number of strains and isolates. This interesting polysaccharide specificity may be used in practice to detect bacteria on seeds (Coleno, 1968b) if the serum is adequately prepared (Coleno, 1970), which means that it must have a good anti-polyosidic antibody titre. But in studying a mixture of species, attention has to be paid to various types of antigens reacting in the complement-fixation test which have not been studied in bacterial phytopathology until now. Their existence has been shown in a previous paper (Coleno, 1970).

The purpose of this paper is to provide some details on these antigens in three species of phytopathogenic pseudomonads: *P. phaseolicola, P. syringae* and *P. mors-prunorum.*

Material and methods

Table 1 gives the material used and its origin. In addition of these strains we have

Table 1. Species and strains used in the study on complement-fixing.

Species, strains	Origin	Number in the original collection
P. phaseolicola 1 and 6	INRA Angers	22, 302
13, 14, 53, 61 and 87	INRA Rennes	
88, 90 and 91	NCPPB	1321, 605, 1103
P. syringae	INRA Angers	311, 612, 560, 229
P. mors-prunorum	INRA Angers	542, 04
P. fluorescens	INRA Angers	121
P. marginalis 384	INRA Rennes	
P. tolaasii	INRA Angers	1287, 1288
P. cichorii	INRA Angers	1005
P. lachrymans	INRA Angers	70
P. viridiflava	INRA Angers	1142
P. solanacearum 30	INRA Rennes	

INRA collection: domaine de Bois Labbé, Beaucouzé par Angers (M. Ridé, J. Prunier). NCPPB collection: National Collection of Phytogenic Bacteria, Plant Physiology Laboratory, Harpenden, Herts, England (R. A. Elliott).

used about fourty isolates of the same species. All antigens were prepared from aerated cultures. The samples were taken in the logarithmic phase of growth. The bacterial suspensions in 8% saline were washed three times. Afterwards they were diluted either to 2.10^9 cells per ml and killed by formol (suspension S) or centrifuged and directly extracted with a phosphate buffer (pH 7.2 $\pm$ 0.5) by ultra-sonation for 5 min in an MSE ultrasonic grinder. The container was kept in an icetank throughout the operation. After centrifuging, the supernatant was filtered off and dialysed for 48 h against an 8% saline, then reduced to a concentration of 5 mg N protein/ml (extract E). O antigen was prepared by Boivin's technique according to the details given by Staub; they also served as test antigens.

Rabbits were immunized according to two methods.

Method I, extract E Once a week for three weeks an intradermal injection in the fat of the neck with 2 ml of an equal mixture of Freund's complete antigen (difco-certified) and a test antigen (extract E); three times a week for three weeks an injection of antigen alone. This brings about normal immunisation.

For short immunisation, the last three injections were omitted.

For long immunisation, two or more series of three injections were added, at intervals of one month.

Method II (Staub) suspension S Six intravenous injections (0.5, 1, 2, 2, 2 and 2 ml) at a rate of three injections per week. After a month's rest: 9 intravenous injections (0.5, 0.5, 0.5, 1, 1, 1.5, 2, 2 and 2 ml) at a rate of three injections per week. This is the normal immunisation procedure.

For short immunisation the last three injections were omitted.

Table 2. Serological relationships among some phytopathogenic pseudomonads using the complement-fixation test.

Normal immunisation	Antisera against					
	P. phaseo-licola	*P. sy-ringae*	*P. mors-prunorum*	*P. margi-nalis*	*P. fluo-rescens*	*P. solana-cearum*
P. phaseolicola	+++	++	++	−	−	−
P. syringae	++	+++	++	−	−	−
P. mors-prunorum	++	++	+++	−	−	−
P. cichorii	+	±	+	−	−	−
P. lacrymans	+	±	+	−	−	−
P. viridiflava	+	±	+	−	−	−
P. marginalis	±	−	−	+++	+	−
P. fluorescens	±	−	−	+	+++	−
P. tolaasi	±	−	−	+	+	−
P. solanacearum	−	−	−	−	−	+++

+++ = control fixing, ++ = strong fixing (from 10 up to 80 or 90%), + = weak fixing (5–10% of the control), ± = very weak fixing, − = no fixing.

For long immunisation two or more series of six injections of 2 ml were added, each series preceded by a 1 ml subcutaneous injection. The lapse of time between the series was three weeks.

Blood samples were always taken by intracardiac puncture eight days after the last injection. Several animals were immunised each time. Only the best sera, with a titre of at least 1/1280 with the homologous antigen were used. Unless stated otherwise, the results relate to sera prepared following one of the normal immunisation procedures.

Cross-absorptions were performed according to Kauffman (1954) using living cells. They were carried out before the destruction of endogenous complement activity of the test serum.

The complement-fixation test was achieved as described by Levine (1967). The results were clear enough to use the rapid plate technique in routine testing. This technique is not very sensitive but practical (Cours sérologie Institut Pasteur, Paris 1965), particularly in a qualitative study like this. Therefore, the results are presented as indicated under Table 2.

Results

A comparison of the extracts of the three species and their sera (Tables 1 and 2) shows that all have antigens in common. When compared with extracts from other phytopathogenic pseudomonads they behave as a homogeneous group. This is not surprising in view of the serological relationships between the three species (Coleno, 1968a). The same situation occurs with *Pseudomonas fluorescens*, *P. tolaasi* and *P.*

Table 3. Influence of immunisation route on appearance of specific antibodies in the complement-fixation test.

Antigens from	Antisera against								
	P. phaseolicola 14 absorbed by			*P. syringae* 612 absorbed by			*P. mors-prunorum* 542 absorbed by		
	P. phaseolicola 14	*P. syringae* 612	*P. mors-prunorum* 542	*P. phaseolicola* 14	*P. syringae* 612	*P. mors-prunorum* 542	*P. phaseolicola* 14	*P. syringae* 612	*P. mors-prunorum* 542
Intravenous route (type 2 sera), normal immunisation									
P. phaseolicola	−	+	+	−	−	−	−	−	−
P. syringae	−	−	−	−	−	+	−	−	−
P. mors-prunorum	−	−	−	−	−	−	+	+	−
Intradermal route (type 1 sera), normal immunisation									
P. phaseolicola	−	+	−	−	−	−	−	−	−
P. syringae	−	−	−	+	−	+	−	−	−
P. mors-prunorum	−	−	−	−	−	−	+	+	−

Table 4. Influence of the length of immunisation (unsorbed sera).

	Antisera against *P. phaseolicola*			Antisera against *P. syringae*			Antisera against *P. mors-prunorum*		
	P. phaseolicola	*P. syringae*	*P. mors-prunorum*	*P. phaseolicola*	*P. syringae*	*P. mors-prunorum*	*P. phaseolicola*	*P. syringae*	*P. mors-prunorum*
Short immunisation	+	+	+	.	.	.	+	+	+
Normal immunisation	+	+	+	+	+	+	+	+	+
Long immunisation	.	.	.	+	+	+	.	.	.

marginalis. Similarly, *P. lachrymans*, *P. cichorii* and *P. viridiflava* react uniformly with the antisera, but the absence of homologous sera makes a further study impossible here. Finally *P. solanacearum* shows no relationship with any species tested.

The study of *P. phaseolicola*, *P. syringae* and *P. mors-prunorum* may be continued, however. By absorbing the respective sera with the heterologous antigens, two types of results appear:

1. Depending on the immunisation method (Tables 1 and 3) absorption leaves antibodies which are specific for each of the three species. However, in antisera against *P. phaseolicola* prepared following the intradermal route of injection, the specific antibodies are not detected when the sera are absorbed with *P. mors-prunorum*. The same is true for antisera against *P. syringae* prepared following the intravenous route of injection and absorbed with *P. phaseolicola*. With cross-absorbed antisera against *P. mors-prunorum* the reactions are specific, independent of the immunisation route.
2. Depending on the duration of immunisation, the common antibodies appear very early (Table 4) since a positive result is obtained independent of extract and serum. On the other hand, as apparent from Table 5, in certain cases the specific antibodies sometimes appear rather late, though, generally speaking, still in time if immunisation is continued long enough.

Discussion

Taken as a whole the results show that various antigens reacting in the complement fixation test do exist among the phytopathogenic pseudomonads. The distribution of these antigens is variable. It is possible, however, to assume that they allow serological grouping. In addition, *P. phaseolicola*, *P. syringae* and *P. mors-prunorum* have at least one antigen in common and others specific for each species, which is more or less easily shown.

Furthermore, as shown by Table 3, absorbing the type 2 anti-syringae serum with *P. mors-prunorum* does not remove the specific antibodies against *P. syringae*. With this absorbed serum, antigens of *P. phaseolicola* are not detected. This seems rather paradoxical, as absorption with *P. phaseolicola* does eliminate the antibodies against *P. syringae*. Thus it might be supposed that *P. phaseolicola* and *P. syringae* possess a common precipitating or agglutinating antigen (since Kauffmann's absorption technique has been applied), and that in the case of *P. syringae* this antigen is involved in the antigen-antibody complex capable of fixing the complement. A long immunisation, or immunisation carried out by a favourable route, reveals precipitating, agglutinating (or complement-fixing) antibodies with a narrow specificity and directed against this structure. A specific serum is thus obtained by saturation.

On the other hand, a short immunisation, or immunisation carried out by an unfavourable route, reveals antibodies with a wider specificity and with a double vocation directed against common determinants but also capable of fixing the complement. This type of antibody is eliminated in a saturation by *P. phaseolicola*, even if, in the presence of extracts of the same species, they cannot be fixed because of the

Table 5. Influence of the length of immunisation (absorbed sera).[1]

	Antiserum against *P. phaseolicola* (ID route) saturated by			Antiserum against *P. phaseolicola* (IV route) saturated by			Antiserum against *P. syringae* (IV route) saturated by		
	P. phaseo-licola	*P. syrin-gae*	*P. mors-prunorum*	*P. phaseo-licola*	*P. syrin-gae*	*P. mors-prunorum*	*P. phaseo-licola*	*P. syrin-gae*	*P. mors-prunorum*
Short immunisation	−	−	−	−	+	−	.	.	.
Normal immunisation	−	+	−	−	+	+	−	−	+
Long immunisation	.	.	.	.	.	.	+	−	+

1. The homologous antigen was used for all tests.

Table 6. A hypothetic scheme.

	Antibodies obtained against *P. syringae*	Removed by
Short immunisation	Ac (C C_I)	C C_I and C_I *(P. syringae – P. phaseolicola)*
(or unfavourable route)	Ac (C_I)	C C_I and C_I *(P. syringae – P. phaseolicola)*
Normal long immunisation	Ac (C C_I)	C C_I and C_I *(P. syringae – P. phaseolicola)*
(or favourable route)	Ac (C_I)	C C_I and C_I *(P. syringae – P. phaseolicola)*
	Ac (C)	C C_I and C *(P. syringae)*

Hypothesis C represents antigenic determinant involved in the complement fixation test (specific antigen)
Hypothesis C_I represents antigenic determinant able to precipitate or to agglutinate *P. syringae*: C C_I; *P. phaseolicola*: C_I; *P. mors-prunorum*: d.

absence of a corresponding antigen in *P. phaseolicola.* This is a question therefore of the quality of the antibodies whose specificity increases with time (Table 6).

In the same way it is possible to explain the interference between *P. phaseolicola* and *P. mors-prunorum* (Table 3). Probably the specific antigen of *P. mors-prunorum* acts in a similar way, but it has not been possible to demonstrate this, either because the interferring species is neither *P. phaseolicola* nor *P. syringae*, or because the corresponding antibodies appear much too early.

In a previous paper (Coleno, 1970) it has been shown that for *P. phaseolicola* and *P. syringae* these antigens are connected with the thermolabile part of the antigen 0. It may be used to estimate the contamination rate of seeds and plants (Coleno & Trigalet, 1969).

References

Coleno, A., 1968a. Intérêt des motifs antigéniques communs entre *Pseudomonas phaseolicola* (Burk.) Dowson et *Pseudomonas syringae* van Hall pour le typage de cette dernière espèce. C.r. hebd. Séanc. Acad. Sci., Paris 266: 2516–2518.

Coleno, A., 1968b. Utilisation de la technique d'immunofluorescence pour le dépistage de *Pseudomonas phaseolicola* (Burk.) Dowson dans les lots de semences de haricots contaminés. C.r. hebd. Séanc. Acad. Agric. Fr. 1968: 1016–1020.

Coleno, A. & A. Trigalet, 1969. À propos de la détection des maladies bactériennes transmises par la semence. Sci. Agron. Rennes, 1969 (1).

Coleno, A., L. Hingand & M. R. Barzic, 1970. Contribution à l'étude sérologique de *Pseudomonas phaseolicola* (Burk.) Dowson. Ann. Phytopath. 2(1): 199–207.

Digat, B. & A. Escudie, 1967. Reconnaissance du flétrissement bactérien des Solanées aux Antilles Françaises. Phytiat.-Phytopharm. 16: 187–197.

Kauffmann, F., 1954. Enterobacteriaceae. 2nd ed., Copenhagen.

Köhler, W., 1957. Zur Serologie der *Pseudomonas aeruginosa*. Z. ImmunForsch. exp. Ther. 114: 282.

Lucas, L. T. & R. C. Grogan, 1969a. Serological variation and identification of *Pseudomonas lachrymans* and other *Pseudomonas* nomen species. Phytopathology 59: 1808–1912.

Lucas, L. T. & R. C. Grogan, 1969b. Some properties of specific antigen of *Pseudomonas lachrymans* and other *Pseudomonas* nomen species. Phytopathology 59: 1913–1917.

Müller, H. & H. Kleinmaier, 1958. Zur Sero-struktur der O-Antigene von *Pseudomonas aeruginosa*. Z. Hyg. InfektKrankh. 145: 85–91.

Lovrekovich, L. & Z. Klement, 1961. Species-specific antigens of *Pseudomonas tabaci*. Acta microbiol. hung. 8: 303–310.

Lovrekovich, L., Z. Klement & W. J. Dowson, 1963. Serological investigation of *Pseudomonas syringae* and *Pseudomonas mors-prunorum* strains. Phytopath. Z. 47: 19–24.

Levine, L., 1967. In: D. M. Weir-Blackwell's Handbook of experimental immunology. Blackwell's Scientific Publications, Oxford.

Sandvick, O., 1960. Serological comparison between strains of *Pseudomonas aeruginosa* from human and animal sources. Acta path. microbiol. scand. 48: 56–60.

Staub, A. M., 1965. In: R. L. Whiter, ed.: Bacterial lipido-proteino-polysaccharides ('0 somatic antigen') extraction with trichloroacetic acid. Methods in carbohydrate chemistry, vol. 5, p. 92–95.

Comparative serological studies between Xanthomonas citri and a bacterium causing canker on Mexican lime

T. Namekata
Secção de Micología Fitopatológica, Instituto Biológico, São Paulo, Brasil

A. R. de Oliveira
Departamento de Bioquímica da Universidade de Campinas, Estado de São Paulo, Brasil

Several methods are used to identify closely related bacteria. Serology is one of them, but we have to be careful in interpreting its results.

This paper gives the outcome of the investigations on antigenic properties of *Xanthomonas citri*, the cause of citrus canker, and a bacterial isolate causing canker (in the field) only on Mexican lime. These strains cannot be distinguished by the usual identification tests for *Xanthomonas* species so that serological tests were developed. To obtain more data on antigenicity, *Xanthomonas carotae*, *X. manihotis*, *X. phaseoli* var. *fuscans*, *X. vesicatoria* and *Pseudomonas solanacearum* were included.

Of both isolates (*X. citri* and the bacterial isolate of Mexican lime canker), 48 h old cultures were suspended in a saline solution (0.85 %) in a concentration of 10^8 cells per ml. The suspension was divided into two parts, one of which was heated during 10 min at 60 °C whereas the other part remained untreated.

From these suspensions, rabbits received eight intramuscular injections at 10 days' intervals of the bacterial suspensions (3 ml) emulsified in Freund's incomplete adjuvant (Difco).

For serological tests two types of antigens were prepared: with living bacteria ('0' antigen) and with bacterial extracts. Agglutination, precipitation, gel-diffusion and immuno-electrophoresis tests were performed. The antigens had a titer of 1 : 5120 in homologous tests. From the 7 bacterial strains only 4 showed a positive reaction (*X. carotae*, *X. manihotis*, *X. citri* and the Mexican lime isolate). In heterologous tests titers were always much lower than in homologous ones. By cross-reactions, absorption tests and immuno-electrophoresis it was possible to determine the specific antigenic components of *X. citri* and the Mexican lime isolates. Bacterial extracts obtained by heating the suspension during 45–60 min at 100 °C did not disturb the antigenic specificity.

The advantages of using bacterial extracts are:

a. they increase the speed of reaction in the gel diffusion tests;
b. the precipitation lines in the gel diffusion become very clear;
c. a distinct difference between heterologous and homologous titers is obtained.

It is important to adjust the antigen-antiserum concentration ratio in the serological absorbtion test to avoid misinterpretation of the results.

These serological studies, combined with physiological, morphological, pathological and bacteriological techniques demonstrated that the strain isolated from the

Mexican lime canker differs from *X. citri* on an intraspecific level. Therefore the name *Xanthomonas citri* (Hasse) Dowson forma specialis *aurantifolia* is suggested.

Serological identification of bacteria in bulbs

G. J. Saaltink, D. H. M. van Slogteren and W. Kamerman[1]

Flowerbulbs Research Centre, Lisse, The Netherlands

Abstract

Antisera to different plant pathogenic bacteria, listed in a table, have been prepared for the primary purpose of clinical diagnosis.

Immunization procedure

Bacteria were washed, with saline, from the surface of the nutrient broth agar on which they had been grown, killed by heating the suspension at 60 °C for 1 h, and sedimented by centrifuging at 5 °C for 20 min at 10,000 *g*. The sediment was resuspended in saline and emulsified in Freund's adjuvant oil incomplete. Rabbits were injected intramuscularly with 2 ml of the emulsion (containing some 4.0×10^9 bacteria) in each hindleg. Blood samples were taken about 3 weeks afterwards and when the titres of the antisera were too low, the injections were repeated once or twice with intervals of about 4 weeks. The homologous titre for each bacterium with the best antiserum from a series of rabbits is given in the third column of the table.

The titres vary greatly between species of the same genus: for the antiserum against *Erwinia chrysanthemi* the homologous titre was 1:2560 whereas against *E. carotovora* the highest titre was only 1:320. For antisera against *Xanthomonas hyacinthi* and *X. gummisudans* the homologous titres were 1:2560 and 1:320, respectively.

Diagnostic techniques

The microdroplet agglutination method under liquid paraffin, described for indexing plant viruses (van Slogteren, 1955), was mainly used. Small droplets of antiserum dilutions were pipetted on the polyvinyl-coated inner surface of a petri dish, stirred with small amounts of bacterial material and covered with liquid paraffin. To some extent this method resembles that described by Nováková (1957) for indexing

1. The authors are indebted to Miss M. H. Bunt and Miss N. P. de Vos for their technical assistance. The present address of the last author is: Rijksinstituut voor Natuurbeheer, Leersum, The Netherlands.

Antisera and their serological reactions with other isolates: congeneric species not mutually cross-reacting separated by dotted lines.

Antisera against	Isolated from	Homologous titre	Reaction with other isolates: positive	Reaction with other isolates: negative
Erwinia chrysanthemi	dahlia chrysanthemum	1:2560 1:2560	cross-reacting with their mutual antigens and reacting with isolates from begonia, carnation and with *E. cytolytica* from dahlia	*E. carotovora* isolates from both hyacinth and potato
Erwinia carotovora	hyacinth with 'white slime' disease	1:320	none	*E. chrysanthemi*, *E. carotovora* from potato
Erwinia carotovora	potato (Kew Sa312)	1:320	an isolate causing flower blight in hyacinth and isolates from iris and tulip	*E. chrysanthemi*, *E. carotovora* from hyacinth with 'white slime'
Corynebacterium oortii *Corynebacterium fascians*	tulip gladiolus	1:160 1:160	cross-reacting with their mutual antigens and reacting with *C. betae* and *C. tritici*	*C. michiganense* *C. poinsettiae*
Xanthomonas hyacinthi	hyacinth	1:2560	none	*X. gummisudans*
Xanthomonas gummisudans	gladiolus	1:320	none	*X. hyacinthi*

Antisera prepared by injecting bacteria emulsified in Freund's adjuvant listed in first column.

Erwinia atroseptica. Occasionally Ouchterlony's gel-diffusion method was applied; it yielded positive results with *Erwinia* and *Corynebacterium* species. In tissues of leaves or bulb scales of infected hyacinths, *Xanthomonas hyacinthi* could be directly identified with the microdroplet agglutination method as follows. The homogenate, obtained by homogenizing 3 g infected tissue in saline, was pressed through cloth and the extract centrifuged at 1000 *g* (the sediment was discarded). To prevent spontaneous flocculation, the supernatant was heated for 15 min at 85 °C whereupon the bacteria were sedimented by centrifuging at 10,000 *g* for 20 min. After resuspension in 3 ml saline, clarification by centrifuging at 1500 rpm for 5 min followed and the supernatant was used for agglutination reactions.

All other bacteria listed were indirectly identified by plating extracts from plant tissues on nutrient broth agar. After 2 to 3 days incubation at the corresponding most appropriate temperature, a small amount of material from single colonies was stirred with droplets of antiserum with a needle. Agglutination took place rapidly (within a few seconds between *Erwinia chrysanthemi* and its corresponding antiserum). No reactions were ever observed between the antisera and material from single colonies of saprophytic bacteria present in the extracts as contaminants.

As compared to the previously applied diagnostic procedures (such as physiological tests) requiring periods from 3 to 5 weeks, the serological method requires only 2 hours for a reliable diagnosis (direct method) or 2 to 3 days (indirect method).

Serological relationships

Within each bacterial genus, the presence or absence of serological cross-reactions as well as reactions between the available antisera and other isolates of plant pathogenic bacteria are also in the table.

The data suggest that sometimes isolates hitherto considered to belong to the same species (such as the *Erwinia carotovora* isolated from hyacinth and potato) show no serological relation. On the other hand, *Erwinia cytolytica* could be identified serologically with *E. chrysanthemi.* Cross-reactions were observed between antisera to *Corynebacterum oortii* (Saaltink & Maas Geesteranus, 1969) and *C. fascians* and their mutual heterologous antigens, and serological relationships could be determined between these isolates and *C. betae* and *C. tritici.*

Conclusion

Antisera prepared against plant pathogenic bacteria, the majority of which had relatively low titres, proved useful for clinical diagnosis and for establishing close serological relationships. For determining more distant relationships, other methods of preparing the antigens to be injected should be tried.

References

Nováková, J., 1957. A new method of isolation of blackleg-pathogens from diseased plants. Phytopath. Z. 29: 72–74.

Saaltink, G. J. & H. P. Maas Geesteranus, 1969. A new disease in tulip caused by *Corynebacterium oortii* nov. spec. Neth. J. Pl. Path. 75: 123–128.

Slogteren, D. H. M. van, 1955. Serological micro-reactions with plant viruses under paraffin oil. Proc. 2nd Conf. Potato Virus Dis. Lisse-Wageningen, 1954: 51–54.

Development of the hypersensitivity reaction induced by plant pathogenic bacteria

Z. Klement

Research Institute for Plant Protection, Budapest, Hungary

Abstract

The paper supplies general information on defence mechanisms of plants against bacteria. The most important in all incompatible combinations is the hypersensitivity reaction. During its development three main phases are distinguished: the induction period, the latent period, and cell collapse. The role and the importance of these phases are summarised and discussed.

Introduction

Pathogenic bacteria are known to enter the plant through the stomata or through wounds and to multiply in its intercellular spaces or vessels. The fluid in these spaces and vessels contains in abundance all nutrients necessary for most pathogenic and saprophytic bacteria when cultivated *in vitro*. But inside the living plant only the host specific bacteria are able to grow, involving a compatible host-pathogen combination (Fig. 1).

The question of host specificity, i.e. why the occurrence of plant pathogenic bacteria is limited to special hosts, comes sharply into focus. To solve this problem we must study the defence mechanisms of the hosts. Before discussing the most important defence reaction of plants (the hypersensitive reaction, HR), we should consider the different defence mechanisms.

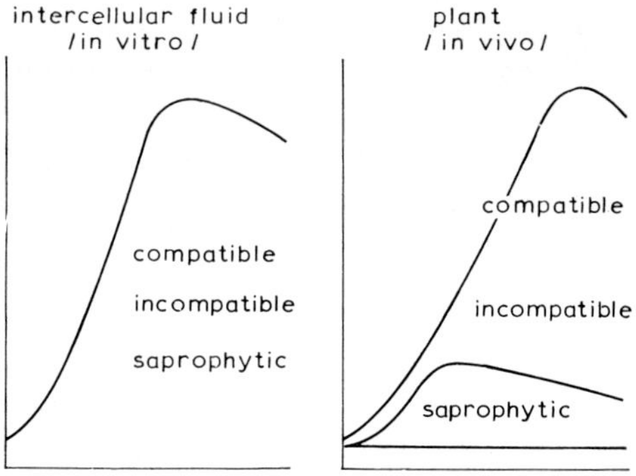

Fig. 1. Multiplication of bacteria in intercellular fluid (in vitro; left) and in the plant (right; in vivo).

Preformed resistance

First let us consider a normosensitive compatible host-pathogen combination, e.g. a sensitive tobacco cultivar infected by virulent *Pseudomonas tabaci*.

The first barrier to be surmounted by the bacteria is the preformed resistance to penetration into intercellular spaces or into the vascular system of the plant. Several kinds of such resistance factors are known. Bacteria are unable to break through an undamaged cuticle: it must be wounded before the pathogen can colonize the plant tissue. If the stomata are closed, bacteria are unable to penetrate into the substomatal cavity, but if the atmosphere becomes humid, the door is opened for infection. Thus, as Sasser *et al.* (1970) have recently shown, the osmotic potential in intercellular leaf fluid is a controlling factor in the development of diseases: dry outside air normally inhibits population increase of pathogenic bacteria, a period of high relative humidity and rainfall reduces the osmotic potential and allows bacterial multiplication.

Other factors play an important role in eliminating the bacteria: the buffering capacity of plant fluids, the lack of certain nutrients for the bacteria, and antibacterial substances in the plant tissue.

When, e.g., the pathogen *Pseudomonas tabaci* has surmounted the entrance barrier and starts to multiply in the host, it forms wildfire toxin, and after a few days the typical disease symptoms develop. Though the pathogen is host specific, the toxins and enzymes are not and these bacterial products do not seem to play a primary role in pathogenesis. They can operate only if the parasite is already established in the plant and their concentration has reached a certain level.

Induced resistance

To gain a better insight into the defense mechanism of plants against bacteria, let us consider what happens with a saprophytic or incompatible pathogenic bacterium in the plant tissue. Next to pre-existent incompatibility, their elimination from the non-host plant may be due to induced resistance. This includes the defense reactions that start in plants as a result of infection. Two types of such induced defense reactions, premunity and hypersensitivity, are known (Fig. 2).

Premunity Lovrekovich & Farkas (1965) have first described this phenomenon; it was termed 'premunity' by Klement & Goodman (1967). They introduced heat-killed cells of *Pseudomonas tabaci* into the intercellular spaces of tobacco leaf, and one day afterwards the leaf was inoculated with living cells of the *Pseudomonas*. In leaves which had been pretreated with a suspension of killed or saprophytic bacteria, no wildfire symptoms occurred and bacterial multiplication was strongly inhibited.

Pretreatment of the plants with saprophytic *Pseudomonas fluorescens* resulted in a protective effect, not only against subsequent infection by bacteria, but also against tobacco mosaic virus infection (Klement *et al.*, 1966). As Király & Klement (unpublished work) have shown, simple proteins such as albumin are also able to induce premunity in plants.

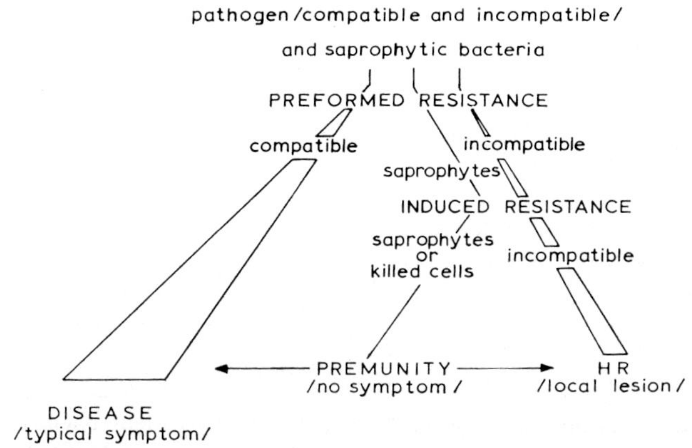

Fig. 2. Defence mechanisms of plants.

Recently, Lozano & Sequeira (1970) were able to nullify the HR by heat-killed *P. solanacearum* cells; details will be given in the paper by Sequeira.

To sum up, premunity is an aquired non-specific immunity manifesting itself in plants pretreated with one bacterium that immunizes or protects the plant from infection by another pathogen. Premunity inhibits both normal disease development and HR. Presumably it is a frequently occurring non-specific reaction of plants. It must be clearly distinguished from the other type of induced defence response, the HR.

Hypersensitive reaction In our laboratory the hypersensitive reaction induced by phytopathogenic bacteria was clearly established in the years 1958–1964. The process is as follows.

When a foreign pathogen, such as *Pseudomonas syringae*, penetrates the intercellular spaces of a tobacco leaf, the bacteria can start to multiply but they stop in a relatively short time (8–10 h). After this penetration the plant cells around the site of infection die, and so the infection is confined.

This hypersensitive reaction works similarly under natural conditions. One or two pathogenic bacterial cells passively penetrate the substomatal cavity through the host's stomata, where they begin to reproduce. Around them a few host cells die, and this confines the infection so that the plant remains practically free from symptoms. The time (8–10 h) is simply too short for the bacteria to increase to such an extent that they induce necrosis in a larger area of tissue. But if high numbers of bacteria are used, the many little local lesions, each invisible to the naked eye, may be so close together that the whole tissue dies. Thus, after the introduction of, say, 10^7–10^8 or more bacteria/ml inoculum into the tissue, the HR (necrosis) can always be seen.

In various laboratories different host-pathogen combinations have been studied in relation to the HR. Klement *et al.* (1964) and Lelliott & Billing (1966) have proved that all pathogenic pseudomonads, except *P. marginata*, induce an HR in foreign hosts. According to Schroth & Hildebrand (1967), at 28 °C xanthomonads also induce an atypical necrosis in the non-host plants.

Table 1. Host-parasite relationships.

Combinations	Hypersensitive reaction	Typical disease symptom
Virulent bacteria – sensitive host plant	−	+
Avirulent bacteria – normally sensitive host plant	+	−
Normally virulent bacteria – resistant host plant	+	−
Pathogenic bacteria – non-hostplant	+	−
Saprophytic bacteria – all plants	−	−

Exactly the same type of reaction occurs when of the normally virulent bacteria meets its resistant host plant. Such cases have been studied with pseudomonads, xanthomonads and *Erwinia* by Stall & Cook (1966), Kennedy & Crosse (1966), Klement & Goodman (1967), Lozano & Sequeira (1970), and others.

If the originally sensitive host comes into contact with the pathogen which has lost its virulence, an HR develops (Klement & Goodman, 1967; Lozano & Sequeira, 1970).

Saprophytic bacteria do not induce an HR, though Lovrekovich & Lovrekovich (1970) have recently reported an extreme case (tobacco plants infiltrated with *P. fluorescens*, kept at 100% relative air humidity in the dark) where the saprophytic bacterium was able to grow in the tobacco leaf and produce necrosis. It seems that such necroses were due to the metabolites of the bacteria, an effect quite different from the HR.

Table 1 shows the types of symptoms and the appearance of the HR in the different host-pathogen combinations.

It seems that the HR is the most important defence reaction of plants in all incompatible host-pathogen combinations. Among plants, no other defence mechanisms are known that are only induced by the HR in incompatible combinations.

As yet there are no answers to the basic questions: why can bacteria only induce an HR in incompatible hosts, and what kind of physiological and biochemical mechanisms are induced in the host by bacteria in the early part of the development of the HR?

If the mechanism of the HR was known, this would not only inform us on the biochemical and physiological mechanisms of this defence reaction, but would also lead to a basic concept of pathogenesis as well as resistance.

In our studies (Süle & Klement, 1970) tobacco was used as host and *P. phaseolicola* as pathogen incompatible to tobacco. Here the HR developed 8–10 h after infiltration of bacteria into the plant's intercellular spaces. Three phases could be distinguished: an induction phase, a latent phase, and the collapse of the host cells (Fig. 3).

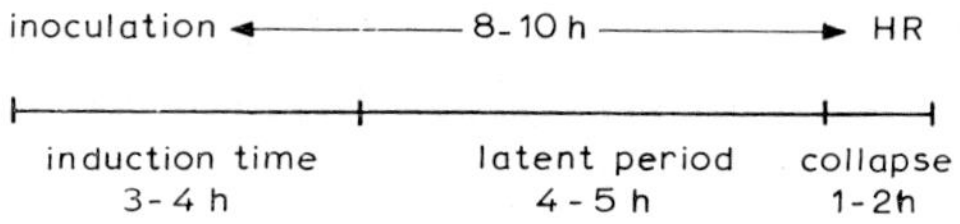

Fig. 3. Phases in the development of the hypersensitive reaction.

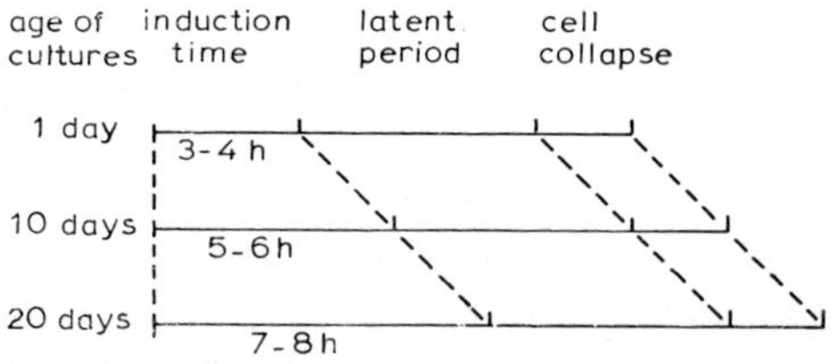

Fig. 4. Development of the hypersensitive reaction.

Induction period

The induction time is the period necessary for the bacterium to start the HR. To measure it, streptomycin was infiltrated into the intercellular spaces to kill the bacterium at different times after inoculation. When the bacteria were killed during the first 3–4 h after inoculation, the HR did not appear; when streptomycin was injected after 3–4 h, the HR appeared 8–9 h after the infiltration of the bacterium. Thus the length of the induction period was about 3–4 h. After this period, during the latent phase, living bacteria are not necessary for the development of the HR.

The concentration of the inoculum does not influence the time of appearance of the HR (Lozano & Sequeira, 1970), but it depends on the metabolical activity of the bacterium.

The age of bacterial cultures also influences the time of appearance of the HR: with 1, 10 and 20 days old cultures it appeared after 8–9 h, 10–11 h and 12–13 h, respectively. Thus the HR came 3–4 h later if the bacterial culture was an old one.

To study the effect of bacterial age for the first period, streptomycin was infiltrated into the intercellulars at different intervals after bacterial inoculation with young or old cultures. As seen in Fig. 4, the induction time was influenced by the age of the bacteria, but latent period and cell collaps period did not change.

Since, *in vitro*, the generation time of *P. phaseolicola* is about 4–5 h, it seems that the multiplication of the bacterium may be an important factor in the induction of the HR. It was impossible, however, to detect any multiplication of *P. phaseolicola* in tobacco tissue during this period. The factor which induces the HR is unknown. The only thing we know is that young cultures, which are metabolically more active, induce the HR earlier than old ones.

These experiments seem to indicate that the hypersensitive response can be induced only by living, metabolically active bacteria, though Sequeira & Ainslie (1969) and Lozano & Sequeira (1970) have prepared bacterial cell-free extracts which were able to produce necrosis. Unfortunately, there are not enough details available about these very interesting experiments. As Dr. Sequeira informed me, it is possible that the cell-free extract consists of toxic substances which might give an HR-like reaction. If it were possible to establish the induction factor of the HR, we would be much closer to solving the problem of pathogenicity and host-specificity of bacteria.

Latent period

The latent period is the time between the start of plant reaction and tissue collapse. This period is independent of the presence of living bacteria: when the bacterial cells are killed by antibiotics during the latent period, the HR will still appear. Therefore, as soon as the controlling factors are activated in the host tissue at the end of the induction period, the defense reaction continues by itself.

During this period the host is symptomless, but some physiological and biochemical changes occur, such as an increase in respiration rate, its maximum occurring just before tissue collapse (Németh & Klement, 1967; Fig. 5). The latent period is too short to leave time for the development of the 'slow' biochemical reactions. Most probably this fast development is why an increase in polyphenoloxidase or peroxidase activity, the most widespread biochemical symptom of many other hypersensitive host-parasite complexes (fungi and viruses) is not observed in the bacterial system (Németh *et al.*, 1969).

Klement & Goodman (1967) reported that the HR was inhibited by high temperature. If their inoculated hypersensitive tobacco plants were kept in an incubator at 37 °C, no HR developed. My co-worker Dr. Süle tried to find out which HR development phase was inhibited by the high temperature. For this purpose tobacco plants were infiltrated with a suspension of *P. phaseolicola* at 28 °C (low temperature) and put into an incubator at 37 °C (high temperature) for one hour intervals during the 8 h incubation period (Table 2). No HR developed when the plants were kept in an

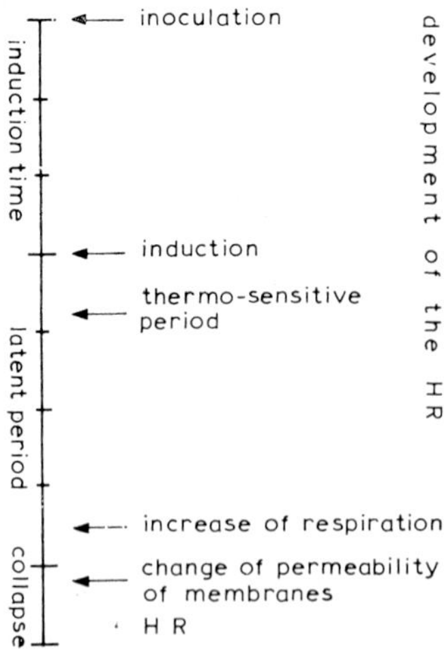

Fig. 5. Effect of bacterial age on length of development of hypersensitive reaction.

Table 2. Hypersensitivity reaction (HR) after various periods at low temperature followed by high temperature.

	Induction time				Latent period				
Period of low temperature (25 °C)	0 h	1 h	2 h	3 h	4 h	5 h	6 h	7 h	8 h
Period of high temperature (37 °C)	8 h	7 h	6 h	5 h	4 h	3 h	2 h	1 h	0 h
Hypersensitivity reaction	−	−	−	−	−	+	+	+	+

incubator at 37 °C for at least 5 h after inoculation. Afterwards the high temperature had no effect on the appearance of the HR.

In another experiment bacteria in tobacco leaves were killed by streptomycin four hours after inoculation, which was one hour after the induction time. Afterwards the plants were kept at 37 °C. Here the HR did not develop either, showing that the high temperature (37 °C) has no effect on the bacteria or on the induction of the HR. The conclusion is that the thermosensitive period is in the early part of the latent period (Fig. 4). The high temperature has no effect on the induction factors but does influence the biochemical and physiological process of the HR induced in the plant.

From the experiment of Lozano & Sequeira (1970) it seems that light is important for the development of the HR, because the synthesis of substances that induce the HR appears to be light dependent.

As mentioned before, the other induced defence reaction of plants, called premunity, is induced by heat-killed or saprophytic bacteria. Lozano & Sequeira (1970) established that the typical disease symptom, and also the HR, were prevented by heat-killed bacterial cells.

As said earlier, albumin also induces a protective effect against the development of the HR. In tobacco leaves pretreated with albumin the HR was inhibited and the necrosis appeared 3 h later than in the non-pretreated leaves. It was established by Süle & Klement (unpublished work) that inhibition was influenced only during the latent period, but that the induction time did not change.

Tissue collapse

After the latent period, tissue collapse is rapid (1–2h) because the permeability of the host cell membrane is changed (Goodman, 1968). Electrolytic measurements of the change in permeability of host tissue has revealed a rapid increase in permeability at the end of the latent period with a maximum at the beginning of the period of tissue collapse (Fig. 4). During this process different ions leave the cell plasma, so that probably the inactive phenols also pass the membranes of the vacuoles. These oxidized phenol compounds may cause cell death and finally necrosis.

The conclusion is that none of the investigated physiological and biochemical changes in the host tissue during the development of the HR are initial causes. They all are the consequences of the HR, as shown by the increase in ammonia and ethylene content recently detected in necrotized tissues by Lovrekovich *et al.* (1970, 1971).

Though many basic questions are still unanswered, the results presented here are fragments which bring us closer to the final solution. And if we succeed in supplying the missing parts, the picture of the HR will become complete and clear.

There are two reasons for optimism. First in the study of HR bacteria are more reasonably managed by simple methods than fungi and viruses. And, secondly, an increasing number of scientists realize the importance of research in this field.

References

Goodman, R. N., 1968. The hypersensitive reaction in tobacco: A reflection of changes in host cell permeability. Phytopathology 58: 872–873.

Kennedy, B. W. & J. E. Crosse, 1966. Inoculation procedures for comparing reaction of soybeans to bacterial blight. Pl. Dis. Reptr. 50: 560–565.

Klement, Z., G. L. Farkas & L. Lovrekovich, 1964. Hypersensitive reaction induced by phytopathogenic bacteria in the tobacco leaf. Phytopathology 54: 474–477.

Klement, Z. & R. N. Goodman, 1967. The hypersensitive reaction to infection by bacterial plant pathogens. A. Rev. Phytopath. 5: 17–44.

Klement, Z., Z. Király and B. J. Pozsár, 1966. Suppression of virus multiplication and local lesion production in tobacco following inoculation with a saprophytic bacterium. Acta Phytopathl. hung. 2: 303–308.

Lelliott, R. A., E. Billing & A. C. Hayward, 1966. A determinative scheme for the fluorescent plant pathogenic pseudomonads. J. appl. Bact., 29: 470–489.

Lovrekovich, L. & G. L. Farkas, 1965. Induced protection against wildfire disease in tobacco leaves treated with heat-killed bacteria. Nature, Lond. 205: 823–824.

Lovrekovich, L. & H. Lovrekovich, 1970. Tissue necrosis in tobacco caused by a saprophytic bacterium. Phytopathology 60: 1279–1280.

Lovrekovich, L., H. Lovrekovich & R. N. Goodman, 1970. Ammonia as a necrotoxin in the hypersensitive reaction caused by bacteria in tobacco leaves. Can. J. Bot. 48: 167–171.

Lozano, J. C. & L. Sequeira, 1970. Differentiation of races of *Pseudomonas solanacearum* by a leaf infiltration technique. Phytopathology 60: 833–838.

Lozano, J. C. & L. Sequeira, 1970. Prevention of the hypersensitive reaction in tobacco leaves by heat-killed bacterial cells. Phytopathology 60: 875–879.

Németh, J. & Z. Klement, 1967. Changes in respiration rate of tobacco leaves infected with bacteria in relation to the hypersensitive reaction. Acta phytopath. hung. 2: 303–308.

Németh, J., Z. Klement & G. L. Farkas, 1969. An enzymological study of the hypersensitive reaction induced by *Pseudomonas syringae* in tobacco leaf tissue. Phytopath. Z. 65: 267–278.

Sasser, J. M., R. W. Miller & D. J. Fieldhouse, 1970. Osmotic potential a controlling factor in the development of bacterial spot diseases. Phytopathology 60: 1311 (Abstr.).

Sequeira, L. & V. Ainslie, 1969. Bacterial cell-free preparations that induce or precent the hypersensitive reaction in tobacco. Int. Bot. Congress Abstr. p. 195.

Stall, R. E. & A. A. Cook, 1966. Multiplication of *Xanthomonas vesicatoria* and lesion development in resistant and susceptible pepper. Phytopathology 56: 1152–1154.

Süle, S. & Z. Klement, 1971. Effect of high temperature and the age of bacteria on the hypersensitive reaction of tobacco. Acta phytopath. hung. 6 (in press).

Prevention of the hypersensitive reaction

L. Sequeira

Department of Plant Pathology, University of Wisconsin, Madison, Wis., USA

Abstract

The hypersensitive reaction (HR) induced in tobacco leaves by the introduction of many incompatible bacteria can be delayed or prevented by prior infiltration with heat-killed bacteria. This protective phenomenon is systemic and is both time and light dependent. The factor that induces protection is produced by many plant pathogenic bacteria and is non-specific in its action. A soluble fraction from disintegrated cells of *Pseudomonas solanacearum* induced protection and the active component was precipitated with ethanol or $(NH_4)_2SO_4$. Its activity was destroyed by pronase, trypsin, or hot trichloroacetic acid. Separation of proteinaceous constituents of the ethanol ppt on Sephadex G-200 yielded an active fraction consisting of several proteins, separable by gel electrophoresis.

Introduction

The hypersensitive reaction (HR) induced in tobacco leaves by the introduction of many incompatible bacteria can be delayed or prevented by a variety of treatments, including high temperature (Klement & Goodman, 1967), long periods of darkness (Lozano *et al.*, 1968), application of antibiotics (Klement & Goodman, 1967) and calcium (Cook & Stall, 1970). One of the most effective treatments consists of prior (18 h) infiltration with heat-killed bacteria (Lozano & Sequeira, 1970). In another section of this symposium, Dr A. A. Cook discusses in detail the various physical and chemical factors that effect the HR. It would serve no useful purpose, therefore, to review here the importance of these factors in the prevention of the HR. This discussion will be limited to the influence of heat-killed cells. This is a phenomenon which has been studied in some detail in our laboratories, and particularly by J. C. Lozano, Sheila Aist, and Virginia Ainslie. It is the purpose of this paper to summarize the available information on this phenomenon and to relate this information to similar phenomena in other host-parasite systems.

The phenomenon of induced protection

Incompatible strains of *Pseudomonas solanacearum* induce a rapid HR when infiltrated in tobacco leaves at a concentration of at least 3.5×10^7 cells/ml (Lozano & Sequeira, 1970). At 28 °C the HR is evident 10–12 h after infiltration. If similar cell suspensions are killed by heating them in a water bath at 90–95 °C for 20 min and these are infiltrated on tobacco leaves prior to infiltrating with live cells, the HR obtained 24 h later may be modified substantially (Lozano & Sequeira, 1970).

Leaves challenged immediately after infiltrating with heat-killed cells will show a normal HR, but as the period allowed between treatment and challenge infiltration is increased, the HR is delayed or completely prevented. A series of intermediate reactions may be obtained, in which green islands remaining within the challenged area increase in number and size as the interval between treatment and challenge increases. Under appropriate conditions of light and adequate bacterial concentration, the treated area may be fully protected if challenged 18 h after treatment with heat-killed cells. The phenomenon is conveniently demonstrated in tobacco leaves in which alternate intercostal areas (panels) are infiltrated with heat-killed cells. If a small area (2 cm^2) in the centre of all panels is challenged 24 h later, the HR appears only on those panels which were not treated with heat-killed cells.

An interesting property of the protective effect is that it becomes systemic. If sufficient time is allowed, and a high concentration of bacteria is used, untreated leaf panels become fully protected, and those of untreated leaves above and below the infiltrated leaf will show a partial or delayed HR (Lozano & Sequeira, 1970).

Conditions that affect protection

Prevention of the HR depends on (a) time allowed between treatment and challenge, (b) concentration of heat-killed and live bacteria used for infiltration, (c) temperature and (d) light regime (Lozano & Sequeira, 1970). If the concentration of bacteria used in the challenge infiltration is much higher than that in the original treatment, or if the plants are subjected to continuous high temperature (36 °C), only partial protection will be obtained.

Perhaps the most striking effect is that of light. If the leaves are covered with aluminium foil (with provisions for adequate gas exchange), no protection or only partial protection will be obtained when leaves are subsequently challenged. The extent of the HR will depend on the period of exposure to darkness. Conversely, treated leaves exposed to continuous light for 24 h show complete protection in both treated and untreated panels (Lozano & Sequeira, *op. cit.*)

Specificity of the reaction

The protective effect of heat-killed cells is non-specific, both in terms of the bacteria that induce it and of the host reaction to different pathogens. Prevention of the HR in tobacco leaves was obtained with heat-killed cells of *Pseudomonas lachrymans* and *Xanthomonas axonopodis*, in addition to *P. solanacearum*, but not by those of *Escherichia coli* (Lozano & Sequeira, *op. cit.*). It seems evident that many, but not all, plant pathogenic bacteria produce the factor responsible for the protective response.

The protective response is, of course, not limited to the HR. This reaction merely provides a convenient reproducible method of measuring the host response to heat-killed bacterial cells. The protective effect of heat-killed cells against compatible pathogenic bacteria was first reported by Lovrekovich & Farkas in 1965. They indicated that heat-killed cells of *P. syringae*, *P. tabaci*, and *Corynebacterium flaccumfaciens* protected tobacco against attack by *P. tabaci*. Sleesman *et al.* (1970) have recently reported that *P. glycinea* cells also protect against the tobacco wildfire bacterium. That the protective response is not limited to bacteria was shown by Loebenstein & Lovrekovich (1966) who reported that TMV local lesion formation was interfered with in tobacco leaves infiltrated with heat-killed cells of *P. syringae*. Heat-killed cells of *P. solanacearum* will affect both the number and size of TMV local lesions in *Nicotiana tabacum* and *N. glutinosa* (unpublished results).

Non-specific protective responses have been induced in tobacco by inoculation with live cells of incompatible bacteria (Averre & Kelman, 1964), *Peronospora tabacina* (Mandryk, 1962), *Thielaviopsis basicola* (Hecht & Bateman, 1964), and as a result of previous systemic or localized viral infections (Ross, 1965). It is unlikely that all of these responses, and those to heat-killed cells, are due to a common factor. The response of tobacco to infection may result in the production of an 'interferon'-like substance, akin to that produced by animal cells as a response to viral RNA, but less specific in its action.

Introduction into protected leaf areas of high populations of a strain of *Pseudomonas solanacearum* that normally induces the HR, results in an immediate and precipitous drop in numbers of bacterial cells (Lozano & Sequeira, 1970). In one instance, populations dropped from 10^8 to 10^2 cells/ml in 36 h, a decrease which was much more rapid than that found during the course of the normal HR.

Nature of the substance that induces protection

Initial studies suggested that the substance in heat-killed *P. solanacearum* cells that induces protection against the HR is proteinaceous (Sequeira & Ainslie, 1969). More recent work has confirmed these original conclusions and has provided information on the general properties of the substance, or substances, involved (Sequeira *et al.*, 1971).

Disintegration of *P. solanacearum* cells by high-speed grinding with glass beads, or by sonication, yields soluble and insoluble (cell wall) fractions, both capable of preventing the HR in a manner identical to that induced by heat-killed cells.

Because most of the protective activity was associated with the cell walls, initial efforts were directed at purifying cell wall components. A standard assay was devised to determine the relative protective properties of each fraction. In the standard assay, the fifth and sixth (from the base) fully developed leaves of one month old tobacco plants are infiltrated with each fraction, or with dilutions from each. To reduce variability, only alternate panels at the centre of each leaf are used. After 24 and/or 48 h, all panels are challenged with live B-1 cells and the reactions are noted 24 h later in accordance with a standard HR index. Dilution end points can be used to determine the relative biological activity of each fraction.

Purification of bacterial cell wall polysaccharides by extraction of cells with hot trichloroacetic acid (TCA) and digestion of the residue with trypsin, followed by washing with ammonium bicarbonate and ammonia, yielded protein-free cell wall polysaccharides which were not active in the standard assay.

The methods for cell disruption used apparently solubilized a considerable amount of the cell wall-bound proteins and these could be precipitated from the soluble fraction by addition of ethanol, $(NH_4)_2SO_4$, or cold TCA. Ethanol was the most effective agent; a suspension of the ethanol ppt in buffer was nearly as effective (on a protein basis) as the crude extract in preventing the HR. When compared with heat-killed cells, the ethanol ppt was more active; it induced full protection within 7 h, as compared with 18 h for cells.

The active substance in the ethanol ppt was destroyed by trypsin or pronase, and its activity was reduced by extensive sonication or prolonged autoclaving. It appeared to be a protein (or proteins) of remarkable heat stability. Exposure to 95 °C for 10 min reduces, but does not destroy, the protective effect of the ethanol ppt and other fractions.

Fractionation and purification of the proteins in the ethanol ppt has been accomplished by chromatography on Sephadex columns (Sequeira *et al.*, 1971). Elution through Sephadex G-200 yielded three major protein peaks, one of which was coincident with a high peak of protection in the standard assay. The protection afforded by these fractions exhibited the same lability when treated with trypsin, and the same stability when subjected to heat, as the ethanol ppt. These fractions were contaminated with nucleic acids and these were removed by chromatography on DEAE-Sephadex. Active fractions were then separated by gel electrophoresis with carrier ampholytes at various pH ranges. A major protein band with an isoelectric point at pH 7.2 gave partial, but inconsistent protection after elution in citrate buffer. Because of difficulties in the separation of proteins from the carrier ampholytes and problems associated with toxicity of the carriers to tobacco tissues, it is not possible at present to conclude whether one or several proteins constitute the protective factor from bacterial cells.

Nature of the protective effect - a look into the future

Present evidence suggests that the factor from *Pseudomonas solanacearum* that prevents the HR in tobacco is a protein of fairly high molecular weight (probably in the range 150,000 to 200,000). We can only speculate as to how this large protein is released in the intercellular spaces of the leaf tissues and is able to exert its remarkable effect on tobacco cells. The effect is not due to the inherent toxicity of foreign proteins, since infiltration with casein, trypsin, ribonuclease, etc. did not result in a protective effect (Sequeira *et al.*, 1971). Neither did toxic compounds, growth regulators, several salts of heavy metals, chelating substances, etc. Active fractions from bacterial cells are not highly toxic and, unless very high protein concentrations are used, protected panels show only slight chlorosis by 96 h after infiltration.

The bacterial protein (or proteins) apparently trigger a sequence of metabolic changes which require several hours, and these changes, or their products, become systemic. It is tempting to suggest that a new protein is synthesized which either inactivates bacteria directly or acts as a catalyst for the release of toxic compounds of wide antimicrobial spectra. Recent evidence, obtained by Mr. Thomas Wacek (unpublished), suggests that the latter alternative may be the correct one. Lovrekovich *et al.* (1968) have suggested that the factor from heat-killed bacteria that induces protection against the wildfire disease of tobacco is a peroxidase. At concentrations reported to be effective, however, direct injection of horseradish peroxidase did not result in protection against the HR in tobacco leaves (Sequeira *et al.*, 1971).

It is perhaps dangerous to speculate as to the broad application of a phenomenon which heretofore has been demonstrated only in tobacco leaves. Merely for the purpose of stimulating discussion, I would like to suggest that the phenomenon does represent a generalized method of resistance in plants to foreign organisms, i.e. to certain specific proteins produced by these organisms. Pathogenic, compatible organisms (e.g. race 1 of *Pseudomonas solanacearum*) presumably multiply too rapidly and spread out before the protective response becomes effective. Other incompatible organisms (e.g. race 2) induce a violent HR at very high inoculum levels, which would mask any possible protective response. Protection is effective, however, against those incompatible pathogens (e.g. race 3) or saprophytes which do not induce the HR but are gradually eliminated from plant tissues. At least as far as plant pathogenic bacteria are concerned, the HR may seem only a useful artifact resulting from massive inoculation with some bacteria at the hands of the plant pathologist. The tobacco plant, at least, seems to have less violent but equally effective means of ridding itself of unwanted invaders.

References

Averre, C. W. & A. Kelman, 1964. Severity of bacterial wilt as influenced by ratio of virulent to avirulent cells of *Pseudomonas solanacearum* in inoculum. Phytopathology 54: 779–783.

Cook, A. A. & R. E. Stall, 1970. Effect of calcium on loss of electrolytes from pepper leaves inoculated with *Xanthomonas vesicatoria*. Phytopathology 60: 1288.

Hecht, E. I. & D. F. Bateman, 1964. Nonspecific acquired resistance to pathogens resulting from localized infections by *Thielaviopsis basicola* or viruses in tobacco plants. Phytopathology 54: 523–530.

Klement, Z. & R. N. Goodman, 1967. The role of the living bacterial cells and the induction time in the hypersensitive reaction of tobacco plants. Phytopathology 57: 322–323.

Klement, Z. & R. N. Goodman, 1967. The hypersensitive reaction to infection by bacterial plant pathogens. A. Rev. Phytopathol. 5: 17–44.

Loebenstein, G. & L. Lovrekovich, 1966. Interference with tobacco mosaic virus local lesion formation in tobacco by injecting heat-killed cells of *Pseudomonas syringae*. Virology 30: 587–591.

Lovrekovich, L. & G. L. Farkas, 1965. Induced protection against wildfire disease in tobacco leaves treated with heat-killed bacteria. Nature, Lond. 205: 823–824.

Lovrekovich, L., H. Lovrekovich & M. A. Stahmann, 1968. The importance of peroxidase in the wildfire disease. Phytopathology 58: 193–198.

Lozano, J. C. & L. Sequeira, 1970. Differentiation of races of *Pseudomonas solanacearum* by a leaf infiltration technique. Phytopathology 60: 833–838.

Lozano, J. C. & L. Sequeira, 1970. Prevention of the hypersensitive reaction in tobacco leaves by heat-killed bacterial cells. Phytopathology 60: 875–879.
Mandryk, M., 1962. The relationship between acquired resistance to *Peronospora tobacina* in *Nicotiana tabacum* and soil nitrogen levels. Aust. J. agr. Res. 13: 10–16.
Ross, A. F., 1965. Systemic effects of local lesion formation. Viruses of Plants. Proc. Conf. Plant Viruses, Wageningen.
Sequeira, L. & V. Ainslie, 1969. Bacterial cell-free preparations that induce or prevent the hypersensitive reaction in tobacco. Abstr. 9th Int. Bot. Congr.: 195.
Sequeira, L., S. Aist & V. Ainslie, 1971. Prevention of the hypersensitive reaction in tobacco leaves by proteinaceous constituents of *Pseudomonas solanacearum*. Submitted to Phytopathology 61.
Sleesman, H. C., J. E. Perley & H. A. J. Hoitink, 1970. Susceptible and hypersensitive reactions in tobacco and their prevention by cell-free extracts of *Pseudomonas tabaci* and *Pseudomonas glycinea*. Phytopathology 60: 1314.

Alteration of hypersensitivity in plants to bacterial infection

A. A. Cook

Department of Plant Pathology, University of Florida, Gainesville, Florida

Abstract

Hypersensitivity caused by phytopathogenic bacteria is dependent on some prerequisite conditions and is subject to environmental, biological and chemical alteration. Despite the similarities of the basic syndromes induced, differences detected with pseudomonad and xanthomonad species now require elaboration of original theories proposed. The utility of the hypersensitive host responses to establish the chronology of biological mechanisms associated with pathogenicity by bacteria remains to be fully appreciated.

Hypersensitivity as a phytopathological phenomenon was originally associated with the response of some plants to virus or fungus infection (Müller, 1959). That phytopathogenic bacteria can induce a hypersensitive response in plants was demonstrated by Klement and his co-workers (Klement, 1963; Klement & Lovrekovich, 1962; Klement *et al.*, 1964). The early visual detection of resistance to bacterial infection in plants obviously is of practical importance (Cook & Stall, 1969; Klement & Lovrekovich, 1961; Lozano & Sequeira, 1970). In addition, hypersensitivity caused by phytopathogenic bacteria may provide a tool to study and elucidate the biology of bacterial pathogenesis (Starr, 1959).

It should be realized that the term 'hypersensitive response' has been used casually in describing bacterial disease symptoms. Certainly not all resistance to bacterial disease in plants results from related and specific conditions that conform to the definitive criteria for hypersensitivity. Distinction between forms of susceptibility, which may include effective practical resistance, and hypersensitivity necessitate a critical consideration of the host-pathogen relations.

The causal biology of bacterial disease symptoms, particularly leafspots, has not been established. No consistent association between pathotoxins and symptom has been demonstrated and attempts to isolate from whole bacteria or culture media materials that consistently cause hypersensitive or susceptible necrosis have not generally met with success. Some knowledge of the causation of plant disease symptoms by phytopathogenic bacteria has been obtained from information derived from or directed to alteration of the hypersensitive response.

Temperature has been recognized as a major factor that influences bacterial disease development in nature, i.e., symptoms in susceptible plants. Variation in length of incubation is generally associated with temperature differences, but this criterion is more likely a reflection of bacterial multiplication rate *in vivo* following infection. As determined in recent studies (Klements & Goodman, 1967; Stall & Cook, 1966), symptoms in susceptible plants begin to appear after the *in vivo* bacterial population reaches a minimum level that is approximately the same for a number of different diseases. Physical means to introduce directly such minimum numbers of bacteria into plant tissues have not been devised, thus necessitating incubation and multiplication *in vivo* before symptoms appear. The hypersensitive response results from *in vivo* bacterial population approximately 10^2 lower than that for susceptible plant symptoms, and this concentration of bacteria can be readily introduced. In preliminary (unpublished) studies with *Xanthomonas vesicatoria*, we found that post-inoculation temperature directly influences the length of incubation in pepper before hypersensitive symptoms develop. Through the range of 20–30°C, a 5° decrease in temperature approximately doubled the incubation period. The same results were obtained by Lozano & Sequeira (1970) with *Pseudomonas solanacearum* in tobacco, where at 16°C hypersensitivity was delayed for 72 hours. It is quite possible that the hypersensitive response can be completely prevented by post-inoculation exposure of plants to a temperature lower than 16°C, even though an adequate inoculum concentration (i.e. greater than the minimum) is introduced. In a recent study by Kennedy & Crosse (1966) there was evidence that pre-inoculation temperature influenced symptom expression, but this is not an acknowledged general occurrence.

For inoculation, bacteria are usually suspended in water or in saline. Under normal laboratory or greenhouse conditions, the visible water congestion created by (hypodermic) infiltration inoculation disappears after some minutes with no obvious ill effects to leaf tissues. It has not been established whether the associated water is lost by transpiration or absorbed by host cells. However, as unpublished data have shown, leaves placed immediately after inoculation under conditions that do not permit infiltrated water or saline to dissipate, fail to develop the hypersensitive response in the usual time. In this manner the appearance of symptoms can be delayed, but the influence of such delay on bacterial viability or multiplication *in vivo* has not been determined. Post-inoculation exposure of infiltrated leaves to relative humidities of 50 or 87–92% did not influence the appearance of either susceptible or hypersensitive symptoms (Lozano & Sequeira, 1970).

Light has not been generally recognized as a critical environmental factor in the development of bacterial plant disease symptoms besides influencing plant growth. In two recent studies, however, light has been found to change symptom expression. In the first, by Smith & Kennedy (1970), susceptible lesions were progressively retarded in size and number by pre- and post-inoculation light periods (4–5 days) of decreasing intensity. In the second study, by Lozano & Sequeira (1970), the influence of light was related to the prevention of hypersensitive response by prior infiltration with heat-killed bacteria. Effective prevention of the hypersensitive response was achieved only if plants injected with heat-killed bacteria were incubated in light (1800 ft-c) before being challenge-inoculated with live bacteria. In an earlier report, by Starr (1959), susceptibility of two varieties of sesame to two bacterial diseases was reported to be enhanced by a short light period (12 h) and a high nitrogen regime, while resistance was manifest under a long (16 h) or short light period (12 h) and both high and low nitrogen fertilization. No information is available relating nutrition of test plants to production of the hypersensitive response. However, there is considerable evidence of *in vivo* nutrition influencing the virulence of vertebrate pathogens (Weinberg, 1966).

Altered semipermeability of host tissues, a characteristic of plant disease (Wheeler & Hanchey, 1968), appears earlier and increases more rapidly in hypersensitive (parabiotic) than in susceptible (eusymbiotic) host-parasite combinations. Host cell membrane destruction characteristically begins a few hours (4 or more) after inoculation, certainly before early hypersensitive symptoms are apparent, and becomes more extensive throughout the period of developing necrosis (Sasser *et al.*, 1968). Complete membrane disorganization ultimately results. The same kind of cellular effects occur in susceptible plants but are usually not apparent within 24 hours after inoculation. Membrane deterioration becomes progressively more extensive during incubation; it attains a level comparable to hypersensitive tissues by the time symptoms appear. Disruption of cellular membranes results in electrolyte loss (by leaching) from inoculated tissues; and it has also been correlated quantitatively with the appearance of a bacterial inhibitor *in vivo*.

A sharp increase in electrolyte loss from hypersensitive leaf tissue within hours after inoculation has been demonstrated by Cook & Stall (1968) and Goodman (1968). Recent analysis of the chemical elements leached from leaves has revealed more potassium than calcium, phosphorus, nitrogen or magnesium. Both magnesium and potassium were found to increase in the intercellular fluid of susceptible leaves (Sinclair *et al.*, 1970). Neither the processes that contribute to membrane deterioration and electrolyte loss, nor the source of the electrolytes has been determined. The knowledge that host cell membranes start to show deterioration when electrolyte loss first begins and that this condition becomes rapidly more severe as electrolyte loss increases, has been used as a basis for assessing cellular effects in host tissues.

Infiltration of calcium into leaves, prior to inoculation or simultaneously with bacteria, serves to suppress electrolyte loss from leaf tissues (Cook & Stall, 1970). Loss of electrolytes is inversely correlated with the infiltrated calcium concentration. Barium chloride and uranyl acetate similarly reduce electrolyte loss from inoculated leaf tissue but magnesium, manganese, nitrogen and potassium are ineffective. Calcium does not afford any consistent influence on electrolyte loss from healthy leaves exposed to various concentrations of ammonia (Lovrekovich *et al.*, 1969, 1970; Stall *et al.*, 1970) or volatile products from bacteria *in vitro* (Cook & Stall, 1969; Sasser *et al.*, 1968).

Of the prerequisites for inducing the hypersensitive response (Klement & Goodman, 1967), probably the most important (and certainly one of the most intriguing) is a minimum concentration of bacteria in the inoculum. Inocula containing 10^7 or more viable pathogenic cells/ml readily induce a hypersensitive response in leaves of plants insusceptible to the inoculated bacterial species. This level is relatively constant for all bacterial species and host plant combinations tested. The short incubation time that typifies hypersensitivity to bacterial infection is generally inadequate for a significant multiplication of the bacteria *in vivo*. The numbers of viable bacteria in hypersensitive tissues, which rapidly decrease after the onset of visible symptoms, have been related to the presence of inhibitory material. Inoculum containing less than 10^7 cells/ml is generally ineffective; it produces no confluent necrosis, irrespective of incubation time, but it has been reported to cause death of isolated host cells. Increase of bacteria in the inoculum progressively shortens incubation time. With susceptible plants, inoculum concentration is much less critical for the production of necrosis. During the extended incubation period, the bacteria multiply and, prior to symptom production, reach a population level approximately 10^2 greater than attained in plant tissues that react in a hypersensitive manner. Reduction of the concentration of viable bacteria in the inoculum merely lengthens the incubation period necessary for multiplication *in vivo*. Conversely, it has been reported that inoculation with more than 1×10^9 cells/ml will induce susceptible necrosis in 24 hours.

Two recent studies have confirmed the earlier report by Lovrekovich & Farkas (1965) that pre-inoculation infiltration of tobacco leaves with heat-killed bacteria alters symptoms following challenge inoculation with live bacteria. In one study (Sleesman *et al.*, 1970) such pre-treatment proved to induce both hypersensitive and susceptible responses in tobacco, before inoculation with live bacteria retarded development of susceptible symptoms for 60 hours. Pre-treatment with heat-killed cells prevented development of the hypersensitive response. A heat-stable fraction from whole or sonicated cells of compatible and incompatible bacteria as well as insoluble cell wall materials also provided protection. No systemic effects were noted and no indication was found that inoculum concentrations for pretreatment or challenge inoculations were critical.

In the other study (Lozano & Sequeira, 1970), the hypersensitive response was reported to be prevented by similar pretreatment, but it was more complete after infiltration with higher concentrations of dead bacteria than live bacteria. Prevention was more efficient if pre-treatment was with the same species or pathotype as that used in the challenge inoculation, if the live bacteria were inoculated 18 hours or more after infiltration of the heat-killed bacteria, some systemic protection being noted. It seemed that the temperature used to kill bacteria influenced the efficacy of protection against hypersensitivity.

Soluble and insoluble fractions capable of preventing hypersensitivity, obtained from sonicated heat-killed cells have been reported earlier from this laboratory (Sequeira & Ainslie, 1969). Our recent attempts, as yet unpublished, to prevent hypersensitivity in pepper and tobacco by pre-inoculation injection of heat-killed xanthomonad bacteria met with little success: in neither host was hypersensitivity prevented, irrespective of the temperature (60–120 °C) used to kill the bacteria (all plants were incubated in continuous fluorescent light).

Goodman (1967) has reported protection against infection of susceptible plants with a virulent organism to result from previous inoculation (30 minutes) with an avirulent isolate of the same organism in a 10^2 greater concentration. A nonpathogenic (parabiotic) and a saprophytic bacterium also gave protection against infection, but a pathogenic (eusymbiotic) bacterium did not. Dilution of virulent bacterial cells in an inoculum with avirulent cells (Goodman, 1965) of the same organism, at a ratio of

1:19 or greater, caused lower disease indices (for susceptible plants), although there was no evidence of inhibition of the former by the latter when grown together *in vitro* (Averre & Kelman, 1964).

The theories originally developed about induction of the hypersensitive and susceptible plant responses (Klement, 1968; Klement & Goodman, 1967), and the variety of plant and bacterial species available for study gave promise of an early solution for both symptom types. Up till now the biology of bacterial disease symptoms has been limited but some general information about the two distinct responses has been established. Symptoms related to the hypersensitive response are confined to the inoculated tissues and the infection resulting from the original introduction of bacteria does not expand further.

The population of viable bacteria in inoculated tissues declines significantly following the appearance of hypersensitive symptoms, possibly as a result of a bacterial inhibitor. At least with some (hypersensitive) bacterium/host combinations, the necrotic host tissue becomes almost milk-white, apart from a narrow dark brown or black margin. These characteristics readily distinguish the hypersensitive response from the susceptible plants wherein the area of infection in host tissue continues to expand and the bacteria to multiply *in vivo* even after necrosis has become evident. The necrotic host tissue commonly assumes some approximately uniform shade of brown and may appear water-soaked and translucent.

In our studies of xanthomonads, and particularly *Xanthomonas vesicatoria*, a second type of hypersensitive response has been encountered. Although we have not yet completed the full characterization of this symptom type, the visible effects first appear after about the same incubation time as those of the typical hypersensitive response. So far as we have been able to determine, an inoculum concentration of approximately 10^7 is prerequisite for this symptom. In appearance, the resulting necrosis is dark brown and, in typical cases, more distinctly confined to the lower surface of the leaf and to the tissues originally inoculated. Isolates that induce this symptom type on pepper *(Capsicum annuum)* have been found in *X. phaseoli*, *X. oryzae* and *X. vesicatoria* (tomato strain). An atypical yellowing symptom induced in tobacco by virulent race 3 isolates and some avirulent mutants of *Pseudomonas solanacearum* (Lozano & Sequeira, 1970) resembles the hypersensitive response in that the affected area does not enlarge. However, multiplication *in vivo* during incubation precludes immediate consideration of yellowing as a form of hypersensitivity.

Both hypersensitive and susceptible symptom expression in plants have been studied more extensively using pseudomonad species. The causation of these symptom types by xanthomonad species is generally similar, with major exceptions such as the influence of light on development of hypersensitivity by some pseudomonads which has not been noted with xanthomonad species. The routines, mentioned previously for prevention of the hypersensitive response, have been devised using tobacco as a host plant and pseudomonads as the challenge bacteria. Attempts to duplicate these results with isolates of xanthomonads, both for heat-killed cells and for challenge bacteria, resulted in partial protection and not complete prevention of hypersensitivity development. This permits the tentative conclusion that pathogenesis by pseudomonads and xanthomonads involve different biological processes. The relation of the pathogenic mechanisms of species in these two groups to those of bacterial pathogens in other genera remains to be established.

The utility of the hypersensitive response caused by phytopathogenic bacteria has been established both for practical and academic purposes (Cook & Stall, 1969; Lozano & Sequeira, 1970). That the response is a pathological process amenable to alteration by a range of environmental, biophysical and chemical conditions has been repeatedly demonstrated. The possibility of exacting control of many of the factors contributing to symptom expression in plants offers the opportunity for examination of these factors separately. It appears obvious from these considerations, and from the recognition of more than two basic symptom types, that the concept of bacterial infection resulting only in a single form of hypersensitive or susceptible symptom, must be examined more critically. That avirulent cultures of bacteria can induce hypersensitivity (Klement & Goodman, 1966), emphasizes the need for a more detailed study of the mechanisms of defense reactions.

Cell-free fractions capable of inducing symptoms resembling hypersensitivity have been isolated from sonicated cells (Sequeira & Ainslie, 1969; Sleesman *et al.*, 1970), but more research is necessary to unravel the biochemical nature of rapid necrosis caused by some bacterial pathogens and the absence of comparable symptoms following infection by other pathotypes (Huang & Goodman, 1970). Certainly further study of these different symptom types and others, now less well defined, will necessarily entail differences at the cellular level. Present evidence leads to the conclusion that host cell membrane integrity is affected and that such effect is contingent upon minimum or greater concentration of bacteria *in vivo*. The concomitant appearance of a bacterial inhibitor of undetermined origin and inconsistent nature and specificity (Lozano & Sequeira, 1970; Stall & Cook, 1968) may be a general phenomenon, while various other aspects of symptom expression may be more directly related to the inoculated species or pathotype.

The wide range of bacterial pathogens and host plants available for comparative study offers ample material for a critical study of the causative nature of bacterial pathogenesis. The similarity of some aspects of hypersensitivity in plants to vertebrate pathologic phenomena (Mann & Collier, 1969) adds impetus to a further study of this symptom type.

References

Averre, C. W. & A. Kelman, 1964. Severity of bacterial wilt as influenced by ratio of virulent to avirulent cells of *Pseudomonas solanacearum* in inoculum. Phytopathology 54: 779–783.

Cook, A. A. & R. E. Stall, 1968. Effect of *Xanthomonas vesicatoria* on loss of electrolytes from leaves of *Capsicum annuum*. Phytopathology 58: 617–619.

Cook, A. A. & R. E. Stall, 1969. Necrosis of leaves induced by volatile materials produced in vitro by bacteria. Phytopathology 59: 259–260.

Cook, A. A. & R. E. Stall, 1969. Differentiation of pathotypes among isolates of *Xanthomonas vesicatoria*. Pl. Dis. Reptr 53: 617–619.

Cook, A. A. & R. E. Stall, 1970. Effect of calcium on loss of electrolytes from pepper leaves inoculated with *Xanthomonas vesicatoria*. Phytopathology 60: 1288.

Goodman, R. N., 1965. In vitro and in vivo interactions between components of mixed bacterial cultures isolated from apple buds. Phytopathology 55: 217–221.

Goodman, R. N., 1967. The protection of apple stem tissue against *Erwinia amylovora* infection by avirulent strains and three other bacterial species. Phytopathology 57: 22–24.

Goodman, R. N., 1968. The hypersensitive reaction in tobacco: a reflection of changes in host cell permeability. Phytopathology 58: 872–873.

Huang, J. & R. N. Goodman, 1970. The relationship of phosphatidase activity to the hypersensitive reaction in tobacco induced by bacteria. Phytopathology 60: 1020–1021.

Kennedy, B. W. & J. E. Crosse, 1966. Inoculation procedure for comparing reaction of soybeans to bacterial blight. Pl. Dis. Reptr 50: 560–565.

Klement, Z., 1963. Rapid detection of the pathogenicity of phytopathogenic pseudomonads. Nature, Lond. 199: 299–300.

Klement, Z., 1968. Pathogenicity factors in regards to relationships of phytopathogenic bacteria. Phytopathology 58: 1281–1221.

Klement, Z., G. L. Farkas & L. Lovrekovich, 1964. Hypersensitive reaction induced by phytopathogenic bacteria in the tobacco leaf. Phytopathology 54: 474–477.

Klement, Z. & R. N. Goodman, 1966. Hypersensitive reaction induced in apple shoots by an avirulent form of *Erwinia amylovora*. Acta phytopath. hung. 1: 177–184.

Klement, A. & R. N. Goodman, 1967. The hypersensitive reaction to infection by bacterial plant pathogens. A. Rev. Phytopath. 5: 17–44.

Klement, Z. & R. N. Goodman, 1967. The role of the living bacterial cell and the induction time in the hypersensitive reaction of tobacco plants. Phytopathology 57: 322–323.

Klement, Z. & L. Lovrekovich, 1961. Defense reactions induced by phytopathogenic bacteria in bean pods. Phytopath. Z. 41: 217–227.

Klement, Z. & L. Lovrekovich, 1962. Studies on host-parasite relations in bean pods infected with bacteria. Phytopath. Z. 45: 81–88.

Lovrekovich, L. & G. L. Farkas, 1965. Induced protection against wildfire disease in tobacco leaves treated with heat-killed bacteria. Nature, Lond. 205: 823–824.

Lovrekovich, L., H. Lovrekovich & R. N. Goodman, 1969. The role of ammonia in the wildfire disease of tobacco caused by *Pseudomonas tabaci*. Phytopathology 59: 1713–1716.

Lovrekovich, L., H. Lovrekovich & R. N. Goodman, 1970. Ammonia as a necrotoxin in the hypersensitive reaction caused by bacteria in tobacco leaves. Can. J. Bot. 48: 167–171.

Lozano, J. C. & L. Sequeira, 1970. Differentiation of races of *Pseudomonas solanacearum* by a leaf infiltration technique. Phytopathology 60: 834–840.

Lozano, J. C. & L. Sequeira, 1970. Prevention of the hypersensitive reaction in tobacco leaves by heat-killed bacterial cells. Phytopathology 60: 875–879.

Mann, J. & H. B. Collier, 1969. The effect of calcium ion on hypotonic hemolysis of human erythrocytes. Life Sci. 8: 1343–1349.

Müller, K. O., 1959. Hypersensitivity. In: J. G. Horsfall & A. E. Dimond, eds. Plant pathology, An advanced treatise, I, p. 469–519. New York.

Sasser, J. M., R. E. Stall & A. A. Cook, 1968. Membrane disruption in pepper cells induced by *Xanthomonas vesicatoria* and by volatile products of the bacterium. Phytopathology 58: 1066.

Sequeira, L. & V. Ainslie, 1969. Bacterial cell-free preparations that induce or prevent the hypersensitive reaction in tobacco. Abstr. XI Intern. bot. Congr. p. 195.

Sinclair, M. G., J. M. Sasser & T. J. Gulya, 1970. Pepper leaf intercellular fluid composition after inoculation with *Xanthomonas vesicatoria*. Phytopathology 60: 1314.

Sleesman, H. C., J. E. Perley & H. A. J. Hoitink, 1970. Susceptible and hypersensitive reactions in tobacco and their prevention by cell-free extracts of *Pseudomonas tabaci* and *Pseudomonas glycinea*. Phytopathology 60: 1314.

Smith, M. A. & B. W. Kennedy, 1970. Effect of light on reactions of soybean to *Pseudomonas glycinea*. Phytopathology 60: 723–725.

Stall, R. E. & A. A. Cook, 1966. Multiplication of *Xanthomonas vesicatoria* and lesion development in resistant and susceptible pepper. Phytopathology 56: 1152–1154.

Stall, R. E. & A. A. Cook, 1968. Inhibition of *Xanthomonas vesicatoria* in extracts from hypersensitive and susceptible pepper leaves. Phytopathology 58: 1584–1587.

Stall, R. E., A. A. Cook & C. B. Hall, 1970. Association of ammonia with electrolyte leakage from leaves of *Capsicum annuum*. Phytopathology 60: 1315.

Starr, M. P., 1959. Bacteria as plant pathogens. A. Rev. Microbiol. 13: 211–238.

Thomas, C. A., 1965. Effect of photoperiod and nitrogen on reaction of sesame to *Pseudomonas sesami* and *Xanthomonas sesami*. Pl. Dis. Reptr 49: 119–120.

Weinberg, E. D., 1966. Roles of metallic ions in host-parasite interactions. Bact. Rev. 30: 136–151.

Wheeler, H. & P. Hanchey, 1968. Permeability phenomena in plant disease. A. Rev. Phytopath. 6: 331–350.

Reactions of the host plant to specific pathogens

K. Rudolph

Institut für Pflanzenpathologie, 34 Göttingen, Germany

Abstract

Two phenomena in bacterial infected plants are typical for the compatible or normosensitive reaction: supply of nutrients for the pathogen, and repression of resistance inducing processes.

Accumulation of compounds at infection sites, increased permeability of cell membranes, leakage of cell components into the intercellular fluid, and increased water congestion are often observed in plant tissues infected by bacteria. The resistant or hypersensitive reaction normally occurring after massive invasion of foreign organisms seems to be inhibited during the compatible combination. The underlying mechanisms are as yet unknown, but marked changes in host metabolism during the susceptible reaction have been reported.

Investigations on soluble bean leaf proteins after infection by *Pseudomonas phaseolicola* have revealed changes in the isoenzyme patterns for several dehydrogenases (substrates: malate, iso-citrate, glutamate, and glucose-6-phosphate).

Introduction

Whereas many phytopathogenic bacteria can induce a hypersensitive reaction in non-host plants, typical disease symptoms are only caused by specific pathogens in their hosts. The following contribution will deal with reactions which occur with susceptible or compatible combinations.

Observations on the hypersensitive reaction may lead to the definition of susceptibility as 'the non-occurrence of a resistant reaction'. This definition is not quite sufficient, since it may allow the conclusion that infected susceptible plants do not basically differ from uninfected healthy plants, and that only with incompatibility additional reactions do occur. In fact, several observations indicate that also in a compatible combination, various reactions of the host plant are induced. Many of these reactions are probably secondary processes, their nature depending on the type of disease. The question is, however, whether the very different types of bacterial diseases have some reactions in common, which are typical and necessary for the compatible or normosensitive reaction.

Based on the study of the literature and on some experiments, two aspects of such reactions will be dealt with here: the supply of nutrients for the pathogen, and the repression of resistance inducing processes.

Methods and Materials

Bean plants were grown in vermiculite in growth chambers, irrigated with a Hoagland solution, and infected with *Pseudomonas phaseolicola* as described by Rudolph & Stahmann (1966). The bean varieties were: Red Mexican 3 (resistant to race 1) and Red Kidney 35 (susceptible to race 1). The profile of the soluble leaf proteins was obtained as described by the same authors (1964a).

Leaf extracts and sonic bacterial extracts (Rudolph & Stahmann, 1966) were separated by polyacrylamide disc electrophoresis (Davis, 1964). It was found that separation of isoenzymes could be improved by coating the glass tubes with dimethyldichlorosilane. For comparison of different extracts the same amounts of soluble protein (300 μg) were applied on the gel. Larger amounts of protein gave rise to artificial bands of dehydrogenases. Protein concentration was estimated by micro-Kjeldahl analysis following precipitation by trichloroacetic acid. In extracts from heavily infected leaves, or in infection centers, bacterial polysaccharides inhibited protein precipitation, but by repeatedly freezing and thawing extracts with trichloroacetic acid precipitation was achieved.

The reaction mixture for development of dehydrogenase activity (modification of van der Helm, 1962) contained 32 ml 0.1 *M* Tris (pH 7.5), 12 ml 1 % malate, glutamate, iso-citrate, or glucose-6-phosphate (pH 7.5), 12 ml 0.1 % NAD or NADP, 12 ml 0.1 *M* NaCN, 8 ml 0.05 *M* $MgCl_2$, 20 ml 0.1 % nitroblue tetrazolium, and 2,5 ml 0.1 % phenazinemethosulphate. Gels were incubated at room temperature and then fixed in 7 % acetic acid. Control mixtures without NAD (NADP) or substrate gave no reaction. Incubation mixtures without NaCN stained very slowly. The bands were numbered according to the recommendations of the Standard Committee on Enzymes (Webb, 1964), so that the isoenzyme with the greatest (anodic) mobility was assigned number 1.

Results and discussion

Supply of nutrients for the pathogen The nutritional requirements of phytopathogenic bacteria are quite simple (Starr, 1964a, b, 1949; Patel & Walker, 1963; Rickard & Walker, 1965a, b; Kamenova & Shchelkunova, 1970), although it has been shown that saprophytic pseudomonads are 'more versatile nutritionally and more reactive biochemically' than plant pathogenic pseudomonads (Misaghi & Grogan, 1969). In some cases it has been supposed that susceptibility depends on the supply of specific organic compounds in a certain balance necessary for bacterial growth (Thomas & Orellana, 1963; Watanabe *et al.*, 1965).

Some studies, especially those of Walker and co-workers, have shown the importance of the nutrition of the host plant for bacterial diseases (Hedges & Fisher, 1946; van Gundy & Walker, 1957; Nayudu & Walker, 1960, 1961; Patel & Walker, 1963; Rickard & Walker, 1965c). Usually, however, the nutritional state of the host plant only modifies the reaction of the plant after inoculation.

Auxotrophic mutants have been used in attempts to show a correlation between susceptibility and availability of certain organic compounds (Garber *et al.*, 1956; Garber & Schaeffer, 1957). However, naturally occurring phytopathogenic bacteria do not reveal properties of auxotrophic mutants.

In general, the total amount of components which are available to the parasite seems to be decisive. As with other diseases, accumulation of substances has been observed at bacterial infection sites, acting as metabolic sinks (Shaw & Samborski, 1956). Sinclair *et al.* (1970) analysed the intercellular fluid of susceptible pepper leaves after inoculation with *Xanthomonas vesicatoria*. They report a 10-fold increase of carbohydrates, amino acids, and proteins. From the steadily increasing levels of some of the constituents of the intercellular fluid they conclude that the infected leaf accumulates materials normally available to the other plant parts.

Tumours, induced by *Agrobacterium tumefaciens*, mobilize essential nutrients. Apart from a greater ability to synthesize protein, tumour tissue also mobilizes nitrogen from other parts of the plant (Link & Goddard, 1951), because tumours and adjacent tissues contain more nitrogen than healthy tissues further away. Increased levels of growth hormones in tumours (Kulescha, 1949; Henderson & Bonner, 1952; ElKhalifa & Lippincott, 1968; Bouillenne & Gaspar, 1970), and in tissues affected by other bacterial diseases (Sequeira, 1963; Freebairn & Buddenhagen, 1964) may be partly responsible for this effect. Observations of Lovrekovich & Loebenstein (1966) have shown that three phytopathogenic pseudomonads induced gall formation in *Datura stramonium*, but that a saprophytic pseudomonas species did not; this indicates an effect of growth hormones.

An alteration in cellular membrane permeability seems to play an important role in the increased availability of organic compounds. Increased permeability has been demonstrated in apple leaves infected by *Erwinia amylovora* (Burkowicz & Goodman, 1969), in angular leaf spot of cucumber (Williams & Keen, 1967), and in crown gall tumour cells (Wood & Braun, 1965). One out of several possible reasons for increased permeability is a higher activity of phosphatidase, as reported for phytopathogenic bacteria, especially for *Erwinia amylovora* (Tseng & Bateman, 1968; Mount & Bateman, 1969; Huang & Goodman, 1970). As permeability increased so did the size of water congested lesions in angular leaf spot of cucumber (Williams & Keen, 1967). The importance of water congestion for bacterial infection has been emphasized early (Clayton, 1936; Johnson, 1945). Recently, a correlation between water congestion and susceptibility to *Xanthomonas pruni* and *X. vesicatoria* has been reported (Matthee & Daines, 1968, 1969).

Obviously, water-congested tissue is a favourable environment for bacterial growth. Enhanced permeability of plant tissue allows nutrients to leak out of the cells without a concomitant onset of resistance reactions. It may be mentioned here that incubation of resistant bean leaves in air of nearly 100% relative humidity after artificial inoculation renders them susceptible to *Pseudomonas phaseolicola* (Rudolph, unpublished).

Repression of resistance inducing processes Susceptibility has been defined as an abnormal situation in which the pathogen is unable to induce a hypersensitive reaction (Klement, 1968), or to put it in a positive way: the situation in which the pathogen is able to inhibit the resistance reaction. In considering the drastic metabolic changes evoked by the pathogen – such as increases in permeability, accumulation of organic compounds, increased respiration, partial destruction of cell walls and cell organelles – it is remarkable that the final cell collapse occurs so late during the susceptible reaction. For a certain period, an equilibrium is established that allows the pathogen to withdraw nutrients from the host without releasing bacteriostatic compounds or inducing any other resistance reaction.

Sometimes the repression of resistance inducing processes might be due to bacterial toxins causing a partial paralysis of the host metabolism, e.g. by inhibiting certain enzymes (Sinden & Durbin, 1968; Patil, 1970). It has been shown that the genetic resistance of bean varieties to *Pseudomonas phaseolicola* can be broken down by applying the purified toxin (Rudolph, 1970). However, these toxins do not explain specificity and therefore they can not be the sole cause of the susceptible reaction.

It has been postulated by Schnathorst & DeVay (1963) that the susceptible reaction is possible because host and parasite have antigens in common. A comparison of reactions of antisera of different varieties of cotton has indicated that susceptibility to angular leaf spot is reflected by the presence of 2 or 3 additional minor bands and a greater reaction intensity with antigens of pathogenic bacteria than with those of non-pathogens. The authors suggest that a common antigen base is a prerequisite for a compatible relationship and they assume that, during infection, the possible exchange of substances (such as transfer or messenger RNA between a host and a parasite) might have a less disruptive effect on cell metabolism if these substances provide common synthetic information (Schnathorst, 1966; DeVay *et al.*, 1967). Although this hypothesis is very fascinating, until now very little evidence has been obtained to support it. Further studies with bacteriophages may result in new interpretations (Civerolo, 1971).

Another possibility would be that the common antigens are in some way related to cell-wall degrading enzymes of the parasite, or cell-wall synthesizing enzymes of the host. This would mean a link to Albersheim's hypothesis that in many instances of pathogenesis by bacteria or fungi, it is an interaction between the pathogen and the carbohydrates of the host which determines the pathogen's ability to produce enzymes capable of degrading the host cell walls (Albersheim *et al.*, 1969).

For diseases with tissue disintegration, the specificity of the cell-wall degrading enzymes should be of primary importance for disease development. So, the parasitic ability of *Erwinia aroidea* depends on rapid growth and production of macerating enzymes which, acting in advance of the parasite itself, prevents the formation of an effective barrier (Wood, 1967). For a soft rot disease type it is difficult to assume, however, that compatibility involves an active reaction of the host plant. An active role of the host in susceptibility would be a participation of the host in cell-wall degradation, or an induced mechanism that inhibits or restrains a resistance reaction. No evidence supports these possibilities.

With other bacterial diseases the pattern of proteins and enzymes differs between infected susceptible plants and healthy plants. For instance, Rudolph & Stahmann (1964a) have demonstrated such a change in the profile of soluble bean leaf proteins

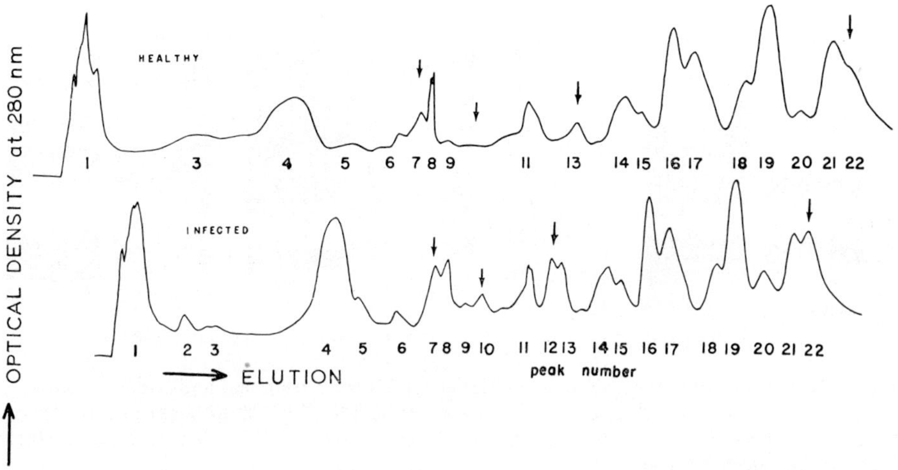

Fig. 1. Optical density profile at 280 nm of effluents from diethylaminoethylcellulose chromatography. The extracts from healthy and infected plants (12 days after infection) contained 120 mg protein. Arrows indicate peaks which changed after infection.

after infection by *Pseudomonas phaseolicola*. After separation of the crude leaf extract on DEAE-cellulose columns some new peaks appeared and some others became more prominent in the infected susceptible leaf (Fig. 1: arrows). The study of different enzymes by polyacrylamide disc electrophoresis also revealed changes after the compatible reaction. In several instances new or increasing isoenzymatic bands appeared to be of bacterial origin, such as a catalase (Rudolph & Stahmann, 1964b) and an acid phosphatase (Rudolph & Stahmann, 1966).

Analysis of dehydrogenases showed that the host also took part in the new appearance or increase of isoenzymes. So a new band for malate dehydrogenase (1.1.1.37) was observed during early disease stages (Fig. 2); it increased so markedly during later stages that the separation of the other bands was impaired. These disease induced changes were confirmed by a second and third re-electrophoresis of the isolated bands. As a sonicated bacterial extract showed a band for malate dehydrogenase at the same position, it seemed obvious that the increasing band was of bacterial origin. From several additional findings, however, it was concluded that this was only partially so because (a) ageing leaves showed a similar increase of this band, (b) the greatest increase of this band was observed in the green tissue surrounding the yellow halos induced by the disease, (c) after mixing of healthy bean leaf and bacterial extracts this band could not be demonstrated after electrophoresis, (d) the bacterial zone with the same Rf value showed a lower stability than the plant band. It was assumed, therefore, that the increase of this band was a reaction of the susceptible intact host tissue following infection by *Pseudomonas phaseolicola*.

NADP-dependent malate dehydrogenase (1.1.1.40) showed only one band in extracts from healthy leaves (Fig. 3). In infected leaves a new zone appeared at the

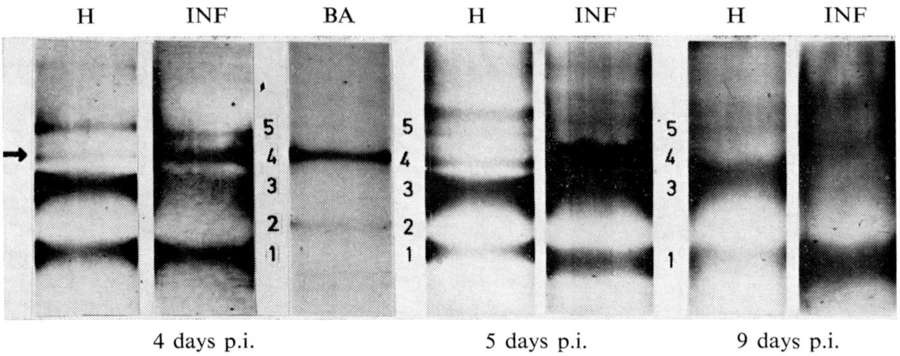

Fig.2. Zymograms of malate dehydrogenases (1.1.1.37) from bean leaf and bacterial extracts, obtained by polyacrylamide disc electrophoresis. H: healthy, INF: halo-blight infected leaves (4, 5, and 9 days after infection), BA: bacterial extract. Arrow indicates isoenzyme 4 which increased in infected leaves.

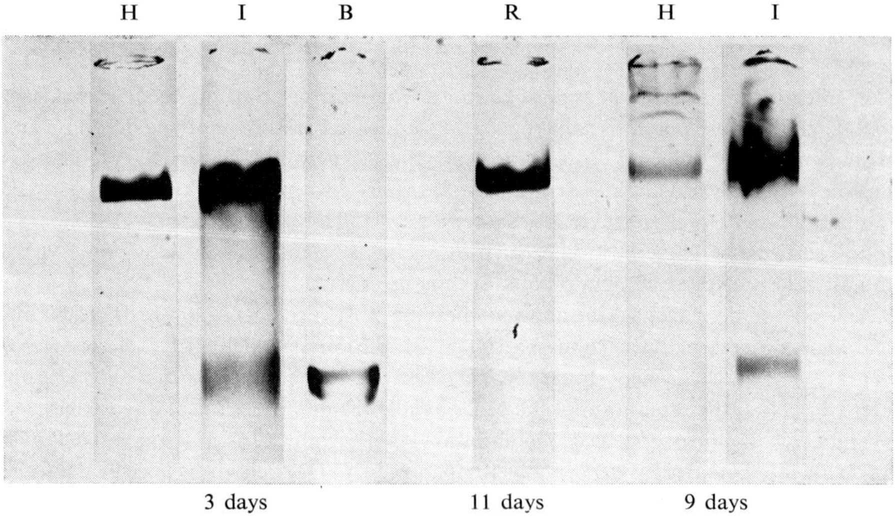

Fig. 3. Zymograms of NADP-dependent malate dehydrogenases (1.1.1.40) from bean leaf and bacterial extracts, obtained by polyacrylamide disc electrophoresis. H: healthy, I: halo-blight infected leaves (3 and 9 days after infection), B: bacterial extract, R: rusted leaves (11 days after infection).

same position as a bacterial band. The total enzyme activity showed a 3-fold increase after infection, indicating that the host enzyme also increased, since the bacterial enzyme contributed only a small portion to the total activity. As no new zone was observed in rust-infected leaves, this case is a different type of bean leaf disease.

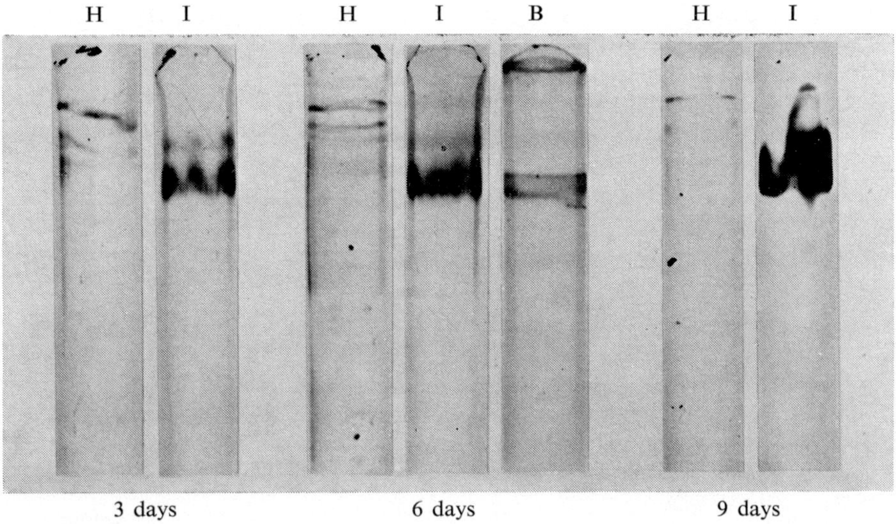

Fig. 4. Zymograms of glutamate dehydrogenase (1.4.1.3) from bean leaf and bacterial extracts, obtained by polyacrylamide disc electrophoresis. H: healthy, I: halo-blight infected leaves (3, 6 and 9 days after infection), B: bacterial extract.

Glutamate dehydrogenase (1.4.1.3) showed a clear band only in infected leaves and at the same position as a bacterial band (Fig. 4). As the total activity increased up to 20 times the original in infected leaves, it seemed unlikely that the new band was only contributed by the bacteria, which gave only a faint band for glutamate dehydrogenase. So it was assumed that bound host glutamate dehydrogenase was released during the disease process. Since the halo blight disease is characterized biochemically by a very high increase of ornithine, it is probable that the very high increase of glutamate dehydrogenase can be explained by the metabolization of ornithine via glutamate.

Iso-citrate dehydrogenase (1.1.1.42) showed a strong band after infection (Fig. 5). Bacterial extracts gave a band at the same position. Healthy leaves showed a faint zone at this position, but only after very long exposure. Therefore, it can not be excluded that a plant enzyme participated in this increase.

Glucose-6-phosphate dehydrogenase (1.1.1.49) showed a new band of bacterial origin in the infected leaves (Fig. 6), while the host band decreased, especially during late infection stages. Rusted leaves did not show a new band.

In evaluation all these observations it has to be kept in mind that changes in dehydrogenases probably represent only a small part of the total reactions occurring during pathogenesis. Other enzymes may be more important. The role of dehydrogenases for the compatible combination can be seen in two ways: supply of metabolites for the pathogen, and regulation of the redox-potential. The redox-potential may determine the degree of resistance. Lovrekovich *et al.* (1967) have suggested that dehydrogenases inhibit resistance inducing processes in *Erwinia*-infected potato tubers.

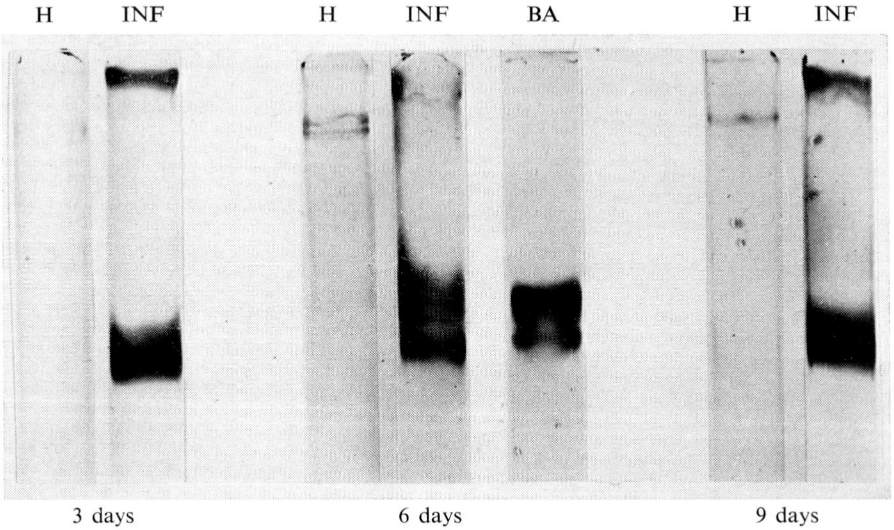

Fig. 5. Zymograms of iso-citrate dehydrogenase (1.1.1.42) from bean leaf and bacterial extracts, obtained by polyacrylamide disc electrophoresis. H: healthy, INF: halo-blight infected leaves (3, 6 and 9 days after infection), BA: bacterial extract.

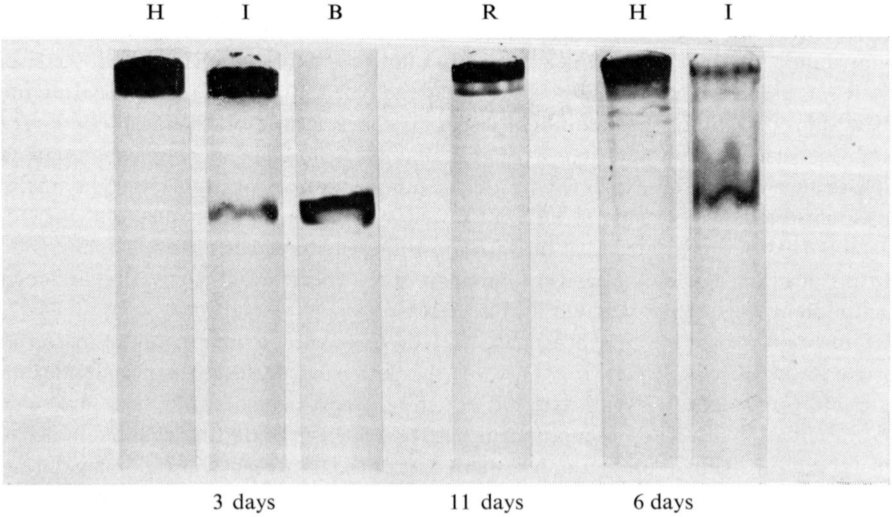

Fig. 6. Zymograms of glucose-6-phosphate dehydrogenase (1.1.1.49) from bean leaf and bacterial extracts, obtained by polyacrylamide disc electrophoresis. H: healthy, I: halo-blight infected leaves (3 and 6 days after infection), B: bacterial extract, R: rusted leaves (11 days after infection).

An increase of dehydrogenases has also been reported in mitochondria from tumour tissue (Lance, 1961). Since tumorous tissue represents a rather exceptional type of bacterial disease, it can not be discussed here in detail, also in regard to nucleic acids (Kupila & Stern, 1961; Srivastava, 1965; Schilperoort *et al.*, 1967; Börner, 1969) or new antigens (DeVay *et al.*, 1970).

The interpretation of the susceptible reaction as a repression of resistance inducing processes does not preclude that synthesis or oxidation of aromatic compounds is concomitantly intensified. This has been shown for *Pseudomonas solanacearum* infected tobacco plants by Sequeira and co-workers (Pegg & Sequeira, 1968; Sequeira, 1969). Peroxidase increases have been reported in susceptible tissue after infection with *Pseudomonas phaseolicola* (Rudolph & Stahmann, 1964b), with *Agrobacterium tumefaciens* (Curtis, 1970), and with *Agrobacterium rhizogenes* (Huisingh & Durbin, 1970). It is still unknown whether the oxidation reactions during the compatible combination proceed along other pathways than during the incompatible reaction.

It was the aim of this contribution to deal with reactions of the host in the compatible combination. A comparison of the two aspects considered – supply of nutrients for the pathogen, and repression of resistance inducing processes – leads to the conclusion that many reactions of the susceptible host can be grouped under the first aspect. On the other hand, only few observations indicate a participation of the host in changing the normal resistant reaction, after invasion by pathogenic bacteria, into a susceptible reaction in the case of specific pathogens. Therefore, we have to assume that the pathogen is mainly responsible for the inhibition of resistance inducing processes. A deeper insight into the basic resistance mechanisms is needed to understand the inhibition of these processes and to explain the fascinating phenomenon of specificity.

References

Albersheim, P., T. M. Jones & P. D. English, 1969. Biochemistry of the cell wall in relation to infective processes. A. Rev. Phytopath. 7: 171–194.

Börner, B., 1969. Über RNS-Veränderungen in Keimpflanzenwurzeln von *Vicia faba* nach Infektion mit *Agrobacterium tumefaciens*. Biol. Zbl. 88: 451–456.

Bouillenne, C. & T. Gaspar, 1970. Auxin catabolism and inhibitors in normal and crown gall tissues of *Impatiens balsamina*. Can. J. Bot. 48: 1159–1163.

Burkowicz, A. & R. N. Goodman, 1969. Permeability alterations induced in apple leaves by virulent and avirulent strains of *Erwinia amylovora*. Phytopathology 59: 314–318.

Civerolo, E. L., 1971. Interaction between bacteria and bacteriophages on plant surfaces and in plant tissues. 3. Intern. Conf. Pl. Pathog. Bact., Wageningen.

Clayton, E. E., 1936. Water soaking of leaves in relation to the development of the wildfire disease of tobacco. J. agr. Res. 52: 239–269.

Curtis, C. R., 1970. Disc electrophoresis of proteins from bean leaves infected with *Agrobacterium tumefaciens*. Phytopathology 60: 1014.

Davis, B. J., 1964. Disc electrophoresis. II. Method and application to human serum proteins. Ann. N.Y. Acad. Sci. 121: 404–427.

DeVay, J. E., R. J. Romani, A. M. Monadjem & M. Etzler, 1970. Induction of phytoprecipitin in sunflower gall tissue in response to infection by *Agrobacterium tumefaciens*. Phytopathology 60: 1289.

DeVay, J. E., W. C. Schnathorst & M. S. Foda, 1967. Common antigens and host-parasite interactions. In: C. I. Mirocha & I. Uritani, eds: The dynamic role of molecular constituents in plant parasite interaction. Am. Phytopath. Soc., Inc.

ElKhalifa, M. D. & J. A. Lippincott, 1968. The influence of plant-growth factors on the initiation and growth of crown-gall tumours on primary pinto bean leaves. J. exp. Bot. 19: 749–759.

Freebairn, H. T. & I. W. Buddenhagen, 1964. Ethylene production by *Pseudomonas solanacearum*. Nature, Lond. 202: 313–314.

Garber, E. D. & S. G. Schaeffer, 1957. Free histidine content of turnip varieties and their resistance to histidine-requiring mutants of *Erwinia aroidea*. J. Bact. 74: 392–395.

Garber, E. D., S. G. Schaeffer & M. Goldman, 1956. The virulence of biochemical mutants of *Erwinia aroidea* for varieties of radish and turnip. J. gen. Microbiol. 14: 261–267.

Gundy, S. D. van & J. C. Walker, 1957. Relation of temperature and host nutrition to angular leaf spot of cucumber. Phytopathology 47: 615–619.

Hedges, F. & H. Fisher, 1946. Association of *Xanthomonas phaseoli* and the common bean-mosaic virus, *Marmor phaseoli*. III. The effect of varying amounts of nitrogen on pathogenicity. Phytopathology 36: 613–623.

Helm, H. J. van der, 1962. L-glutamate dehydrogenase isoenzymes. Nature, Lond. 194: 773.

Henderson, J. H. M. & J. Bruner, 1952. Auxin metabolism in normal and crown gall tissue of sunflower. Am. J. Bot. 39: 444–451.

Huang, J.-S. & R. N. Goodman, 1970. The relationship of phosphatidase activity to the hypersensitive reaction in tobacco induced by bacteria. Phytopathology 60: 1020–1021.

Huisingh, D. & R. D. Durbin, 1970. Some physiological effects of *Agrobacterium rhizogenes* on tomato. Phytopathology 60: 1101–1105.

Johnson, J., 1945. Infection experiments with detached water-congested leaves. Phytopathology 35: 1017–1028.

Kamenova, S. M. & S. A. Shchelkunova, 1970. Ispol'zovanie kislot pervichnogo okisleniya sakharov bakteriyami roda *Pseudomonas* (The utilisation of acids of primary sugar oxidation by bacteria of *Pseudomonas* genus). Vest. leningr. gos. Univ., Ser. Biol. 25: 97–101.

Klement, Z., 1968. Pathogenicity factors in regard to relationships of phytopathogenic bacteria. Phytopathology 58: 1218–1221.

Kulesha, Z., 1949. Relation entre le pouvoir de prolifération spontanée des tissus de topinambour et leur teneur en substance de croissance. C.r. Soc. Biol. Paris 143: 354–355.

Kupila, S. & H. Stern, 1961. DNA content of broad bean (*Vicia faba*) internodes in connection with tumour induction by *Agrobacterium tumefaciens*. Pl. Physiol. 36: 216–219.

Lance, C., 1961. Modification de la répartition des oxydases terminales licé à la transformation tumorale induite par l'*Agrobacterium tumefaciens* dans les tissus vegetaux. C.r. Acad. Sci. Paris 251: 933–935.

Link, G. K. K. & D. R. Goddard, 1951. Studies on the metabolism of plant neoplasms. I. Oxygen uptake of tomato crown-gall tissue. II. The terminal oxidase patterns of crown-gall and auxin tumours of tomato. Bot. Gaz. 113: 185–190.

Lovrekovich, L. & G. Loebenstein, 1966. Induction of non-hyperplastic galls by living and heat-killed phytopathogenic bacteria. Phytopathology 56: 578–579.

Lovrekovich, L., H. Lovrekovich & M. A. Stahmann, 1967. Inhibition of phenol oxidation by *Erwinia carotovora* in potato tuber tissue and its significance in disease resistance. Phytopathology 57: 737–742.

Matthee, F. N. & R. H. Daines, 1968. Effects of soil types and substrate aeration on stomatal activity, water diffusion, pressure deficit, water congestion, and bacterial infection of peach and pepper foliage. Phytopathology 58: 1298–1301.

Matthee, F. N. & R. H. Daines, 1969. The influence of nutrition on susceptibility of peach foliage to water congestion and infection by *Xanthomonas pruni*. Phytopathology 59: 285–287.

Misaghi, I. & R. G. Grogan, 1969. Nutritional and biochemical comparisons of plant-pathogenic and saprophytic fluorescent pseudomonads. Phytopathology 59: 1436–1450.

Mount, M. S. & D. F. Bateman, 1969. Production of phosphatidase C by *Erwinia carotovora*. Phytopathology 59: 401.

Nayudu, M. V. & J. C. Walker, 1960. Bacterial spot of tomato as influenced by temperature and by age and nutrition of the host. Phytopathology 50: 360–364.

Nayudu, M. V. & J. C. Walker, 1961. Tomato leaf composition in relation to bacterial spot. Phytopathology 51: 368–372.

Patel, P. N. & J. C. Walker, 1963. Relation of air temperature and age and nutrition of the host to the development of halo and common bacterial blights of bean. Phytopathology 53: 407–411.

Patil, S. S., 1970. Inhibition of ornithine carbamyl transferase from bean plants by the toxin of *Pseudomonas phaseolicola*. Pl. Physiol. 46: 752–753.

Pegg, G. F. & L. Sequeira, 1968. Stimulation of aromatic biosynthesis in tobacco plants infected by *Pseudomonas solanacearum*. Phytopathology 58: 476–483.

Rickard, S. F. & J. C. Walker, 1965a. Carbon nutrition of the bean wilt bacterium. Phytopath. Z. 52: 19–28.

Rickard, S. F. & J. C. Walker, 1965b. Nitrogen nutrition of the bean wilt bacterium. Phytopath. Z. 52: 131–144.

Rickard, S. F. & J. C. Walker, 1965c. Mode of inoculation and host nutrition in relation to bacterial wilt of bean. Phytopathology 55: 174–178.

Rudolph, K., 1970. The halo-blight toxin of *Pseudomonas phaseolicola:* Influence on host parasite relationship and counter effect of metabolites. Nato advanced study institute on phytotoxins in plant diseases, Pugnochiuso, June 7–21.

Rudolph, K. & M. A. Stahmann, 1964a. Some investigations on the proteins of bean leaves infected by *Pseudomonas phaseolicola*. Z. PflKrankh. PflSchutz 71: 107–112.

Rudolph, K. & M. A. Stahmann, 1964b. Interactions of peroxidases and catalases between *Phaseolus vulgaris* and *Pseudomonas phaseolicola* (halo blight of bean). Nature, Lond. 204: 474–475.

Rudolph, K. & M. A. Stahmann, 1966. Multiple hydrolases in bean leaves (*Phaseolus vulgaris* L.) and the effect of the halo blight disease caused by *Pseudomonas phaseolicola* (Burkh.) Dowson. Pl. Physiol. 41: 389–394.

Schilperoort, R. D., H. Veldstra, S. O. Warnaar, G. Mulder & J. A. Cohen, 1967. Formation of complexes between DNA isolated from tobacco crown gall tumours and RNA complementary to *Agrobacterium tumefaciens*. Biochim. biophys. Acta 145: 523–525.

Schnathorst, W. C., 1966. Unaltered specificity in several xanthomonads after repeated passage through *Phaseolus vulgaris*. Phytopathology 56: 58–60.

Schnathorst, W. C. & J. E. DeVay, 1963. Common antigens in *Xanthomonas malvacearum* and *Gossypium hirsutum* and their possible relationship to host specificity and disease resistance. Phytopathology 53: 1142.

Sequeira, L., 1963. Growth regulators in plant disease. A. Rev. Phytopath. 1: 5–30.

Sequeira, L., 1969. Synthesis of scopolin and scopoletin in tobacco plants infected by *Pseudomonas solanacearum*. Phytopathology 59: 473–478.

Shaw, M. & D. J. Samborski, 1956. The physiology of the host-parasite relations. I. The accumulation of radioactive substances at infections of facultative and obligate parasites including tobacco mosaic virus. Can. J. Bot. 34: 389–405.

Sinclair, M. G., J. M. Sasser & T. J. Gulya, 1970. Pepper leaf intercellular fluid composition after inoculation with *Xanthomonas vesicatoria*. Phytopathology 60: 1314.

Sinden, S. L. & R. D. Durbin, 1968. Glutamine synthetase inhibition: Possible mode of action of wildfire toxin from *Pseudomonas tabaci*. Nature, Lond. 219: 379–380.

Srivastava, B. I. S., 1965. Studies on nucleic acids from healthy and tumorous tomato stem tissue. Biochim. biophys. Acta 103: 349–352.

Starr, M. P., 1946a. The nutrition of phytopathogenic bacteria. I. Minimal nutritive requirements of the genus *Xanthomonas*. J. Bact. 51: 131–143.

Starr, M. P., 1946b. The nutrition of phytopathogenic bacteria. II. The genus *Agrobacterium*. J. Bact. 52: 187–194.

Starr, M. P., 1949. The nutrition of phytopathogenic bacteria. III. The Gram-positive *Corynebacterium* species. J. Bact. 57: 253–259.
Thomas, C. A. & R. G. Orellana, 1963. Amino compounds and reducing sugar in relation to resistance of sesame to *Pseudomonas sesami*. Phytopath. Z. 46: 101–104.
Tseng, T.-C. & D. F. Bateman, 1968. Production of phosphatidases by phytopathogens. Phytopathology 58: 1437–1438.
Watanabe, T., Y. Sekizawa & M. Oda, 1965. Biochemical studies of the host-parasite relationship between *Xanthomonas oryzae* and rice plant. I. Sulfur containing amino acids as an essential growth factor of the bacterium. Ann. phytopath. Soc. Japan 30: 1–7.
Webb, E. C., 1964. Nomenclature of multiple enzyme forms. Nature, Lond. 203: 821.
Williams, P. H. & N. T. Keen, 1967. Relation of cell permeability alterations to water congestion in cucumber angular leaf spot. Phytopathology 57: 1378–1385.
Wood, H. N. & A. C. Braun, 1965. Studies on the net uptake of solutes by normal and crown-gall tumour cells. Proc. natn. Acad. Sci. 54: 1532–1538.
Wood, R. K. S., 1967. Physiological plant pathology. Botanical Monograph Vol. 6, eds. W. O. James & J. A. Burnett. Oxford.

Resistance of potato tubers to infection by Erwinia carotovora var. atroseptica

L. E. Webb and R. K. S. Wood

Food Research Institute, Norwich, Great Britain

Abstract

The occurrence of *Erwinia carotovora* var. *atroseptica* in potatoes samples from a growing crop was investigated. Before mother tuber breakdown the bacterium could be recovered from *c*. 20 % of the mother tubers although they appeared healthy and only 0.3 % of the plants in the crop developed symptoms of blackleg. The bacterium could be isolated from 10 % of the daughter tubers but only after breakdown of mother tubers. Following inoculation of seed tubers, development of blackleg symptoms in the growing plant was found to depend on the method of inoculation and on the growing conditions.

Mother and daughter tubers were sampled from plants during the growing season and susceptibility to soft rot was assessed. Mother tubers were found to be more susceptible than daughter tubers and susceptibility was correlated with higher sugar levels in the tubers. Susceptibility of the daughter tubers increased during the storage period after harvest. Sugar levels and susceptibility to soft rot increased in tubers following treatment with ethylene chlorhydrin vapour.

Crude enzyme extracts from soft rotted potato tissue caused a rapid loss of water from potato discs.

Introduction

The development of blackleg symptoms in a potato crop grown from apparently healthy tubers has posed many questions on the etiology of the disease caused by soft rotting bacteria of the *Erwinia carotovora* group. Outbreaks of symptoms could be explained if the pathogen were soil borne, but Graham (1958) and Logan (1968) have shown that it does not overwinter in the soil and it does not appear, therefore, that crops are infected from year to year from contaminated soil. It is probable, however, that tubers are latently infected by soft rot bacteria and that such tubers do not necessarily produce plants with blackleg symptoms. There is good evidence for this. Bonde (1950) showed that, when tubers from plants with blackleg symptoms were planted the following year, only 1 % of the plants in the crop developed blackleg. Leach (1930) found that blackleg did not develop in plants grown from infected and rotting seed potatoes; as the sprouting tuber became established, development of the soft rot was arrested by the formation of a cork barrier. Van den Boom (1967) showed that watering potato plants with suspensions of *E. atroseptica* did not cause blackleg to develop. The extent to which the tubers in these experiments were infected was not determined. However, as early as 1923, Jennison reported that soft rotting bacteria could be recovered from the heel end of apparently healthy tubers and the development of soft rot from the lenticels of mother tubers which led to the development of blackleg in plants was demonstrated by Davidson (1948) to depend on water-logged conditions in the soil. Rudd Jones & Dowson (1950) and Logan (1964) have shown that the development or arrest of soft rot lesions in potato tubers is reversible according to conditions of incubation and method of inoculation. More recently, Pérombelon (1968) has shown that all apparently healthy mother tubers in a growing crop can be latently infected with soft-rotting *Erwinia* spp. and that daughter tubers became infected after mother tuber breakdown although blackleg did not develop in the crop. This showed that the development of blackleg symptoms bears little relation to the transmission of the bacterium from one crop to the next.

The main objects of the work to be summarized below were to trace the infection of mother and daughter tubers by the bacterium and to assess the susceptibility of tubers to soft rot at different times during the growing season and after harvesting.

Materials and methods

Potato plants and tubers used throughout the experiment were *Solanum tuberosum* cv. Majestic.

The proportion of tubers in a crop at Silwood Park, Berks., infected by the bacterium was determined as follows. From the middle of June, 1970, potato plants were sampled every 10 days. The mother tuber and one daughter tuber from each plant were carefully removed from the soil to prevent damage, put into separate beakers and taken to the laboratory. The tubers were washed in hypochlorite (2% available chlorine) for 30 sec followed by running tap water for 30 min and then allowed to dry. Parts of the periderm were comminuted in 4.5 ml of 0.9% sterile saline; 1 ml was added to 5 ml of 1.5% sodium polypectate and poured on to a basal layer of Logan's medium (1963) modified from Stewart (1962), and dilutions of 1/10 and 1/100 were also plated out in a similar way. Plates were incubated at 25 °C until pectolytic bacteria could be detected as depressions in the pectate layer. Bacteria from these depressions were streaked on to a medium modified from Blackhurst (1961) which contained sodium polypectate (1.0%), cas-amino acids (vitamin free, 0.46%), agar (1.5%) and calcium carbonate (0.1%) to keep the medium at approximately neutral pH. The rapid growth of the pectolytic *Erwinia* species made them easily distinguishable from other bacteria; their identity was confirmed by the use of *E. carotovora* var. *atroseptica* antiserum and by the ability to produce blackleg when inoculated into potato stems.

The bacterium used to test the susceptibility of tubers to soft rot was *Erwinia atroseptica* No. 1277 from the NCPPB, Harpenden, Herts. A standard suspension used for inoculation was prepared from an 18 h nutrient broth shake culture at 25 °C. The culture was centrifuged and the bacteria resuspended in sterile 0.9% saline and the concentration of cells adjusted to 1×10^9/ml by means of a nephelometer.

The usual procedure for inoculating potatoes was as follows. A wound was made in the tuber by pushing a sterile 0.5 cm diam. cork borer into the tuber to a depth of

2 cm; 0.03 ml of the bacterial suspension was run into the wound and the wound sealed with a vaseline/wax mixture. Tubers were incubated for 4 days at 25 °C and the volume of rot was measured by removing the rot from the tuber and running water into the cavity from a burette.

Samples of tuber tissue for sugar analysis were prepared as follows. Ten discs (1 mm thick) were cut from a cylinder of tissue 1 cm diam. removed from the tuber. The discs were washed for 5 min in running water, weighed, then transferred to 100 ml of distilled water which was boiled for 30 min to extract and macerate the tissue, allowed to cool and then frozen at −20 °C to precipitate high molecular weight carbohydrates. After thawing, volumes of liquid were adjusted to the original and filtered; 1 volume of filtrate was diluted with 3 volumes of distilled water before analysis.

Reducing sugars in 1 ml samples of extract were determined by the method of Nelson (1944) modified by Somogyi (1951), and total soluble sugars were determined in 0.3 ml samples using anthrone reagent (Hewitt, 1958). Standard curves were prepared against glucose.

Potato tubers were treated with 'rindite' (ethylene chlorhydrin, ethylene dichloride, carbon tetrachloride, 7:3:1 v/v) in closed containers at the rate of 0.5 ml/kg of tubers for 4 days, or with ethylene chlorhydrin at the rate of 0.8 ml/kg for 4 days (Denny, 1945). The liquid was applied to a cotton wool wad and allowed to evaporate in the container.

Results

The occurrence of E. carotovora var. atroseptica on potato tubers The numbers of tubers from which *E. carotovora* var. *atroseptica* was isolated during the growing season are shown in Table 1. The presence of pectolytic bacteria in each sample from the tubers as indicated by the formation of depressions on Logan's medium, and the number of positive identifications of *E. carotovora* var. *atroseptica* out of each sample of 20 tubers are given. It can be seen that *E. carotovora* var. *atroseptica* was readily isolated from mother tubers before breakdown, and at a time when there were no obvious signs of rotting. Furthermore, very little blackleg developed in the crop. Out of *c.* 8,400 plants only 25 showed symptoms. The bacterium was not recovered from daughter tubers until the mother tubers had begun to break down. Most probably, therefore, the source of contamination of the daughter tubers was in the mother tubers. The significance of the other unidentified pectolytic bacteria was not evaluated.

The parts played by mode of tuber infection and growing conditions were studied in the following experiment. The bacterium was placed in superficial wounds in young potato tubers still attached to the parent plant. Soft rot did not develop in any of the tubers. The tubers were stored for 8 months and then sampled for the bacterium before planting: about 75% were infected. Other tubers were inoculated shortly before planting with a suspension of the bacterium washed from a nutrient agar slope. Both sets of tubers were planted in 12″ pots under the following conditions in June 1969:

Table 1. Isolation of bacteria and identification of *Erwinia carotovora* var. *atroseptica* in samples of 20 tubers taken from potato plants during June and August 1970.

Sampling time (days)	Number of mother tubers senescing or decayed	Mother tubers		Daughter tubers	
		presence of pectolytic bacteria	positive identification of *E. carotovora* var. *atroseptica*	presence of pectolytic bacteria	positive identification of *E. carotovora* var. *atroseptica*
0 (June 6)	0	20	16[1]	—	—
10	0	16	7	0	0
20	0	20	9	0	0
30	1	20	1	7	0
40	7	17	5	0	0
50	7	18	11	4	2
60	8	18	6	2	0
70	19	19	8[1]	1	1
80 (August 26)	19	17	3	0	0

1. A further pectolytic colony did not agglutinate in *E. carotovora* var. *atroseptica* antiserum but caused blackleg symptoms at 25 °C when inoculated into potato stems.

1. unshaded greenhouse, floor lined with polythene sheeting so that the plant pots stood in water; warm humid conditions were maintained;
2. area outside greenhouse roofed with green netting to give shade and with the plant pots standing in water so that cooler, wet conditions were obtained;
3. unshaded greenhouse with infrequent watering so that warm, dry conditions were obtained;
4. area outside greenhouse roofed with green netting for shade and polythene for protection from the rain; plants were watered infrequently so that cooler dry conditions were obtained.

Table 2 shows the percentage of plants which developed blackleg under each growing condition, and the relative yield of tubers per plant. As would be expected, the plants grew best and yielded most under warm growing conditions, humid or dry; yields and growth were less under the cooler wet conditions and lowest under the cooler dry conditions. Blackleg developed only in plants from tubers inoculated immediately before planting and symptoms only started to develop a month after inoculation. All tubers produced plants. Unexpectedly, blackleg did not develop in yields and growth were less under the cooler wet conditions and lowest under the warmer drier conditions and there was more blackleg in plants grown under the cooler conditions. It seemed, therefore, that there was least blackleg in conditions which supported best growth. Tubers, in which an arrested lesion had formed following inoculation 8 months earlier, did not produce plants with blackleg symptoms although most of them were infected with the bacterium.

Table 2. Relative yields of tubers and percentage of plants developing blackleg under different environmental conditions, grown from tubers inoculated with *Erwinia carotovora* var. *atroseptica.*

Treatment of tubers	Environment I warm, humid		Environment II cooler, wet		Environment III warm, dry		Environment IV cooler, dry	
	a	b	a	b	a	b	a	b
1. Inoculated	70	0	29	0	60	0	9	0
Uninoculated control	65	0	29	0	73	0	13	0
2. Inoculated	49	0	12	64	36	32	4	77
Uninoculated control	87	0	60	0	100	0	39	0

a = relative yield of tubers.
b = percentage of plants which developed blackleg.
1. Tubers inoculated 8 months before planting and in which it was estimated that 75% carried viable *E. carotovora* var. *atroseptica.*
2. Tubers inoculated with *E. carotovora* var. *atroseptica* immediately before planting.

Susceptibility of tubers to soft rot The susceptibility of mother and daughter tubers to soft rot during the growing season was assessed by removing tubers from the growing crop and inoculating them with the bacterium; in 4 days this caused the rots shown in Table 3. It was difficult to inoculate mother tubers successfully because they developed large cavities in the medulla soon after sampling began. However, the results in Table 3 show that more soft rot developed in mother than in daughter tubers and that the amounts of rot in daughter tubers remained at relatively low levels. Total

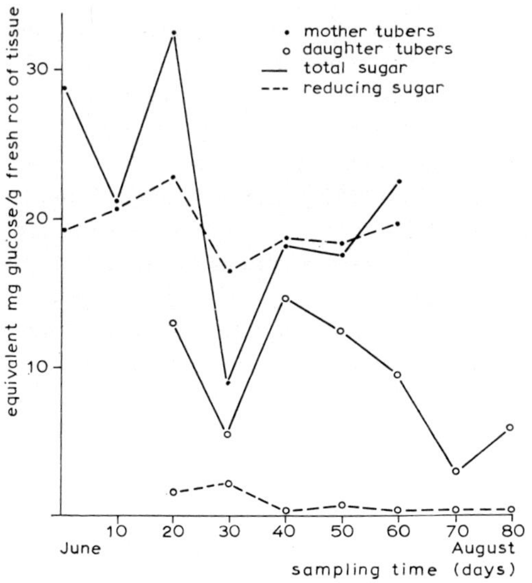

Fig. 1. Total and reducing sugar content of mother and daughter tubers sampled at 10 days intervals.

Table 3. Mean volume of rot produced in tubers after *Erwinia carotovora* var. *atroseptica* was inoculated into samples of 20 mother and daughter tubers taken periodically from the growing crop.

Sampling time (days)	Number of sampled mother tubers	Number of inoculations producing soft rot in mother tubers	Mean vol. of rot in mother tubers (ml)	Mean volume of rot in daughter tubers (ml)
0 (June 6)	20	20	1.14	—[3]
10	20	20	2.81	—
20	17	13	1.65[1]	0.54[4]
30	18	12	—[2]	0.67
40	17	15	—	0.59
50	15	13	—	0.67
60	7	7	—	0.61
70	6	6	—	0.57
80 (August 26)	4	3	—	0.64

1. Average of 11 inoculations.
2. Volume of rot not measured as tubers were senescing and contained cavities.
3. Daughter tubers too small to inoculate.
4. Average of 5 inoculations.

sugars and reducing sugar in samples from tubers are given in Fig. 1, which shows that the level of reducing sugar in the susceptible mother tubers was some 800% higher than that in the less susceptible daughter tubers and some 100% higher than the total sugars of the daughter tubers.

The susceptibility of tubers to soft rot during storage was investigated by sampling and inoculating tubers every 10 days. The regression line for mean volumes of rot per tuber against time (Fig. 2) shows that susceptibility of tubers increased with length of storage.

Treatment of tubers with commercial sprout suppressants, Fusarex[1] (containing tecnazene) at the rate of 1/2 lb/cwt of tubers and Mirvigran 5G[2] (containing *iso* propyl chloro-phenyl carbamate) at the rate of 1 lb/cwt of tubers for 14 weeks prevented sprouting of the tubers but had no effect on the susceptibility of tubers to soft rot during this period.

The susceptibility of tubers to soft rot following treatment with 'rindite' or storage at room temperature is given in Fig. 3, which shows the volumes of rots 3 and 6 days after inoculation. For 3 to 12 days after 'rindite' treatment the 6 day rots were about double the 3 day rots. These differences were less consistent for the untreated tubers.

Three days after 'rindite' treatment tubers were considerably more susceptible to

1. Plant Protection Ltd.
2. CIBA Agrochemicals Ltd.

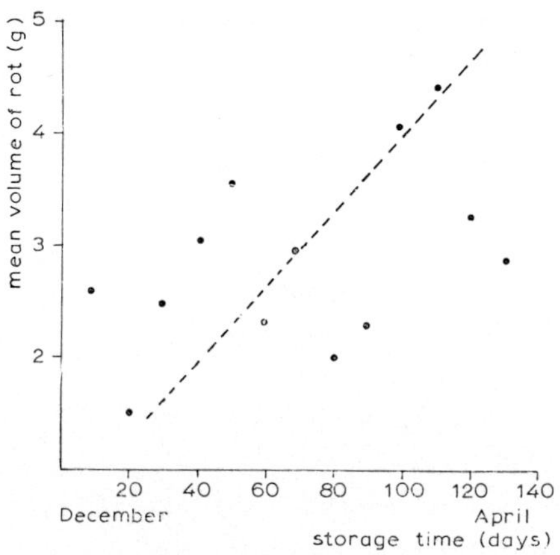

Fig. 2. Mean volume of soft rot developing in tubers inoculated with *Erwinia carotovora* var. *atroseptica* after different storage times.

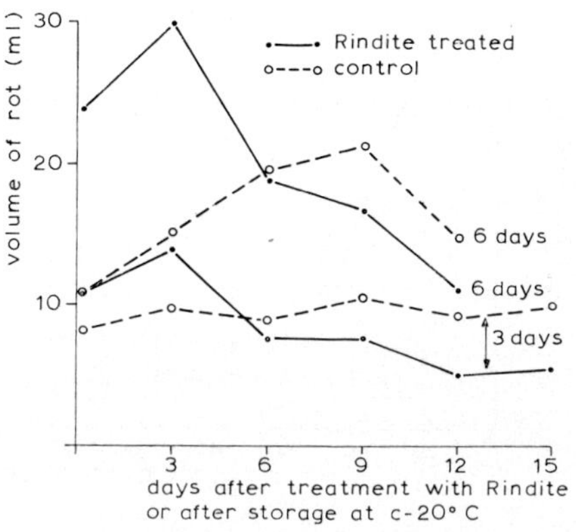

Fig. 3. Effects of 'Rindite' on rots produced by *Erwinia carotovora* var. *atroseptica*.

rot. Afterwards susceptibility decreased and was about the same as in the controls. Susceptibility of control tubers increased with storage at room temperature so that after 9 days they were somewhat more susceptible than 'rindite' treated tubers. Thereafter, they became less susceptible. A striking difference was that rots in 'rindite' treated tubers were usually white, whereas those in untreated tubers were usually dark coloured.

Potato tubers removed from the growing crop in mid August 1970 were treated with ethylene chlorhydrin and inoculated with the bacterium to assess their susceptibility to soft rot and samples of the unrotted potato tissue were analysed for total and reducing sugars. The procedure was repeated 5 weeks later with tubers shortly after harvest. Volumes of rot and the results of the sugar analyses are given in Table 4,

Table 4. Mean concentration of sugars in potatoes treated with ethylene chlorhydrin and untreated controls.

	Total sugar (equivalent mg)	Reducing sugar (glucose/g fresh weight)	Mean volume of rot after inoculation with *Erwinia atroseptica* (ml)
Tubers from growing crop, August 16, 1970, Rindite			
treated	9.3	—	0.9
untreated	7.0	—	0.6
Tubers after lifting, September 20, 1970, Rindite treated	13.8	—	1.1
untreated	7.9	—	0.6

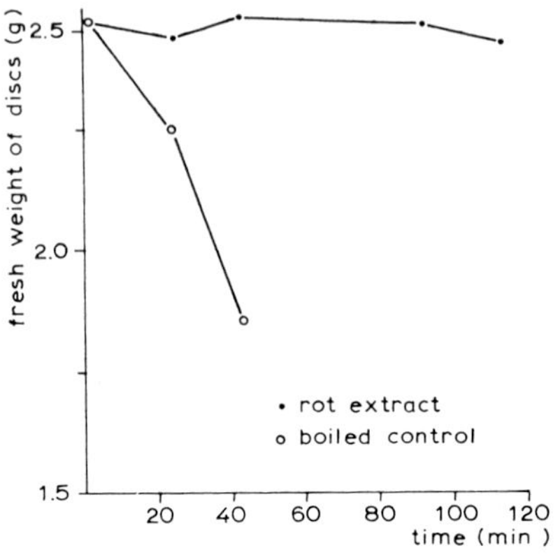

Fig. 4. Effect of water extract of soft rot on fresh weight of potato discs.

which shows that the susceptibility of ethylene chlorhydrin treated tubers to soft rot was greater in the second sample which also contained about twice the concentration of total sugars.

Effect of extracts of soft rot on potato tissue Tissue of white spreading soft rots which had been stored at $-20\,^{\circ}$C was thawed, mixed with water, filtered and a cell free extract was obtained by centrifugation. A portion of this extract was kept at $100\,^{\circ}$C for 5 minutes. Weighed potato discs 2.7 cm diam., 1 mm thick, were placed in untreated and heated extract and periodically removed from the solutions, blotted, weighed and returned to the solutions. Losses in weight are shown in Fig. 4. The untreated extract had caused a weight loss of about a third at a time when the discs were beginning to lose coherence. There was little change in fresh weight of discs in the heated extract.

Conclusions

During the isolation of bacteria from apparently healthy tubers it was usually impractical to sample from the entire surface of the tuber and the growth of saprophytes on Logan's medium often decreased the number of depressions from which bacterial colonies could be picked and identified. Although sampling from a rot has the advantage over sampling from apparently healthy tissue in that the pathogen has already been selected and is present in large numbers, the non-uniformity of the results in Table 1 and the fact that less than 100% of the tubers were found to be infected with *Erwinia* spp. suggests that the sampling technique which was used for apparently healthy tissue was not entirely satisfactory. Nevertheless it is clear that a large number of the mother tubers were infected with *Erwinia* spp. where very few gave rise to plants with blackleg.

The influence of dormancy on the susceptibility of tubers to soft rot was demonstrated by van den Boom (1967) who found that a sprout inhibitor which prolonged dormancy also prevented the increase in susceptibility to soft rot which occurred in untreated controls. Similar results were not obtained following treatment with 'Fusarex' and 'Mirvigran 5G'. The increased susceptibility of tubers after treatment with ethylene chlorhydrin in parts confirms that dormant tubers are more resistant to soft rot.

The increase in susceptibility of mother tubers during the growing period was associated with increases in reducing sugar, whereas the increase in susceptibility of tubers during storage and of ethylene chlorhydrin treated tubers was accompanied by increases in non-reducing sugars. The role of these sugars in susceptibility probably merits further investigation, particularly because an early and dramatic effect of the cell separating enzymes produced by *Erwinia* spp. is a striking decrease in the permeability of protoplast membranes leading to rapid loss of electrolytes. (Mount *et al.* 1970; Hall & Wood, 1970). The sugars and other metabolites that will also be released under the influence of these enzymes will undoubtedly affect growth of the bacteria and the secretion of enzymes and, therefore, the rate at which rots develop. Substances such as ethylene chlorhydrin and endogenous factors in tubers may well affect susceptibility, both by altering the permeability of protoplasts so that the early growth of the bacterium in the tuber is affected, and by causing changes in the amounts of metabolites within protoplasts of types that would be released by cell separating enzymes and that would materially affect the growth and activity of the pathogen. These possibilities are now being investigated.

References

Blackhurst, F. M., 1961. Physiological studies of *Verticillium* wilt of tomatoes. Ph.D. Thesis, University of London.

Bonde, R., 1950. Factors affecting potato blackleg and seed piece decay. Bull. Mi. agric. Exp. Stn 482.

Boon, T. van den, 1967. Untersuchungen über die Voraussetzungen für das Auftreten der Schwarzbeinigkeit der Kartoffel. Phytopath. Z. 58: 239–276.

Davidson, R. S., 1948. Factors affecting the development of bacterial soft-rot of potato tuber initials. Phytopathology 38: 673–687.

Denny, F. E., 1945. Synergistic effects of three chemicals in the treatment of dormant potato tubers to hasten germination. Contr. Boyce Thompson Inst. 14: 1–14.

Graham, D. C., 1958. Occurrence of soft rot bacteria in Scottish soils. Nature, Lond. 181: 61.

Hall, J. A. & R. K. S. Wood, 1970. Plant cells killed by soft rot parasites. Nature, Lond., 227: 1266–1267.

Hewitt, B. R., 1958. Spectrophotometric determination of total carbohydrate. Nature, Lond., 182: 246–247.

Jennison, H. M., 1923. Potato blackleg with special reference to the etiological agent. Ann. Mo. bot. Gdn 10: 1–72.

Leach, J. G., 1930. The identity of the potato blackleg pathogen. Phytopathology 20: 743–751.

Logan, C., 1963. A selective medium for the isolation of soft rot coliforms from soil. Nature, Lond. 199: 623.

Logan, C., 1964. Bacterial hard rot of potatoes. Eur. Pot. J. 1: 45–46.

Logan, C., 1968. Survival of the potato blackleg pathogen over winter. 1st. International Congress of Plant Pathology, London, 14–26th July.

Mount, M. S., D. F. Bateman & H. G. Basham, 1970. Induction of electrolyte loss, tissue maceration and cellular death of potato tissue by an endopolygalacturonate trans-eliminase. Phytopathology 60: 924–931.

Nelson, N., 1944. A photometric adaptation of the somogyi method for the determination of glucose. J. biol. Chem. 153: 375.
Pérombelon, M., 1968. The Scottish Horticultural Research Institute, 15th Annual Report: 34.
Rudd Jones, D. and W. J. Dowson, 1950. On the bacteria responsible for soft rot in stored potatoes and the relation of the tuber to invasion by *Bacterium carotovorum* (Jones) Lehmann and Neumann. Ann. appl. Biol. 37: 563–569.
Somogyi, M., 1952. Notes on sugar determination. J. biol. Chem. 195: 19.
Stewart, D. J., 1962. A selective-diagnostic medium for the isolation of pectinolytic organisms in the Enterobacteriaceae. Nature, Lond. 195: 1023.

Reaction of soybean tissue culture to pathogenic and saprophytic bacteria

B. W. Kennedy, T. W. Mew and L. Olson[1]

Department of Plant Pathology, University of Minnesota, St. Paul, Minnesota, U.S.A.

Abstract

Callus tissues of five soybean varieties were inoculated with a total of 57 isolates of bacteria representing 26 species. Based upon either a color or a population standard, we could not show with certainty that callus distinguished between saprophytic and plant pathogenic bacteria. Death of callus by freezing or heat resulted in indiscriminate invasion by a wide variety of bacteria and a general absence of characteristic colour changes. Reaction of callus tissues to bacteria was not easily changed via manipulation of the physical and chemical environment where callus was grown.

We thought that tissue cultures might be used to elucidate specificity of host-parasite associations. Our commitment was to begin with conventional methods and innovate as judgment directed when data became available. Considerable preliminary efforts indicated that conclusive interpretations would be difficult. Variability in callus lots was an unexpected problem, in view of the chemical and physical environment control we were able to provide. Specific combinations of soybean genotype and bacterial cultures was an easy choice since we knew from experience a great many of the natural attributes of these combinations. Thus, availability of proper biological materials for testing was excellent.

Callus cultures were established from hypocotyls arising from germinated seeds. A modified medium according to Miller (1961), but with the addition of 0.134 mg/l of cytokinin, 5 mg/l NAA, and a vitamin mix substitute was used in all studies to produce callus. On occasion, various chemicals were incorporated such as streptomycin, tetrazolium, methionine sulfoximine, and cycloheximide in order to measure effects upon callus or its reaction to invasion by bacteria. Also, on some occasions extracts washed from intercellular spaces of soybean cotyledons were adjusted to pH 5.6, sterilized by filtration, and added to media at a rate of 0.1 %–5 % of the total volume.

After extensive preliminary investigations, detailed studies were made on callus of two soybean cultivars and five bacteria. Various ways were used to inoculate with the

1. This research was supported by the Agricultural Research Service, U.S. Department of Agriculture, Cooperative Agreement No. 12-14-100-9373(34) administered by the Crops Research Division, Beltsville, Maryland.

saprophyte *Pseudomonas fluorescens*, a pseudomonad causing soft rot in vegetables, and three races of *P. glycinea*. Isolates of the soybean pathogen *Xanthomonas phaseoli* var. *sojense* and selected bacterial pathogens of other plants were tested from time to time. Data were often inconclusive and individual treatments were frequently incorporated into 5–13 separate experiments involving up to 34 replications in efforts to establish relationships. Bacterial frequency reached a population peak of about 10^8–10^9 cells/callus piece after four days, regardless of treatment, initial concentration of inoculum, species of bacteria, or genotype of callus under test.

Growth of callus was inhibited severely by 100 μg/ml of methionine sulfoximine in the medium; this effected no apparent change in its reaction to virulent races of *P. glycinea* or to a number of saprophytes. Frequently, but inconsistently, introduction of 10^{-2} M methionine nullified the toxic effects of methionine sulfoximine. Cycloheximide at 1 μg/ml in medium on which callus was grown induced multiplication of three species of Pseudomonas.

It was concluded that the effects of light was nil and that callus containing chlorophyll possessed about the same reaction to a variety of bacteria as did callus without chlorophyll. No specificity was imparted to callus produced on media containing washings from intercellular spaces of green soybean cotyledons.

We were unable to demonstrate production of a phytoalexin. Data taken on color change and bacterial population after periods of time ranging from 20 hours to four days revealed no consistent changes effected by applying live avirulent, dead virulent, saprophytic, or mixtures of avirulent bacteria on callus before inoculating with virulent *P. glycinea*. In any given experiment a striking effect might be observed but could not subsequently be repeated at will. Heat (36–39 °C for 24 hours) applied to callus pieces did not affect bacterial multiplication but did cause disappearance of an electrophoretic band (Peroxidase) in callus from one soybean cultivar. Ammonia was produced on any of a variety of callus-bacterial combinations, be they pathogens on soybean, saprophytes, or pathogens to other plant species. There was no correlation between resistance of the host and the amount of phenolic compounds found therein; neither was there a correlation between phenolics in the host and occurrence or these materials in tissue cultures of the same cultivar. In general, calcium deficiency promoted invasion of callus by bacteria.

The conclusion is that cultivation of soybean *in vitro* is accompanied by a loss in specificity in reaction to bacteria; only slight modification is encountered upon manipulation of the chemical and physical environment.

Reference

Miller, C. O., 1961. A kinetin-like compound in maize. Proc. nat. Acad. Sci. (U.S.) 47: 170–174.

Pathogenicity of Erwinia chrysanthemi on tobacco and Dieffenbachia leaves[1]

U. Mazzucchi

Istituto di Patologia Vegetale, Università degli Studi, Bologna, Italy

Abstract

Some pathogenic aspects of *Erwinia chrysanthemi* are studied for: (a) the validity of heterologous leaf inoculations (using Klement's injection method) as pathogenicity tests, (b) the development of the reactions in the test plants, and (c) the eventual pathogenicity mechanism of the bacterium.

The results of (a) suggest that leaf inoculations on tobacco and *Dieffenbachia* can be used as pathogenicity tests, although the ability to induce confluent necrosis may vary among the host plant strains of the Erwinia.

The loss of turgor caused by the Erwinia on tobacco leaves (b) has shown to start at least three hours before that caused by the phytopathogenic control pseudomonads; it develops much more rapidly.

After purification of the polygalacturonate *trans*-eliminase of one strain of *Erwinia chrysanthemi*, and with various concentrations of the pure enzyme, the same symptoms as those induced by the living bacteria appear on both tobacco and *Dieffenbachia* leaves (loss of turgor, necrosis of tissues).

Introduction

The purpose of this paper is to report on the researches carried out on some pathogenic aspects of *Erwinia chrysanthemi* Burk. *et al.* (*E. carotovora* var. *chrysanthemi* (Burk. *et al.*) Dye, 1969) by paying attention to the following points:

(a) the validity of heterologous leaf inoculations as pathological tests for picking isolates of the pathogen from among the bacteria of diseased tissues,

(b) the development of reactions in the test plants,

(c) the involved pathogenicity mechanism.

Materials and methods

To study point a, tests were conducted with 70 strains of *Erwinia chrysanthemi*,

1. The writer is very grateful to Dr D. W. Dye (Auckland, New Zealand), Dr D. C. Graham (Edinburgh, Scotland), R. A. Lelliott (Hatching Green, England), Dr D. Massfeller (Berlin-Dahlem, Germany) and Dr M. P. Starr (Davis, California, U.S.A.) for kindly supplying the cultures used in this research.

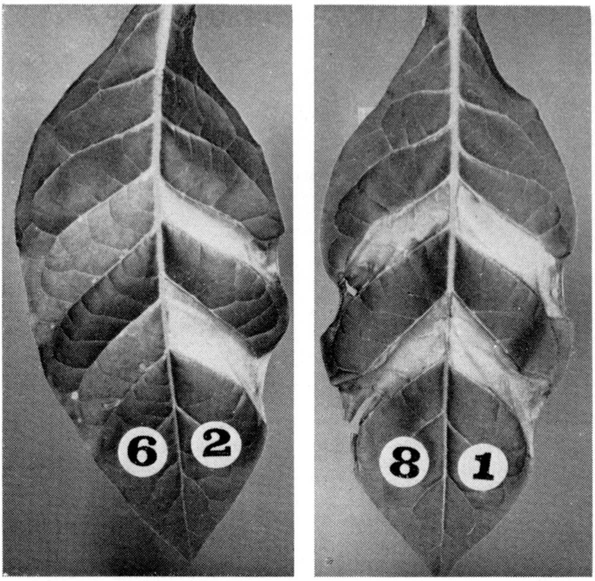

Fig. 1. Internerval zones of tobacco leaves infiltrated with *Erwinia chrysanthemi* (6 and 8), *Pseudomonas syringae* (1) and *P. mors-prunorum* (2) after 5 days at 25 °C. (6 = PDD 2360, from Maize; 8 = NCPPB 2308, from Dieffenbachia).

isolated from carnation (4 strains), Chrysanthemum (4), Dahlia (2), Dieffenbachia (29), maize (14), Parthenium (2), Philodendron (3) and Shasta daisy (12).

The examined strains were isolated in Italy and abroad. The latter came from the NCPPB, the ICPB, the PDD, the BBLB-D, and from the collection of Dr D. C. Graham.[2] Their identity was confirmed serologically (slide agglutination) and with some differential biochemical tests (Graham, personal communication; Dye, 1969; Mazzucchi, 1970). The cultures were grown on YDC-agar at 27 °C, transplanted with intervals of two weeks and maintained at room temperature.

Suspensions made from 24 hour old cultures were injected into the intercellular spaces of *Nicotiana tabacum* (cv. White Burley) leaves and of *Dieffenbachia picta* (cv. *amoena*) leaves. Klement's injection infiltration method was followed (Klement, 1963) using 1 ml syringes and No 20 needles. The concentration of the suspension was about 10^8 cells/ml (more often $2 \times 10^8 - 6 \times 10^8$); it was evaluated turbidimetrically and, for several samples, by the plate count technique on KB agar. For the negative results, the injections were repeated with a concentration of 10^9 cells/ml.

On tobacco plants at the 8–10 leaf stage, only the leaves in the middle of the stem were inoculated. Each strain was injected into alternate internerval zones on one side of each leaf (Fig. 1).

2. NCPPB = National Collection of Plant Pathogenic Bacteria, Harpenden, England; ICPB = International Collection of Phytopathogenic Bacteria, Davis, U.S.A.; PDD = Collection of the Plant Diseases Division, Auckland, New Zealand; BBLB-D = collection of the Biologische Bundesanstalt für Land- und Forstwirtschaft, Berlin-Dahlem, Germany; IPV-BO = collection of the Istituto di Patologia Vegetale, Bologna, Italy.

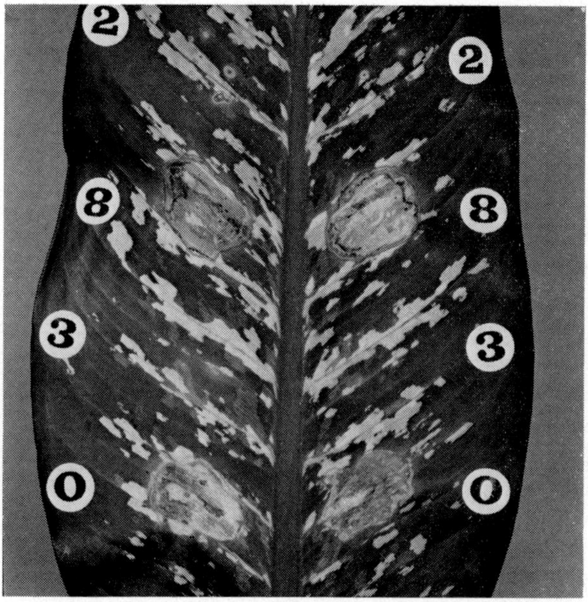

Fig. 2. Young Dieffenbachia leaf infiltrated with *Erwinia chrysanthemi* (8 and 0), *Pseudomonas mors-prunorum* (2) and *P. phaseolicola* (3) after 5 days at 25 °C. (8 = NCPPB 2308, from Dieffenbachia; 0 = NCPPB 2309, from Shasta daisy).

Dieffenbachia plants at the 5–7 leaf stage were kept in a relatively warm (25 °–30 °C) and humid greenhouse. The youngest apical leaves, immediately after opening, were injected in round, aligned areas of about 4 × 3 cm. It was possible to inoculate several cultures on one leaf (Fig. 2).

Pseudomonas syringae (NCPPB 281), *P. mors-prunorum* (NCPPB 560) and an Italian isolate of *P. phaseolicola* (IPV-BO 239/2) were used as controls for all the injections on tobacco and on Dieffenbachia.

After the injections, the plants were put in the light, at 25 °C, with 70–80 % relative humidity and with a photoperiod of 12 hours. The results refer to observations after 2 and 5 days.

For the study of point b, the following cultures were used: 226 IPV-BO, NCPPB 2308 (= IPV-BO 117), NCPPB 2309 (= IPV-BO 145), NCPPB 454, 516, 518, 1956 and the three control pseudomonads. Each tobacco leaf was injected, using the technique described above, on the one side with a strain of *E. chrysanthemi* and on the other with one of the control pseudomonads. Each trial was repeated 4 times for each combination. The development of the symptoms was observed in the light, at 25 °C, with 60–70 % relative humidity, for 13 consecutive hours.

For part c, the polygalacturnonate trans-eliminase of one strain of *E. chrysanthemi* (NCPPB 2308) was purified following the method of Pupillo, Alberghina and Mazzucchi. The pure enzyme, in various concentrations (from 180 to 9 μg/ml) in phosphate buffer KPO_4 0.05 M, pH 8, was injected into the intercellular spaces of tobacco and

Dieffenbachia leaves, using Klement's method.

Following the injections, the plants were kept under the same environmental conditions as those described in part b, and were observed for one week.

Results

The positive reaction of *E. chrysanthemi* on tobacco consisted in confluent necrosis of the infiltrated tissues (Fig. 1). The internerval necroses were similar to those induced by the pseudomonads; nevertheless a greenish coloration, which was always absent in the controls, often appeared in the necrotic tissues. The negative reactions of *E. chrysanthemi* consisted in chlorosis of the infiltrated zones (Fig. 1).

On Dieffenbachia, the pathogenic strains of *E. chrysanthemi* caused wet collapse and, successively, necrosis. This reaction was slower than that of tobacco leaves. The negative reactions consisted of the absence of macroscopic symptoms or, sometimes, in a weak chlorosis or a slight brownish coloration of the tissues.

The control pseudomonads showed negative results on Dieffenbachia; macroscopic symptoms never appeared, apart from temporary water-soaking. The results observed in the negative reactions of *E. chrysanthemi* rarely appeared.

The outcome of the pathological tests are shown in the table. Many strains of *E. chrysanthemi*, especially those from Dieffenbachia and maize, showed negative results.

On tobacco, the development of macroscopic symptoms induced by *E. chrysanthemi* was very characteristic. The first symptom consisted of a slight loss of turgor in the tissues three hours after the injection. This loss increased rapidly in the following hours and the tissue became transparent when observed against the light. After 6 to 7 hours, the loss of turgor was complete and the mesophyll had collapsed. Later the tissues dried and after 12 hours the reactions had finished.

For the three control pseudomonads, the loss of turgor began later, 7 hours after

Results of pathological tests with *Erwinia chrysanthemi* on Dieffenbachia and tobacco leaves (fractions represent positive results/total number of strains examined).

Strain	Dieffenbachia	Tobacco
Carnation	4/ 4	4/ 4
Chrysanthemum	4/ 4	4/ 4
Dahlia	1/ 2	2/ 2
Dieffenbachia	26/29	14/29
Maize	6/14	6/14
Parthenium	2/ 2	2/ 2
Philodendron	3/ 3	3/ 3
Shasta daisy	12/12	12/12
Total	58/70	47/70

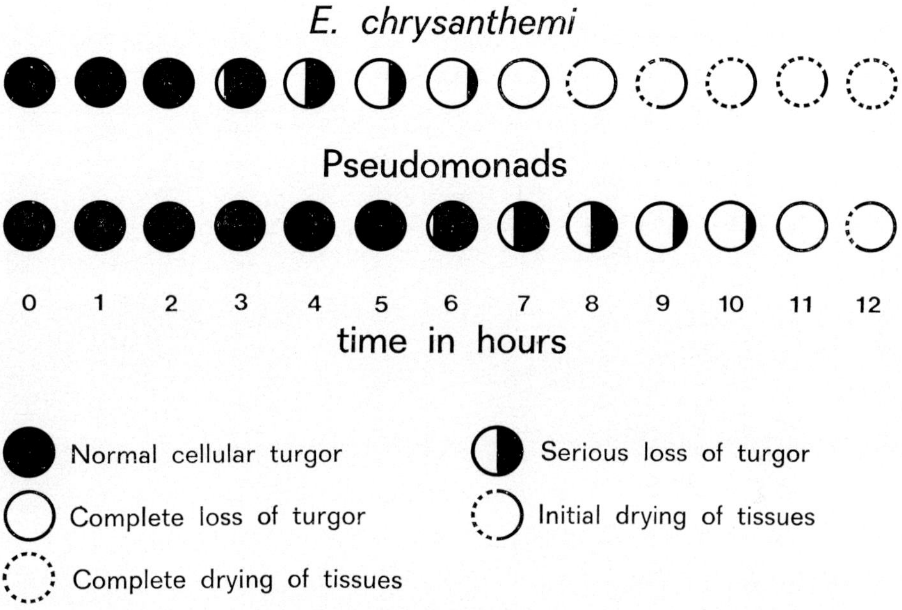

Fig. 3. Development of macroscopic symptoms in tobacco leaves injected with *Erwinia chrysanthemi* and pseudomonads.

the injections, and it became serious after 11 hours. Then the areas were less transparent than those infiltrated with *E. chrysanthemi.*

The enzyme, in the state of homogeneous protein in concentrations of 180, 72, 36, 18 μg/ml, caused loss of turgor and confluent necrosis of the tissues. The lower concentrations of 12 and 9 μg/ml only caused chlorosis of the tissues.

References

Dye, D. W., 1969. A taxonomic study of the genus *Erwinia.* II. The 'carotovora' group. N.Z. Jl Sci. 12: 81–97.

Klement, Z., 1963. Rapid detection of the pathogenicity of phytopathogenic pseudomonads. Nature, Lond. 199: 299–300.

Mazzucchi, U., 1970. L'avvizzimento batterico della Margherita (*Chrysanthemum maximum* Ram.). Phytopath. medit. 9: 111–121.

The relationship of Pseudomonas syringae virulence on bean, Phaseolus vulgaris, to bacterial isolate source

S. M. Saad and D. J. Hagedorn

University of Wisconsin, Madison, U.S.A.

Abstract

Virulence of 30 *Pseudomonas syringae* isolates from 10 hosts was compared on bean, cultivar Tenderwhite, in the greenhouse using inoculum levels of 10^3 and 10^8 cells per ml. Isolates from pear, apple, sour and sweet cherry, lilac, peach, plum, walnut and sorghum produced only tiny dark necrotic flecks within 24–48 h with inoculum containing 10^8 cells/ml. No macroscopic symptoms were obtained with the lower level of inoculum. Growth characteristics of several isolates were studied in bean pods. In the compatible combination, the bacteria multiplied logarithmically for 4 days inducing water-soaked lesions without brown necrosis. Logarithmic growth of incompatible isolates from pear, lilac, and sour cherry terminated after 2 days and induced brown sunken necrotic lesions which fluoresced when exposed to UV light.

Virulence of 30 *Pseudomonas syringae* isolates from 11 hosts was compared on bean cultivar Tenderwhite in the greenhouse using inoculum levels of 10^3 and 10^8 cells per ml. All isolates were stored on a medium containing 0.8 % nutrient broth, 2.0 % glycerol, and 2.0 % agar in slant tubes at 4 °C. Inoculum was prepared by growing the bacteria on the same medium for 24 h at 28 °C. Bacteria were washed from the agar slants with sterile distilled water and diluted to the desired concentration measured by a Bausch and Lomb Spectronic 20 Colorimeter. Beans were sown in vermiculite, then transplanted (one per pot) into 3:1 soil-sand mixture in 5-inch pots and grown in a greenhouse at 21 °C. Bacterial suspensions were atomized, under pressure of 15 ψ and from a distance of 10 inches, onto both surfaces of the first set of young trifoliate leaves when they were 1/3 expanded. In one set of trials, plants were placed in a moist chamber at 98–100 % relative humidity for 24 h prior to and after inoculation. No moist chamber treatment was used in the other trials.

All bean and lima bean isolates tested at both inoculum levels induced typical olive-green water-soaked lesions within 3 to 4 days. These lesions soon became necrotic with marginal chlorosis. The number of lesions obtained varied with the inoculum level, while the time of appearance of symptoms was almost the same with both inoculum levels. The isolates of *P. syringae* from pear *(Pyrus communis)*, apple *(Malus sylvestris)*, sour cherry *(Prunus cerasus)*, lilac *(Syringa vulgaris)*, plum *(Prunus cerasifera)*, peach *(Prunus persica)*, walnut *(Juglans californica)*, and sorghum *(Sorghum vulgare)* produced only tiny dark necrotic flecks within 24–48 h with inoculum containing 10^8 cells per ml. No macroscopic symptoms were obtained with the lower level of inoculum. Three isolates from sweet cherry *(Prunus avium)* tested at both levels never induced any symptoms. Different isolates from the same host showed only minor differences in the size and number of lesions produced. There was no significant difference between misting the plants and 'inoculation in situ', i.e. without the use of a moist chamber, in regard to the type and number of lesions produced by the different isolates at both inoculum levels.

Growth characteristics of several *P. syringae* isolates from different hosts and one isolate of *P. phaseolicola* race 1 were studied in bean pods. Detached bean pods of the cultivar Tenderwhite were sterilized in 1% sodium hypochlorite (Klenzade) for 5 min, rinsed in sterile distilled water, and injected in six places with a hypodermic syringe at the rate of approximately 0.01 ml per injection with inoculum containing 10^7 cells per ml. Injected bean pods were placed in 15 cm petri plates with moist filter paper and incubated at room temperature. The measurement of bacterial growth in vivo was accomplished by taking 7 mm diameter discs with a sterile cork borer at 0, 1, 2, 3, 4, and 7 days after injection. Discs were macerated in a mortar containing 10 ml sterile distilled water. One ml samples were taken from the appropriate dilution of the suspension and mixed with 15 ml crystal violet medium at 45 °C. Plates were dried in an oven for 30 min and incubated at room temperature for 3 days after which counts were made.

The population trends of the pathogenic bean isolates of *P. syringae* and *P. phaseolicola* showed the same characteristics as far as multiplication of the bacteria was concerned. The bacteria multiplied logarithmically in the bean pod tissues for 4 days inducing water-soaked lesions without the formation of any brown necrosis. The logarithmic growth of the isolates of *P. syringae* from pear, lilac, and sorghum was different from the bean isolates and terminated after 2 days. These isolates induced atypical symptoms on the bean pods which consisted of light or dark brown sunken necrotic lesions which fluoresced when exposed to ultra-violet light.

The rapid formation of the atypical symptoms in bean pods and young trifoliate leaves, and the abrupt change from the logarithmic multiplication phase, suggest that the uncongenial isolates of *P. syringae* induce a rapid reaction in the host. This results in destruction, dessication and necrosis of the host tissues and in the localization of the pathogen, preventing its further spread. This is sufficient proof to consider it a hypersensitive host response which may be a defense mechanism operating in the incompatible host-parasite relationship. In addition, these studies indicate a positive relationship of high virulence to bean with *P. syringae* isolates from bean and lima bean only.

Qualification of the infiltration technique to prove pathogenicity

D. Knösel and E. Lange

Department of Microbiology and Phytopathology, University of Hohenheim, Stuttgart-Hohenheim, Germany

In 1961 Stolp tested a great number of phytopathogenic pseudomonads and fluorescent isolates from soil on their pathogenicity. Suspensions of the bacteria were infiltrated in the primary leaves of *Phaseolus vulgaris;* the appearance of lesions and necroses indicated a positive reaction.

During our current work this infiltration technique was compared with the common needle inoculations to prove pathogenicity and virulence of strains of *Pseudomonas*, *Xanthomonas* and *Erwinia* species.

The bacteria were cultivated on a nutrient medium containing yeast extract, pectin and carboxymethyl cellulose. After appropriate incubation the bacteria were suspended in tapwater. Test plants were dipped with shoot and leaves into the suspension and then kept for 3 minutes in a desiccator under a vacuum of 70 to 85 mm Hg. Under these conditions the air leaves the intercellular spaces and when the vacuum is released, the suspension of bacteria infiltrates the tissue. After this treatment the plants were kept in a greenhouse at a high humidity. Development of symptoms was observed during about one week.

According to Stolp, bean plants *(Phaseolus vulgaris)* proved to be a suitable test plant for species of *Pseudomonas* and *Xanthomonas;* for *Erwinia* species young plants of *Physalis floridana*, were successfully used.

With suspensions of 10^7 cells/or more per ml, often a strong reaction of the leaf tissue was observed, whereas with 10^4 to 10^6 cells/ml various symptoms, such as chlorotic spots, lesions, rotting, and necroses could be noted dependent on differences in virulence.

Where the natural hosts of the pathogens were available, needle inoculations were carried out for comparative purposes. In most cases the results corresponded with the infiltration tests. Sometimes infiltration proved to be positive where needle inoculation gave no reaction.

As a rule definite symptoms occurred, or failed to appear, when different species or strains of bacteria were infiltrated in the leaves of the test plants. Therefore, this method may be suitable to prove pathogenicity and virulence of plant pathogenic bacteria.

Plant tumors and crown gall, an analysis of autonomous growth

H. Veldstra

Department of Biochemistry, University of Leiden, The Netherlands

Abstract

A survey is given of the present state of the studies on the crown gall problem, with a special accent on the evidence, recently obtained, that the tumor inducing principle (TIP) is (part of) the *Agrobacterium tumefaciens* DNA.

The essential events in the transformation of the normal plant cell into a crown gall tumor cell are discussed, also in connection with their role in the genesis of other types of plant tumors. A parallel is drawn with the malignant transformation of animal cells by oncogenic viruses and it is considered whether generally introduction of an additional (part of a) genome, foreign to the recipient normal cell, is the primary event in tumor formation, viz. an escape from regulating mechanisms.

Interactions between higher plants and micro-organisms can have very different effects; nevertheless they may be classed in two main groups: those leading to a partial or complete destruction of the host plant and those causing non-lethal morphological changes.

In the latter case the host survives in an adapted form, deviating in several respects from the normal one, often recognizable in a localized excessive growth.

Galls are the well-known examples of this type of interaction. Here the plants react to different organisms such as bacteria, fungi, nematodes, mites or insects by overgrowth (hypertrophy) and cell proliferation (hyperplasy).

Many of these overgrowths are self-limiting, which means that the deviations are still regulated by certain mechanisms. Striking examples, particularly among insect galls, often show organized forms, which in their specialization depend on the insect species involved (Mani, 1964).

Of the self-limiting overgrowths caused by bacteria the best known are the root nodules, resulting from an infection of leguminous roots by *Rhizobium* species which fix nitrogen from the air (Raggio & Raggio, 1962). The infected plant as a whole remains unchanged.

The situation with infections by *Agrobacterium tumefaciens* is quite different. When the wound of a susceptible plant (mostly a dicotyledon) is infected, a crown gall develops. This is a non self-limiting unorganized growth, escaping the regulating mechanism of the host. This causative relation was established as early as 1907 by Smith & Townsend and afterwards much research has been devoted to this type of

Fig. 1. Crown galls on stems of *Kalanchoë daigremontiana* inoculated with *Agrobacterium tumefaciens:*

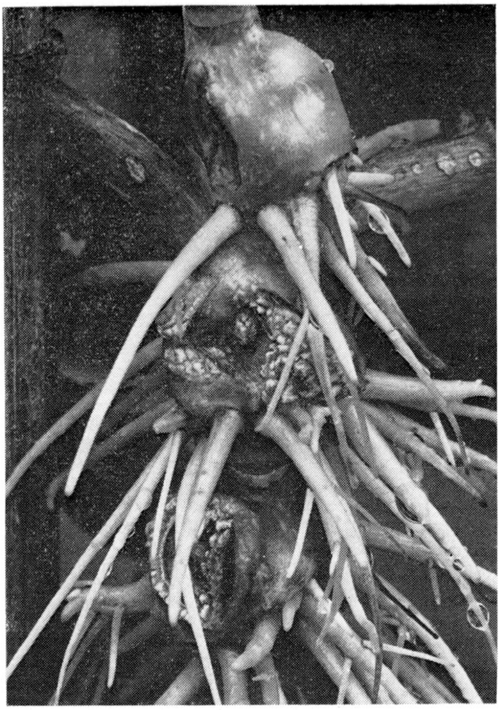

a. Strain B6

b. Strain E III 9.6.1

plant neoplasm (for reviews see Braun & Wood, 1961; Kupila-Ahvenniemi & Therman, 1968; Schilperoort, 1969), also because of its importance for cancer research (Bergmann, 1964; Gautheret, 1964).

Crown gall formation on *Kalanchoë daigremontiana*, the host plant mostly used in our studies, is shown in Fig. 1.

The real tumorous character is evident from chaotic cell division, the tumor that results when sterile crown gall tissue (from tissue culture) is transplanted onto a normal plant, the growth of sterile tissue and the lack of growth of normal host tissue on a simple synthetic medium. Sterile tumor tissue does not require such exact nutritional conditions as normal tissue, as known already for a long time for animal tumor cells.

For an analysis of the transformation process under the influence of the bacteria the following factors are important.

1. Wounding is essential; contact of the bacteria with the intact plant never results in tumor formation. This also applies to two other cases of tumorous growth in plants, viz. a RNA-virus induced wound tumor (Black, 1949, 1952, 1964, 1965) and tumors occurring in interspecific *Nicotiana* hybrids (Kostoff, 1930; Kehr & Smith, 1954). In both cases a tumor always develops after some kind of irritation, such as wounding or, in the latter case, after natural leaf dropping. Apart from the effect wounding (wound sap) in the case of crown gall may have on the bacteria, to be discussed later, the essence of this requirement most probably is initiation of renewed cell division, after which the transformation, already induced in some cells (Kupila-Ahvenniemi & Therman, 1968) advances towards non self-limiting growth. This would imply that wounded cells provide a mitotic stimulant. Indeed, washing the wound may prevent a mitotic response and bacteria in a washed wound do not induce tumor formation, whereas upon addition of wound sap this occurs again (Klein, 1954).
2. Only living bacteria induce tumors. Heat-treated bacteria or cell-free extracts of *Agrobacterium tumefaciens* lack this capacity. However, Kovoor (1967) succeeded in transforming *in vitro* normal cells into tumor cells by *A. tumefaciens* DNA.
3. The bacteria stay intercellular and once the transformation has been accomplished their presence is no longer required. This means that sterile crown gall tissue can be cultured for unlimited periods while preserving its specific properties. Consequently some kind of information, termed the tumor inducing principle (TIP) by Braun (1947), transferred from the bacteria to the sensitive host cell, must be responsible for its transformation into a tumor cell.
4. Bacteria can be effectively introduced in the wound only within a definite interval between wounding and inoculation, showing a distinct optimum (Lipetz, 1966). When the bacteria are transferred to *Kalanchoë* 4–5 days after wounding, no tumor develops, probably because the normal wound healing then has progressed too far. Neither is the presence of the bacteria for 12 h at 25 °C after wounding successful; the maximum effect is obtained after about 27–35 h. This suggests that the bacteria have to stay in the wound (under the influence of wound sap?) to become adjusted (Lipetz, 1965, 1966) to an effective association with a host cell which needs a certain period adjacent to a

damaged cell in order to become what may be called a competent cell (conditioning period, Braun, 1965)). This implies a kind of mutual adaptation leading to a donor-acceptor relationship.

In this connection reference may be made to competition experiments (Schilperoort, 1969) with the non-tumorogenic *Agrobacterium radiobacter* discussed by Bogers in this symposium, suggesting that host cell attachment sites in the wound play a decisive role in tumor initiation, a conclusion arrived at also by Lippincott & Lippincott (1969).

5. In *Kalanchoë* tumors cannot be induced at temperatures above 30 °C (Riker, 1926), though host cells, bacteria and extant tumors are resistant in this sense. Lipetz (1965, 1966), in a detailed analysis of the correlation between the processes of wound healing and conditioning (as indicated under 4), has provided convincing arguments for the following scheme.

At higher temperatures wound healing (cell division) starts earlier. As the conditioning phase is also shortened, this reduces the period of maximal sensitivity to transformation. When the duration of this stage drops below the minimum time required for transformation of a normal cell into a tumor cell, tumors do not appear. This is, indeed, the situation with *Kalanchoë* at temperatures above 30 °C.

6. Once tumor induction has taken place, growth regulators, especially auxins and cytokinins, play an important role in the autonomous proliferation of the tissue (see Fig. 2). The transition to crown gall tissue is a gradual process, which in *Kalanchoë* takes about 4 days to reach an optimum. This process can be interrupted by heat treatment, by killing the bacteria without affecting the host (Braun, 1958, 1962). The growth rate is proportional to the length of the transformation period, which can also be observed when subsequently the tumor tissue is cultivated *in vitro*. In the latter case it appears that the degree of autotrophy, viz. its independence of added growth

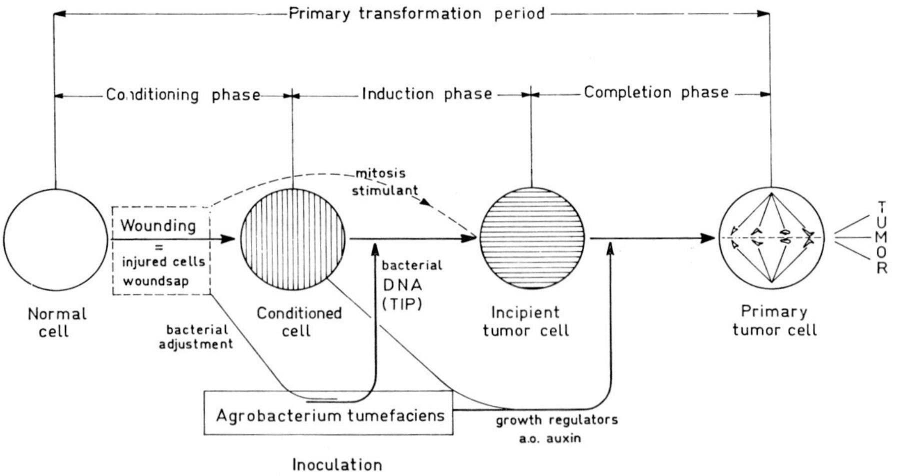

Fig. 2. Scheme of the transformation of a normal plant cell into a primary tumor cell.

regulators, increases with length of transformation period, which led Braun (1960, 1961) to the conclusion that, during the transformation, processes involved in the biosynthesis of growth regulators become progressively activated (derepressed).

Wood & Braun (1961) adduced arguments that in this activation a more efficient ion uptake plays a role, connected with changed membrane properties. This aspect is of more general importance as also with animal cells a difference in topochemistry between the cell surfaces of normal and tumor cells is observed, e.g. after transformation by oncogenic DNA viruses (Cumar *et al.*, 1970). Most probably this is related to the loss of contact inhibition upon transformation.

For a long time the most intriguing problem in crown gall research has been the nature of TIP, about which a number of hypotheses have been put forward (see Kovoor, 1967). However, recent studies indicate rather convincingly that DNA from *Agrobacterium tumefaciens* is the inducing agent.

As mentioned already, Kovoor (1967) has succeeded in transforming *in vitro* normal cells into tumor cells by *A. tumefaciens* DNA. Schilperoort *et al.* (1967), using DNA-RNA hybridization techniques, have collected evidence for the presence of bacterial genetic material in the tumor cell, as partial homology was found between *A. tumefaciens* DNA and tumor DNA but not between the former and normal cell DNA. These results have been corroborated by the work of Quétier *et al.* (1969) and of Milo & Srivastava (1969).

Subsequently Schilperoort *et al.* (1969), by immunodiffusion techniques, have clearly shown that expression of this *A. tumefaciens* DNA occurs in sterile crown gall tissue (see Schilperoort's contribution to this symposium).

Chadha & Srivastava (1971) have confirmed the presence of cross-reacting antigens in sterile tumor tissue, being absent in callus tissue and that, in addition, specific bacterial proteins could be fractionated on a DEAE-Sephadex (A-50) column from the soluble proteins of tumor tissue.

These facts strongly remind us of the course of events with malignant transformation of animal cells by oncogenic viruses, especially by those containing DNA.

Benjamin (1966) and Fujinaga & Green (1966) showed earlier (also with DNA-RNA hybridization) that in animal cells transformed by polyoma virus or oncogenic adenoviruses, RNA complimentary to the respective viral DNA's is present, although no infectious virus particles could be detected. So they concluded that the viral genome persisted in the transformed cells.

Subsequently Westphal & Dulbecco (1968) and Dulbecco (1969), with the same techniques as used by us for crown gall, proved that viral DNA was present in the transformed cell integrated in its genome, whereas partial expression of viral genes takes place.

At first the situation with the RNA containing leucoviruses seems to be different, though it is known that an infection with these viruses requires DNA synthesis to be successful. In this context the spectacular findings of Temin & Mizutani (1970) and Baltimore (1970), extended by the work of Spiegelman *et al.* (1970), should be mentioned. Their studies on the presence of an RNA-dependent DNA polymerase in the virions of the RNA tumor viruses has provided evidence for its essential role in virus replication and transformation. So it is possible to consider all oncogenic viruses from the same point of view. The finally resulting double stranded virus-specific DNA, upon integration in the host cell genome will lead to a final situation quite similar to that reached with the DNA viruses.

In all cases discussed so far, transformation to a tumorous state is linked with the introduction of a genome to a certain extent foreign to the recipient normal cell. This raises the intriguing question whether such a supplementing of the normal genome more generally causes an escape from regulating mechanisms.

In this connection once more reference may be made to the interspecific hybrids of tobacco, where sometimes tumor production is related with the presence of an alien chromosome of one of the parent species on the background of that of the other, in a certain sense recipient, species. Tumor induction can even be caused by a fragment of the alien chromosome (see Ahuja, 1965). Seitz (1965) has supplied a detailed analysis of tumor formation in the hybrid of *Nicotiana langsdorffii* and *Nicotiana glauca*. The first species shows weak callus development at a wound surface, in the second it is very intense, but limited. In the hybrid it runs wild and develops into a tumor.

Tumor formation based on the genetic constitution also occurs in animals; the melanomas developing in certain hybrids of viviparous cyprinodent fishes have been most thoroughly investigated (Gordon, 1969; Anders, 1967). A regulating repression with respect to melanin-forming melanophores, present in the separate genomes, falls out in the combination.

Apart from addition of foreign genetic material from outside, alienation with respect to the normal pattern can also be effected internally by damage of part of the genome. Certain types of human genetic disorders are related to a higher risk of cancer. A good example of this is the hereditary disease xeroderma pigmentosum (XP): the skin of patients with a hereditary taint is highly sensitive to ultraviolet light, often leading to skin tumors. The XP-cells lack the mechanisms for repair of UV-induced single strand DNA-lesions (Cleaver, 1968; Bootsma *et al.*, 1970). Thus a lasting partial damage of a chromosome may have the same effect as the presence of a foreign element, both leading to a derailment of the normal growth pattern. In addition, the frequency of transformation by means of the oncogenic virus SV 40 is much higher with XP-cells than with normal cells (Veldhuisen & Pouwels, 1970).

It may be concluded that often the basic mechanism of non self-limiting growth is an error in the genome, as Burnet (1970) most simply termed it in a recent discussion on the general nature of malignancy.

For many scientists it is the still burning question, whether the transition from normal to tumor cell is basically a somatic mutation or an irregularity in differentiation, or in other words whether genetic or epigenetic modifications are involved.

Though the difference between these two points of view may be smaller than often assumed (see the discussion by Burnet, 1970), each scientist forms an opinion from the facets of the problem which are most striking to him. So for crown gall. Braun

(1969, 1970) clearly prefers a mechanism acting at the epigenetic level, mainly because of the reversal of tumor growth when the teratoma type (see Fig. 1b) of tobacco crown gall tissue is tip-grafted on healthy plants. In this way the abnormal tissue is forced to grow rapidly and becomes totally normal again (Braun, 1959, 1965). (As to the interpretation of this phenomenon, see also Sacristán & Melchers, 1969.)

Further arguments for this interpretation are supplied by similar reversals in animal tumors. Mostly the reversal is effected by bringing the tumor in contact with tissues possessing a high regenerative power, as has been done in Braun's crucial experiment. Now it is just on this type of plant tissue that teratoma type tumors are formed upon inoculation with a less virulent strain of *Agrobacterium tumefaciens*, whereas the same strains induce highly malignant unorganized tumors on tissues with a relatively low regenerative capacity. This suggests that, in the latter case, the ability of the host cells to thrust an effect of its own regulatory mechanism upon the tumor cells is less than with the tissue endowed with a high regenerative tendency.

In this connection one should bear in mind that growth and development of plant tissues is regulated by its phytohormones and that tumor phenotype can be mimicked by addition of auxin and cytokinin in certain proportions to normal tissue (Braun & Stonier, 1958). This imitation remains completely hormone dependent, however. The most important of these hormones are also synthesized by the inducing *A. tumefaciens* (Berthelot & Amoureux, 1938; Locke *et al.*, 1938; Kaper & Veldstra, 1958; Galsky & Lippincott, 1967). Integration of bacterial genetic material in the genome of the host may thus imply introduction of information already known to the cell, and one can imagine that the resulting shift in the pattern of growth regulation (Veldstra, 1968) no longer can be mastered by the regulating mechanisms of the host, the less so when it belongs to the type of low regenerative capacity.

When the basic mechanism of crown gall initiation is considered to take place at the genetic level, an alternative explanation for the result of Braun's reversal experiments is possible. It might be that excision and elimination of the bacterial contribution to the tumor genome occurs under the influence of the high rate of cell division inherent in the tissue on which the grafting is done. Mechanisms for excision of damaged, and thus foreign parts of genomes are known for the repair mechanisms operating after UV-induced lesions, referred to earlier (see Cleaver, 1968; Bootsma *et al.*, 1970; Veldhuisen & Pouwels, 1970).

When at any time it may prove possible to reduce the origins of different types of tumors to the same denominator ('it ain't necessarily so', in view of the complexity of regulatory mechanisms and diverging possibilities of interference transferable to the cell progeny), the studies on the pathogenic interaction between *A. tumefaciens* and higher plants certainly will have contributed a fundamental share.

References

Ahuja, M. R., 1965. Genetic control of tumor formation in higher plants. Quart. Rev. Biol. 40: 329.

Anders, F., 1967. Tumour formation in platyfish-swordtail hybrids as a problem of gene regulation. Experientia 23: 1.

Baltimore, D., 1970. Viral RNA-dependent DNA polymerase. Nature, Lond. 226: 1209.

Benjamin, T. L., 1966. Virus-specific RNA in cells productively infected or transformed by polyoma virus. J. mol. Biol. 16: 359.

Bergmann, L., 1964. Pflanzliche Tumoren und das Krebsproblem. Naturwissenschaften 51: 325.

Berthelot, A. & G. Amoureux, 1938. Sur la formation d'acide indol-3-acétique dans l'action de *Bacterium tumefaciens* sur le tryptophane. C.r. Acad. Sci. Paris 206: 537.

Black, L. M., 1944. Some viruses transmitted by agallian leafhoppers. Proc. Amer. philos. Soc. 88: 132.

Black, L. M., 1949. Virus tumors. Survey biol. Progr. 1: 155.

Black, L. M., 1952. Plant virus tumors. Ann. N.Y. Acad. Sci. 54: 1067.

Black, L. M., 1965. Physiology of virus induced tumors in plants. *In:* W. Ruhland, ed., Encyclopedia of plant physiology 15/2, 236. Springer, Berlin-New York.

Bootsma, D., M. P. Mulder, F. Pot & J. A. Cohen, 1970. Different inherited levels of DNA repair replication in *Xeroderma pigmentosum* cell strains after exposure to ultraviolet light. Mutation Res. 9: 507.

Braun, A. C., 1947. Thermal studies on the factors responsible for tumor initiation in crown gall. Am. J. Bot. 34: 234.

Braun, A. C., 1952. Conditioning of the host cell as a factor in the transformation process in crown gall. Growth 16: 65.

Braun, A. C., 1958. A physiological basis for autonomous growth of the crown gall tumor cell. Proc. natl Acad. Sci. U.S. 44: 344.

Braun, A. C., 1959. A demonstration of the recovery of the crown gall tumor cell with the use of complex tumors of single-cell origin. Proc. natl Acad. Sci. U.S. 45: 932.

Braun, A. C., 1960. The plant tumor cell as an experimental tool for studies on the nature of autonomous growth. Can. Cancer Conf. 4: 89.

Braun, A. C., 1961. Plant tumors as an experimental model. Harvey Lectures 56: 191.

Braun, A. C., 1962. Tumor inception and development in the crown gall disease. Ann. Rev. Plant Physiol. 13: 533.

Braun, A. C., 1965. The reversal of tumor growth. Sci. Am. 213 (5): 75.

Braun, A. C., 1969. The cancer problem. A critical analysis and modern synthesis. Columbia Univ. Press, New York-London.

Braun, A. C., 1970. On the origin of the cancer cell. Am. Scient. 58: 307.

Braun, A. C. & T. Stonier, 1958. Morphology and physiology of plant tumours. Protoplasmatologia 10 (5a): 41.

Braun, A. C. & H. M. Wood, 1961. The plant tumor problem. Adv. Cancer Res. 6: 81.

Burnet, M., 1970. Immunological surveillance. Pergamon Press, Oxford-London, p. 121.

Chadha, K. C. & B. J. S. Srivastava, 1971. Evidence for the presence of bacteria-specific proteins in sterile crown gall tissue. Pl. Physiol. 48: 125.

Cleaver, J., 1968. Defective repair replication of DNA in *Xeroderma pigmentosum.* Nature, Lond. 218: 652.

Cumar, F. A., R. O. Brady, E. H. Kolodny, V. W. McFarland & P. T. Mora, 1970. Enzymatic block in the synthesis of gangliosides in DNA virus-transformed tumorigenic mouse cell lines. Proc. natl Acad. Sci. U.S. 67: 757.

Dulbecco, R., 1969. Cell transformation by viruses. Science 166: 962.

Fujinaga, K. & M. Green, 1966. The mechanism of viral carcinogenesis by DNA mammalian viruses: viral-specific RNA in polyribosomes of adenovirus tumor and transformed cells. Proc. natl Acad. Sci. U.S. 55: 1567.

Galsky, A. G. & J. A. Lippincott, 1967. Production of a gibberellin-like substance by strains of *Agrobacterium tumefaciens*. Pl. Physiol. 42: S29.

Gautheret, R. J., 1964. Les phénomènes tumoraux chez les plantes: les rapports avec le problème du cancer. Annls pharm. fr. 22: 147.

Gordon, M., 1959. Pigment cell biology. Acad. Press, New York, p. 215.

Kaper, J. M. & H. Veldstra, 1958. On the metabolism of tryptophan by *Agrobacterium tumefaciens*. Biochim. Biophys. Acta 30: 401.

Kehr, A. E. & H. H. Smith, 1954. Genetic tumors in *Nicotiana* hybrids. Brookhaven Symp. Biol. 6: 55.

Klein, R. M., 1954. Mechanisms of crown gall induction. Brookhaven Symp. Biol. 6: 97.

Kostoff, D., 1930. Tumors and other malformations on certain *Nicotiana* hybrids. Zentbl. Bakt. ParasitKde 81: 244.

Kovoor, A., 1967. Sur la transformation de tissus normaux de scorsonère provoquée *in vitro* par l'acide désoxyribonucléique d'*Agrobacterium tumefaciens*. C.r. Acad. Sci. Paris 265: 1623.

Kupila-Ahvenniemi, S. & E. Therman, 1968. Morphogenesis of crown gall. Adv. Morphogenesis 7: 45.

Lipetz, J., 1965. Crown gall tumorigenesis: effect of temperature on wound healing and conditioning. Science 149: 865.

Lipetz, J., 1966. Crown gall tumorigenesis. II. Relations between wound healing and the tumorigenic response. Cancer Res. 26: 1597.

Lippincott, B. B. & J. A. Lippincott, 1969. Bacterial attachment to a specific wound site as an essential stage in tumor initiation by *Agrobacterium tumefaciens*. J. Bact. 97: 620.

Locke, S. B., A. J. Riker & B. M. Duggar, 1938. Growth substance and the development of crown gall. J. agric. Res. 57: 21.

Mani, M. S., 1964. The ecology of plant galls. Junk, The Hague.

Milo, G. E. & B. J. S. Srivastava, 1969. RNA-DNA hybridization studies with the crown gall bacteria and the tobacco tumor tissue. Biochem. Biophys. Res. Commun. 34: 196.

Quétier, F., T. Huquet & E. Guillé, 1969. Induction of crown gall: partial homology between tumor-cell DNA, bacterial DNA and the G+C-rich DNA of stressed normal cells. Biochem. Biophys. Res. Commun. 34: 128.

Raggio, M. & N. Raggio, 1962. Root nodules. Ann. Rev. Pl. Physiol. 13: 109.

Riker, A. J., 1926. Influence of environment on development of crown gall. J. agric. Res. 32: 83.

Sacristán, M. D. & G. Melchers, 1969. The caryological analysis of plants regenerated from tumorous and other callus cultures of tobacco. Molec. Genet. 105: 317.

Schilperoort, R. A., H. Veldstra, S. O. Warnaar, G. Mulder & J. A. Cohen, 1967. Formation of complexes between DNA isolated from tobacco crown gall tumours and RNA complementary to *Agrobacterium tumefaciens* DNA. Biochim. Biophys. Acta 145: 523.

Schilperoort, R. A., 1969. Investigations on plant tumors. Crown gall. On the biochemistry of tumor induction by *Agrobacterium tumefaciens*. Thesis Leiden.

Schilperoort, R. A., W. H. Meys, G. M. W. Pippel & H. Veldstra, 1969. *Agrobacterium tumefaciens* cross-reacting antigens in sterile crown gall tumors. FEBS Lett. 3: 173.

Smith, E. F. & Co. Townsend, 1907. A plant-tumor of bacterial origin. Science 25: 671.

Spiegelman, S., A. Burny, M. R. Das, J. Keydar, J. Schlom, M. Trávníček & K. Watson, 1970. Characterization of the products of RNA-directed DNA polymerase in oncogenic RNA viruses. Nature, Lond. 227: 563.

Steitz, E., 1965. Untersuchungen über die Wirkung von Kinetin auf Kallus und Tumoren bei Organkulturen von *Nicotiana glauca*, *N. langsdorffii* und deren amphidiploiden Bastard. Experientia 21: 647.

Temin, H. M. & S. Mizutani, 1970. RNA-dependent DNA polymerase in virons of Rous sarcoma virus. Nature, Lond. 226: 1211.

Veldhuisen, G. & P. H. Pouwels, 1970. Transformation of *Xeroderma pigmentosum* cells by SV40. Lancet p. 529.

Veldstra, H., 1968. On the biochemistry of interference with normal plant development, with special reference to growth regulators. Neth. J. Plant Pathol. 74, suppl. 1: 55.
Westphal, H. & R. Dulbecco, 1968. Viral DNA in polyoma- and SV40-transformed cell lines. Proc. natl. Acad. Sci. U.S. 59: 1158.
Wood, H. N. & A. C. Braun, 1961. Studies on the regulation of certain essential biosynthetic systems in normal and crown gall tumor cells. Proc. natl Acad. Sci. U.S. 47: 1907.

Integration of Agrobacterium tumefaciens DNA in the genome of crown gall tumor cells and its expression

R. A. Schilperoort

Department of Biochemistry, University of Leiden, The Netherlands

Abstract

Experiments have been carried out to obtain evidence for the assumption that the tumor inducing principle (TIP) is identical with *Agrobacterium tumefaciens* DNA.

It was found that c-^{3}H-RNA's synthesized on the DNA's of *A. tumefaciens* strain A_6, B_6 and of *A. radiobacter* form a specific hybrid with tobacco crown gall DNA. In this way it was also shown that the *A. tumefaciens* DNA is located in the nuclei of the tumor cells. By analysing ^{3}H-thymidine labelled and unlabelled crown gall DNA with analytical and preparative CsCl equilibrium density gradients in combination with MAK column chromatography, strong indications were obtained that the *A. tumefaciens* DNA is integrated in the host cell genome.

Ouchterlony immunodiffusion studies revealed that the *A. tumefaciens* DNA in the tumor cells gives rise to the synthesis of *A. tumefaciens* proteins. Moreover, it was observed that tumorogenic *A. tumefaciens* strains showed more precipitation lines in the reaction with antitumor serum than the non-tumorogenic *A. radiobacter*.

The central question in the crown gall problem concerns the character of the tumor inducing principle (TIP, as Braun, 1947, calls it), the information transferred from the organism *(Agrobacterium tumefaciens)* to the normal cell, inducing its transformation into a tumor cell.

During studies on this subject (cf. Veldstra, 1971), and guided by a possible analogy with the transformation of animal cells by oncogenic viruses, it was considered whether introduction of *Agrobacterium tumefaciens* DNA into plant cells, together with its integration into the host genome and subsequent expression in the cells, is responsible for tumors induced by this bacterium. If so, crown gall formation could be seen as a real genetic transformation of normal cells into tumor cells.

Using the sensitive technique of molecular DNA-RNA hybridization it was examined whether there was any specific homology between tumor DNA and *Agrobacterium tumefaciens* DNA. Such a relationship, indicating the presence of *A. tumefaciens* DNA in tumor cells, was indeed found (Schilperoort *et al.*, 1967; Schilperoort, 1969, 1970). Recently these results were confirmed and in certain respects elaborated (Quétier *et al.*, 1969; Milo & Srivastava, 1969; Srivastava, 1970). Furthermore, by a modified Ouchterlony immunodiffusion technique, evidence was obtained for the expression of *Agrobacterium* DNA in sterile crown gall tissue (Schilperoort *et al.*, 1969).

In all experiments the tumor tissue was taken from tumors induced by *A. tume-*

faciens strain A_6 (on *Nicotiana tabacum* cv. White Burley), cultured *in vitro* for more than seven years. The sterility of the tumor tissue was routinely checked by incubating samples of minced tissue on nutrient agar plates, in a pepton liquid medium, in Difco Bacto N.I.H. thioglycollate broth, and in Difco Bacto Sabouraud Dextrose broth. The DNA's used in the hybridization experiments were isolated by SDS-phenol and SDS-pronase-phenol methods (Schilperoort *et al.*, 1967; Schilperoort, 1969). The highly radioactive RNA's (c-RNA's) were synthesized *in vitro* on DNA templates with DNA dependent RNA polymerase. The complementarity tests were carried out according to the liquid hybridization method of Nygaard & Hall (1964). Similar results were obtained with the filter method of Gillespie & Spiegelman (1965).

First of all the complementarity of tumor DNA and tobacco leaf DNA to c-^{3}H-RNA of *A. tumefaciens* A_6 was tested (Table 1). In this series of experiments the DNA's of calf thymus and T_4 phage were included to examine the binding specificity of the RNA's synthesized *in vitro*. To find out more about the specificity of the reaction of the plant DNA's with *A. tumefaciens* c-RNA, these plant DNA's were also incubated with T_4-c-RNA and tobacco leaf c-RNA. The latter was included to make sure that any difference found between tumor DNA and tobacco DNA when they complex with *A. tumefaciens* c-RNA were not due to RNase activity.

The results clearly showed that a highly specific complementarity exists between crown gall DNA and *A. tumefaciens* c-RNA indicating the presence of *A. tumefaciens* A_6-DNA in crown gall tissue initiated by this strain. Almost the same results were found with c-^{3}H-RNA synthesized on strain B_6 DNA, and with c-^{3}H-RNA synthesized on DNA of the non tumorogenic *A. radiobacter* (Schilperoort *et al.*, 1967; Schilperoort, 1969) (Table 2). Nuclear crown gall DNA was used in these experiments

Table 1. Complementarity test with crown gall DNA and *Agrobacterium tumefaciens* A_6 c-^{3}H-RNA.

DNA from	$s_{20,w}$	*A. tumefaciens* c-^{3}H-RNA dpm on filter	Tobacco leaf c-^{3}H-RNA dpm on filter	T_4 phage c-^{3}H-RNA dpm on filter
Input 150 μg:				
crown gall	20	44,400	451,600	3,920
tobacco leaf	22	1,830	350,000	600
calf thymus		2,280	5,210	—
T_4 phage	24	400	112	582,270
A_6 bacterium	19	180,300	—	130
Input 30 μg				
crown gall DNA	20	32,600	—	1,180

Averages of two assays. The c-^{3}H-RNA's had a specific activity of about 10^6 dpm/μg. Input doses: *A. tumefaciens* c-^{3}H-RNA 10^6 dpm, tobacco leaf c-^{3}H-RNA 1.2×10^6 dpm, T_4 c-^{3}H-RNA 10^6 dpm.

Table 2. Complementarity test of nuclear crown gall DNA with c-^{3}H-RNA of *Agrobacterium tumefaciens* A_6 and *Agrobacterium radiobacter* (c-^{3}H-RNA in dpm on filter).

DNA (input 25 μg) from	*A. tumefaciens* A_6 (input 14 × 10^5 dpm)	*A. radiobacter* (input 9 × 10^5 dpm)	Tobacco leaf (input 5 × 10^5 dpm)	T_4 phage (input 5 × 10^5 dpm)
Crown gall nuclei	46,800	60,800	49,100	450
Tobacco leaf nuclei	1,540	1,480	42,500	420
Calf thymus	2,040	—	—	540
T_4 phage	—	350	—	30,800
A_6 bacterium	110,700	—	220	200

Averages of two assays. Specific activity of the c-^{3}H-RNA's 10^6 dpm/μg.

to collect data about the situation of bacterial DNA in tumor cells. In this context it is important to note that this nuclear DNA did not contain any satellite DNA visible by analytical CsCl density gradient ultracentrifugation. Since this nuclear tumor DNA also gave a highly specific hybridization with *Agrobacterium* c-^{3}H-RNA's it was concluded that the *A. tumefaciens* DNA in the tumor cells is partly or completely located in the nucleus.

To establish whether the *A. tumefaciens* DNA in crown gall exists as a free entity or is integrated in the host cell genome, the tumor DNA was examined in analytical and preparative CsCl density gradients. MAK column chromatography was also applied because of its capacity to enrich satellite DNA in a few fractions. Control experiments were carried out with DNA from normal tobacco (cv. White Burley) tissue both from sterile tissue culture material and from tobacco leaves and stems.

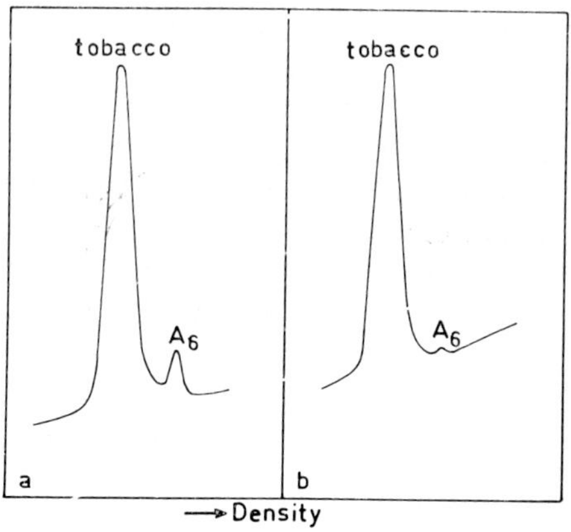

Fig. 1. Densitograms of 5% and 1% *Agrobacterium tumefaciens* A_6 DNA in the presence of tobacco leaf DNA after analytical CsCl density gradient centrifugation.

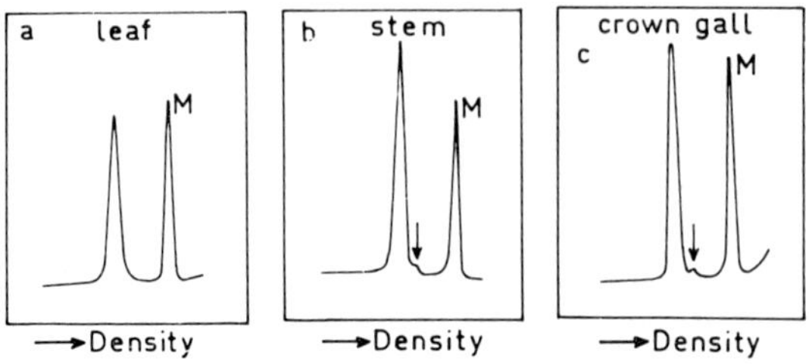

Fig. 2. Densitograms of tobacco leaf DNA, stem DNA and crown gall DNA after analytical CsCl density gradient centrifugation. M: marker M lysodeikticus DNA at $\rho = 1.7310$ g cm^{-3}.
a: tobacco leaf DNA at $\rho = 1.6966$ g cm^{-3}
b: tobacco stem DNA at $\rho = 1.6953$ g cm^{-3}
c: tobacco crown gall DNA at $\rho = 1.6946$ g cm^{-3}

To determine the minimal amount of *A. tumefaciens* A_6 DNA detectable by analytical CsCl density gradient centrifugation, tobacco leaf DNA was mixed with 5% and 1% *A. tumefaciens* A_6 DNA. As is shown in Figs 1a and b, the relative amount of bacterial DNA easily detectable in this way is about 1%.

No satellites were found in CsCl gradients of tumor DNA preparations with a buoyant density of *A. tumefaciens* DNA ($\rho = 1.7177$ g cm^{-3}), even with a large overloading amount of DNA (ca 30 µg instead of the usual 2 µg). Yet a satellite DNA with a density of 1.7059 g cm^{-3} appeared to be present in normal as well as in tumor DNA preparations when the DNA was extracted with the SDS-pronase-phenol method (Schilperoort, 1969). Pronase appeared to be essential for the optimum extraction, since almost no satellite DNA could be extracted when it was omitted, as with the SDS-phenol method.

Tumor DNA preparations always contained more satellite DNA than the other plant DNA's studied. As more of this satellite DNA could be extracted from unpigmented tissue (i.e. tissue culture material and tobacco stems) than from green tobacco leaves (Fig. 2), it was tentatively concluded that this satellite is not extracted from chloroplasts. Its real origin is not yet clear. It may be of cytoplasmic origin (mitochondria, proplastids) since the corresponding nuclear DNA, as mentioned before, did not contain any satellite detectable with UV absorption optics in the analytical ultracentrifuge. To substantiate this supposition no DNA should leak from the nuclei when these are isolated. This problem is under investigation now.

The results summarized in Table 3 show that the ρ value of crown gall DNA did not significantly differ from that of DNA isolated from normal tissue (tissue culture). Unexpectedly the buoyant density of total tobacco leaf DNA ($\rho = 1.6965$) obtained with the SDS-pronase-phenol method was significantly higher. Moreover the density of the tobacco leaf DNA was influenced by the method of extraction: with the SDS-phenol method tobacco leaf DNA with a density of 1.6954 g cm^{-3} was obtained, corresponding with the value found for the leaf-nuclear DNA and stem-nuclear DNA (Table 3).

Such significant differences were not found between normal tissue DNA and tumor

Table 3. Buoyant density of the main fraction of total DNA and nuclear DNA obtained from various kinds of tobacco tissue.

Material from	ρ(g cm⁻³) main fraction DNA	ρ(g cm⁻³) nuclear DNA
Leaves	1.6965	1.6957
	1.6954[1]	
Stems	1.6953	1.6953
Crown gall tissue (culture)	1.6948	1.6946
Normal tissue (culture)	1.6946	

1. DNA isolated without pronase. Total DNA isolated according to SDS-pronase-phenol procedure. Nuclear DNA isolated from nuclei with a high salt method.

tissue DNA, possibly because both tissues lack chloroplasts. As expected, the same buoyant density values were found for the (nuclear) DNA's from tobacco leaves and stems (Table 3). So the fluctuation of density in tobacco leaf DNA may be ascribed to a varying degree of extraction of DNA from chloroplasts, the density of which is slightly higher than that of nuclear DNA.

The other small disparaties in density for the various DNA types might indicate differences in GC content, but also differences in the extent of methylation of cytosine residues in the nuclear DNA (Kirk, 1967).

When DNA preparations (isolated in the presence of pronase) are split up according to GC content by MAK column fractionation (Fig. 3), the buoyant density of the MAK column fractions plotted against fraction number gives no linear relationship for all tobacco tissue DNA's studied. (For convenience, the top fractions of the MAK column elution patterns of the various kinds of DNA were plotted at the same position in the graphs. The buoyant density of a DNA fraction refers to the density of the main peak.)

The elution curve could roughly be divided into left-side DNA fractions (GC content high) and right-side DNA fractions (AT content high). In the first few fractions of the left-side DNA fractions satellite DNA ($\rho = 1.706$ g cm^{-3}) was enriched (Schilperoort, 1969). As with unfractionated DNA, the amount of satellite DNA was always highest in tumor DNA.

In the extreme left-side fractions no satellite with a buoyant density of *A. tumefaciens* A_6 DNA was ever detected. This means that, if a small amount of free bacterial DNA exists in tumor tissue, it must be far below 1%.

The right-side DNA represented more than two thirds of the total DNA and had a uniform buoyant density. This AT-rich component of the DNA sample determines mainly the overall buoyant density of the DNA preparations. Contrary to this, the DNA's in the left-side fractions were all quite different. The GC content of these DNA fractions gradually decreased from left to right. The main peak DNA in these fractions (as far as they were detectable with UV optics in the Spinco E) exhibited densities up to 1.702 g cm^{-3}. Thus DNA fractions with a higher density than the

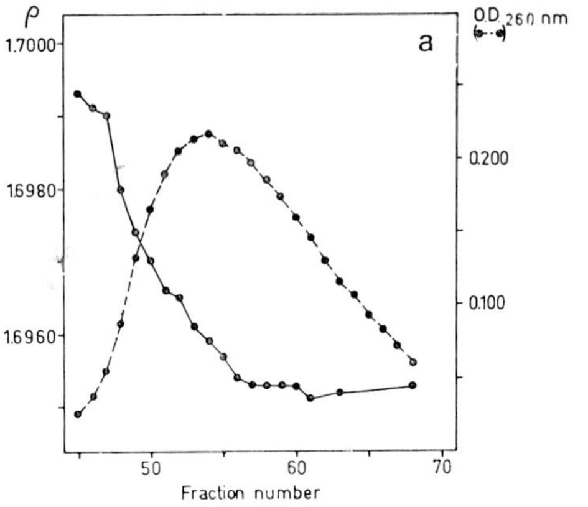

Fig. 3. Buoyant density distribution and corresponding OD_{260} elution pattern
a: of crown gall DNA fractionated by MAK column chromatography;
b: of normal DNA and crown gall DNA fractionated by MAK column chromato graphy (1: DNA isolated from tobacco normal tissue, 2: DNA isolated from tobacco crown gall tissue, both in tissue culture);
c: of tobacco leaf DNA and crown gall DNA fractionated by MAK chromatography (1: DNA isolated from tobacco leaves, 2: DNA isolated from crown gall tissue in tissue culture).

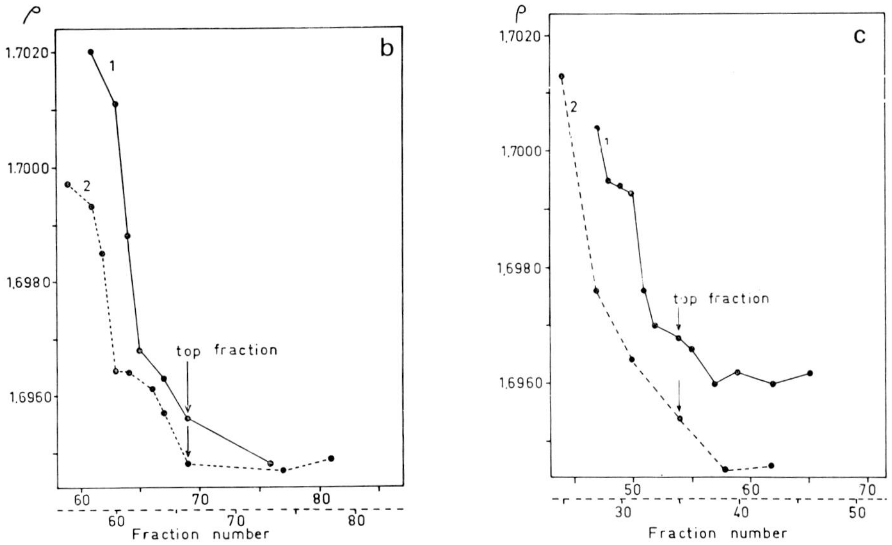

overall density of total DNA were indeed found (as would be expected with *A. tumefaciens* DNA integrated in the plant genome), but these DNA's were not specific to tobacco crown gall tissue, since tobacco normal (tissue culture) DNA and tobacco leaf DNA showed the same density profile on fractionation.

For completeness' sake ^3H-thymidine-labelled crown gall DNA (specific radioactivity 8500 dpm/μg) was also fractionated with MAK-column chromatography. The OD_{260} and radioactivity elution patterns are shown in Fig. 4.

No radioactivity peak was found to the left of that in the figure, which indicates that no special kind of GC-rich DNA (e.g. *A. tumefaciens* DNA) had been eluted before the rest of the ^3H-DNA.

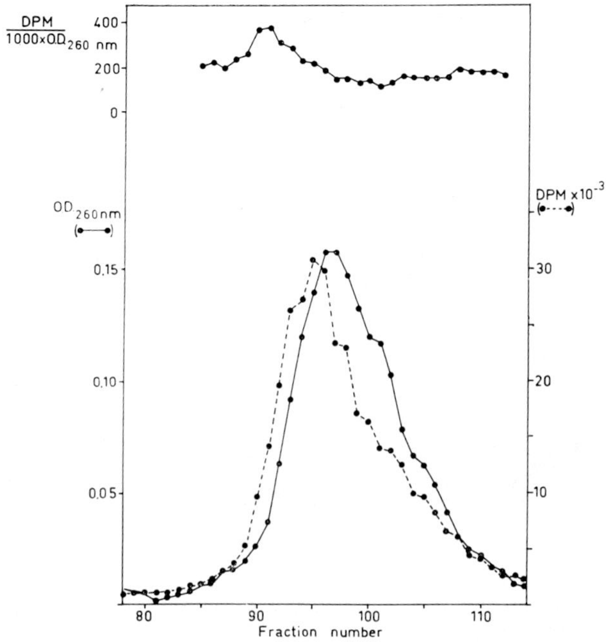

Fig. 4. Crown gall ^{3}H-DNA fractionated by MAK column chromatography.

The same buoyant density distribution and location of satellite DNA (ρ = 1.706 g cm^{-3}) was found as that described for the non-labelled tumor DNA. The enrichment of satellite DNA and its extraordinary high radioactivity, as compared with that of the bulk of the DNA, was shown by the specific radioactivity distribution as indicated in Fig. 4 by the curve of dpm/1000 $\times$ OD_{260}.

Tumor ^{3}H-DNA was also fractionated by preparative CsCl density gradient centrifugation with a fixed angle rotor 40 (Fig. 5). A careful analysis of all fractions of the gradients from top to bottom for three separate ^{3}H-DNA batches was carried out to find *A. tumefaciens* DNA. But a DNA in the buoyant density range of ρ = 1.718 g cm^{-3} never showed up. Thus a separately banding *A. tumefaciens* A_6 DNA in crown gall tissue does not exist.

The unusual radioactivity distribution shown in Fig. 5 suggests the presence of a highly labelled component at the left-side referred to further as left-side radioactive satellite. The presence of this highly labelled fraction was already known from MAK-column fractionation (Fig. 4).

To investigate what causes the unexpected shape of the curve some fractions (marked in Fig. 5 with an arrow) were examined in the analytical ultracentrifuge. The buoyant densities of the DNA's in these fractions are given in Table 4.

The buoyant density distribution over the preparative CsCl density gradient

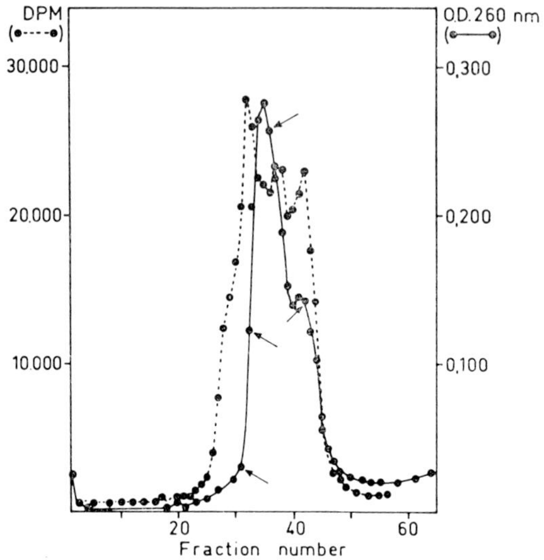

Fig. 5. Crown gall ^{3}H-DNA fractionated by preparative CsCl density gradient centrifugation. Arrows indicate fractions analyzed in Spinco E.

Table 4. Buoyant density of crown gall ^{3}H-DNA's in fractions of preparative CsCl density gradient.

Fraction 31	Fraction 32	Fraction 36	Fraction 42
$\rho(g\ cm^{-3})$ = 1.6962	$\rho(g\ cm^{-3})$ = 1.6948	$\rho(g\ cm^{-3})$ = 1.6947	$\rho(g\ cm^{-3})$ = 1.6946
1.701	1.700		
1.707			

fractions (Fig. 6) and the existence of a satellite DNA with $\rho = 1.707\ g\ cm^{-3}$ were similar to the results obtained after MAK-column fractionation of unlabelled tumor DNA. Both methods showed small amounts of GC-rich DNA and a large amount of AT-rich DNA (with a uniform density of $1.6948\ g\ cm^{-3}$) in tumor DNA.

From the left-side GC-rich fractions of tumor DNA we may conclude that the high GC content is partly the result of the incorporation of a GC-rich *A. tumefaciens* DNA into a relatively AT-rich tobacco DNA.

Fraction 60 of tumor DNA (see Fig. 3b) shifted its density from $1.6964\ g\ cm^{-3}$ to $1.6952\ g\ cm^{-3}$ after the DNA solution had passed through a 16 gauge needle one or three times. Simultaneously a small new peak with a density of $1.7179\ g\ cm^{-3}$ was observed. The density decreased further to $1.6944\ g\ cm^{-3}$ after five passages through the same needle.

The neighbouring fractions 59 and 61 (Fig. 3b) showed no such change upon shearing, nor did a comparable fraction of normal tissue DNA. In other experiments a few GC-rich fractions of tumor DNA gave new peaks with a higher buoyant density

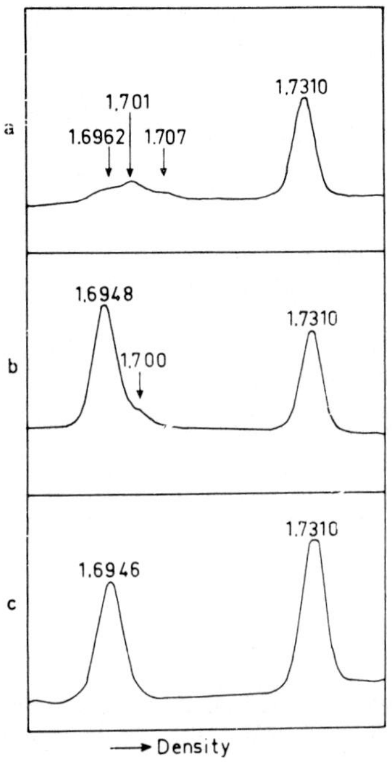

Fig. 6. Densitograms of crown gall ^{3}H-DNA in preparative CsCl density gradient fractions (see Fig. 5 and Table 4) after analytical CsCl density gradient centrifugation. Marker *Micrococcus lysodeikticus* DNA at $\rho = 1.7310$ g $\times$ cm^{-3}.
a: fraction number 31
b: fraction number 32
c: fraction number 42

than the main peak upon shearing, without a significant accompanying drop in density of the main peak (Schilperoort, 1969). A significant drop in density of the DNA in a MAK column fraction and the simultaneous appearance of a peak at the *A. tumefaciens* DNA density were found only for one DNA sample.

From these experiments we may conclude that small quantities of GC-rich tumor DNA produce upon gentle shearing, even smaller amounts of material (never detected without shearing) that might contain bacterial DNA and bands as small bumps at densities in the range of 1.7103 g cm^{-3} to 1.7179 g cm^{-3}. But we have to bear in mind that GC-rich plant DNA, containing e.g. redundant ribosomal-DNA (r-DNA), may also be the cause of these bumps.

As in mixtures of *A. tumefaciens* A_6 DNA and tobacco leaf DNA the amount of the first has to be over 0.5% of the total amount of DNA to be detected by UV-optics in the Spinco E (see Fig. 1), ^{3}H-thymidine labelled crown gall DNA fractions from CsCl density gradients were also sheared. In this way very small amounts (far below 0.5%) should be detected if the molecular weight of the DNA after shearing is still high enough to give a sharp band in the CsCl density gradient.

Fig. 7 shows what happens when the left-side radioactive satellite fractions of several gradients are pooled, concentrated, desalted by dialysis and recycled unsheared and needle-sheared (10 $\times$) in separate CsCl density gradients. As marker DNA's A_6-DNA and a non-labelled crown gall DNA were used. The non-sheared DNA solution was recycled without adding marker DNA. For convenience the graphs of the

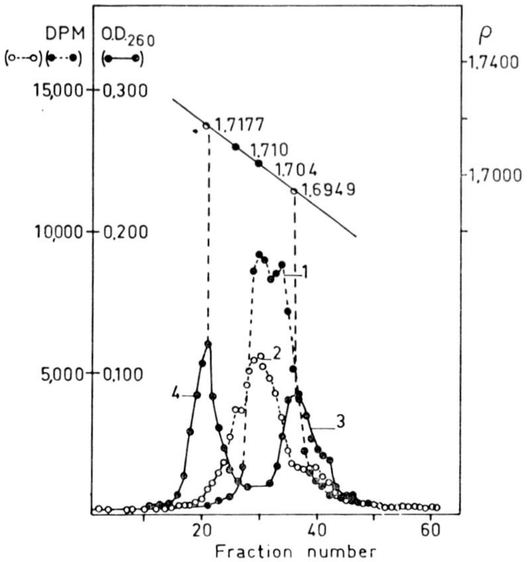

Fig. 7. Unsheared and sheared 'left-side radioactive satellite' ^{3}H-DNA fractionated by preparative CsCl density gradient centrifugation.
1: unsheared ^{3}H-DNA
2: needle-sheared ^{3}H-DNA
3: marker crown gall DNA at $\rho = 1.6949$ g cm^{-3}
4: marker *Agrobacterium tumefaciens* strain A_6 DNA at $\rho = 1.7177$ g cm^{-3}

needle-sheared and unsheared fractionated DNA's were plotted so that the tops of both graphs were at the same fraction number (Fig. 7). Even if this procedure should prove to be not perfect, the deviation can only be very small.

The two fractions (29 and 30 in Fig. 7) of the highest top of the sheared satellite DNA were at $\rho = 1.706$ g cm^{-3} and $\rho = 1.704$ g cm^{-3}. These values remind us of the satellite with $\rho = 1.707$ g cm^{-3} observed when the left side base fraction is recycled in the analytical ultracentrifuge (Fig. 6, fr. 31). As to the unsheared satellite DNA, it is reasonable to consider the left-side top fraction to be at the position of the approximate value of $\rho = 1.704$ g cm^{-3} in analogy to the sheared satellite recycled with added marker DNA's.

Fig. 6 also shows that other satellite DNA's with $\rho = 1.701$ g cm^{-3} (fr. 31) and $\rho = 1.700$ g cm^{-3} (fr. 32) are detected in unsheared DNA by analytical CsCl density gradient centrifugation. Assigning the density value of either 1.707 g cm^{-3} or 1.706 g cm^{-3} (determined from the Spinco E run) to the left-side top fraction of the unsheared DNA in Fig. 7, we find $\rho = 1.701$ g cm^{-3} or 1.700 g cm^{-3}, respectively, on the ρ-value scale of Fig. 7 for the right-side top of the unsheared DNA. Therefore both radioactive satellites detected in the unsheared satellite DNA have already shown up in the analytical ultracentrifuge runs.

Upon shearing (carried out in duplo experiments) the right-side component ($\rho = 1.701 - 1.700$ g cm^{-3}) of the unsheared satellite DNA seems to be at least partly split up into a fraction with a higher buoyant density of approximately 1.710 g cm^{-3} (Fig. 7) and a fraction at the position of the lower density of the non-labelled tumor DNA ($\rho = 1.6949$) added as a marker. This phenomenon is very interesting in connection with the presence of *A. tumefaciens* DNA in tumor DNA and is in agreement with the results obtained after shearing GC-rich MAK column fractions of unlabelled

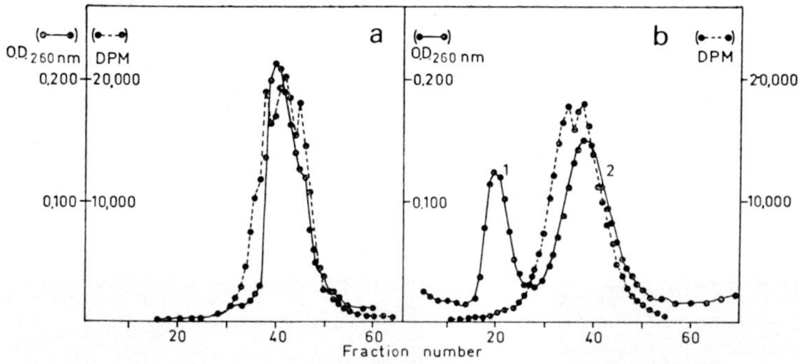

Fig. 8. a: main peak ³H-DNA recycled in a preparative CsCl density gradient;
b: sheared main peak ³H-DNA fractionated by preparative CsCl density gradient centrifugation.

tumor DNA. A definite conclusion at present on its significance would be speculative. No DNA with a density of *A. tumefaciens* DNA has been freed by shear from the GC-rich highly radioactive satellites.

Likewise the main peak DNA fractions (remaining after the left-side radioactive satellite DNA fractions were collected) were pooled and recycled unsheared and needle-sheared in separate CsCl density gradients. The sheared DNA solution was supplemented with *A. tumefaciens* A_6 DNA as a marker. As apparent from Fig. 8b, no *A. tumefaciens* like DNA was freed upon shear. Moreover, the high specific activity of the left-side satellite DNA's still present in the pooled main peak DNA fractions (see Fig. 8a) did not permit the detection of any intermediate DNA if this would have been derived after shear treatment from the relatively low radioactive bulk of the DNA.

Furthermore, it seems that the main peak fractions also contain DNA types with different specific radioactivities. Both the total ^{3}H-DNA (see Fig. 5) and the recycled main peak DNA (Fig. 8a) show three well-defined radioactive peaks; the left can be ascribed to the effect of the left-side radioactive satellites. The top of the OD_{260} patterns coincides with a minimum in the radioactivity profiles indicating that the bulk of the DNA, which is nuclear, has a relatively low specific radioactivity. The whole radioactivity curve, which does not follow the OD_{260} pattern at all, is determined by the presence of small amounts of different strongly labelled DNA types. Their origin and significance is still unknown and is under investigation. Anyway the OD_{260} right-side shoulder (see Figs 5 and 8a) cannot be attributed to a satellite with a buoyant density different from that of the bulk of the DNA, since no such satellite has been found in this fraction when recycled in the analytical ultracentrifuge (Fig. 6, fr. 42).

The DNA in this fraction had the same buoyant density ($\rho = 1.6946$ g cm^{-3}) as found for the bulk of the DNA. So its separation from the main peak DNA cannot be due to a difference in buoyant density. Probably this OD_{260} shoulder is the result of an artifact inherent to the applied technique when large amounts of high molecular weight DNA are used.

What can be concluded from all these experiments is that no free, separately banding *A. tumefaciens* DNA exists in the tumor tissue. So it is supposed that the *A. tumefaciens* DNA detected by the hybridization experiments is integrated in the host genome. Consequently, owing to the buoyant density heterogeneity of all tobacco DNA's

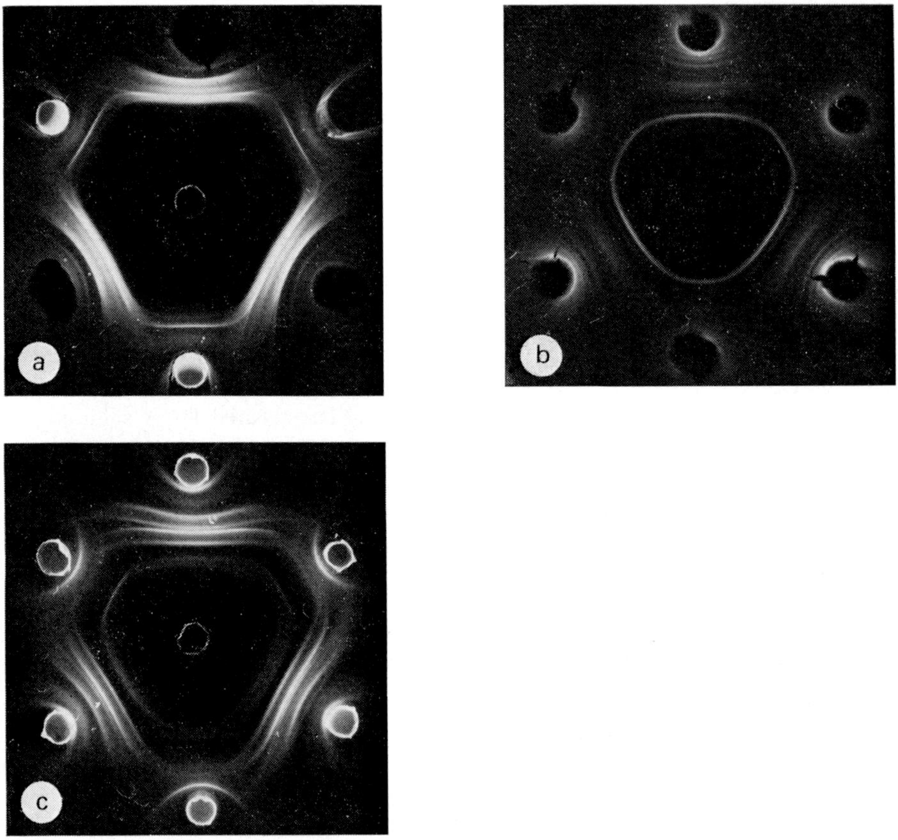

Fig. 9. Micro-immunodiffusion patterns obtained from the reactions of normal tobacco tissue (tissue culture) extract, tobacco leaf extract, and tobacco crown gall (tissue culture) extract versus both normal tissue antiserum and crown gall antiserum.

a: normal tissue extract (central well) versus normal tissue antiserum in the peripheral wells 1, 3, 5 and crown gall antiserum in the peripheral wells 2, 4, 6;

b: leaf extract (central well) versus normal tissue antiserum in the peripheral wells 1, 3, 5 and crown gall antiserum in the peripheral wells 2, 4, 6;

c: crown gall tissue extract (central well) versus normal tissue antiserum in wells 1, 3, 5 and crown gall antiserum in wells 2, 4, 6.

See for well numbering Fig. 12b.

studied (including crown gall DNA), shear experiments alone show inconclusively the presence of *A. tumefaciens* DNA in tumor tissue. Such experiments have to be carried out with nuclear DNA and DNA from well-defined subcellular organelle fractions in combination with hybridization experiments using *A. tumefaciens* c-^{3}H RNA. Perhaps such experiments, now being carried out, will reveal the significance of intermediate DNA's in the crown gall problem.

To see whether the presence of *A. tumefaciens* DNA in the tumor causes synthesis of *A. tumefaciens* like proteins in its cells, Ouchterlony immunodiffusion studies have been carried out (Schilperoort, 1969; Schilperoort *et al.*, 1969). First the antigen-composition of the extracts of normal tissue and crown gall tissue of tobacco were compared in immunodiffusion tests in which the homologous and heterologous reactions were carried out simultaneously. The soluble antigens obtained from tobacco leaves were also studied. The results are shown in Fig. 9.

Despite the differences in the proportion of the antigens in the extracts of tobacco leaves and in those of normal tissue, equal numbers of precipitin lines (about 11) were observed.

The number of detectable precipitin lines in the reaction of normal tissue antiserum with tumor tissue antigens is the same as in the homologous reactions with normal tissue antigens. Nothing indicated that antigens present in normal tissue are absent in tumor tissue.

The precipitin patterns shown in Figs 9a, c and in Fig. 10a also demonstrate that certain antigens are present in a larger amount in tumor tissue than in normal tissue.

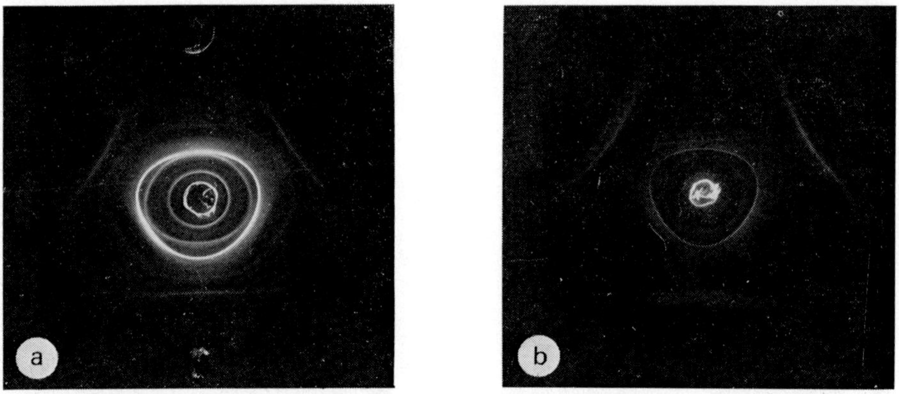

Fig. 10. Micro-immunodiffusion patterns obtained from the reactions of normal tissue antiserum and crown gall antiserum (central wells) versus both normal tissue extract and crown gall extract.

a: normal tissue antiserum versus normal tissue extract in the peripheral wells 1, 3, 5 and crown gall extract in the peripheral wells 2, 4, 6,

b: crown gall antiserum versus normal tissue extract in the peripheral wells 1, 3, 5 and crown gall extract in the peripheral wells 2, 4, 6.

See for well numbering Fig. 12b.

In the tests shown in Fig. 10 the antiserum has been placed in the central well instead of the peripheral wells, so that the inner antigen reactions as shown in Fig. 9 can be seen better in Fig. 10.

The precipitin pattern in Fig. 10b indicates the presence of a few precipitin lines for the tumor antigen preparation which could not be shown in the reaction of normal tissue antigens to tumor tissue antiserum. This indicates that the concentration of these antigens in tumor tissue is higher than in normal tissue. It might even indicate that certain antigens occurring in tumor tissue are not present in normal tissue at all. This is corroborated by crown gall antiserum reacting with *Agrobacterium* soluble antigens, whereas antiserum to normal tissue of tobacco did not react with the *Agrobacterium* extracts (Fig. 11).

Fig. 11a clearly shows that certain antigens of *A. tumefaciens* strain A_6 react with crown gall antiserum and not with normal tissue antiserum (Fig. 11b).

From the precipitation pattern in Fig. 12 at least four precipitin lines are detectable when strain A_6 soluble antigens are tested against tumor tissue antiserum.

The cross-reacting antigens were found in the bacterial extracts of the three virulent strains studied (A_6, B_6 and E III 9.6.1). Two of them seemed to be more concentrated in the extracts of strain A_6.

A. radiobacter antigens also gave precipitates, but the number of precipitin lines differed from that of the virulent strains. This non-tumorogenic *Agrobacterium* gave only two detectable lines which they shared with the A_6 strain (Fig. 12). These cross-reactions were not found when the bacterial antigens were tested against the sera obtained from rabbits before the first injection with the antigen preparations.

The cross-reacting antibodies can be removed from antitumor serum by absorption

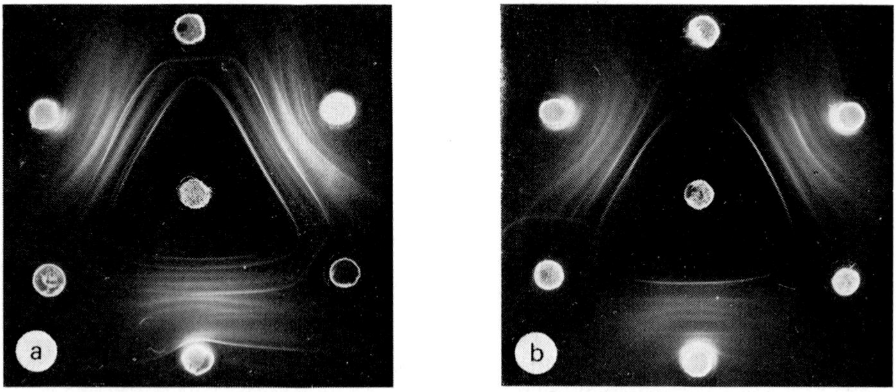

Fig. 11. Micro-immunodiffusion patterns obtained from the reactions of strain A_6 extract (central wells) versus strain A_6 antiserum, crown gall antiserum and normal tissue antiserum.
a: strain A_6 extract versus strain A_6 antiserum in the peripheral wells 2, 4, 6 and crown gall antiserum in the peripheral wells 1, 3, 5;
b: strain A_6 extract versus strain A_6 antiserum in the peripheral wells 2, 4, 6 and normal tissue antiserum in the peripheral wells 1, 3, 5.

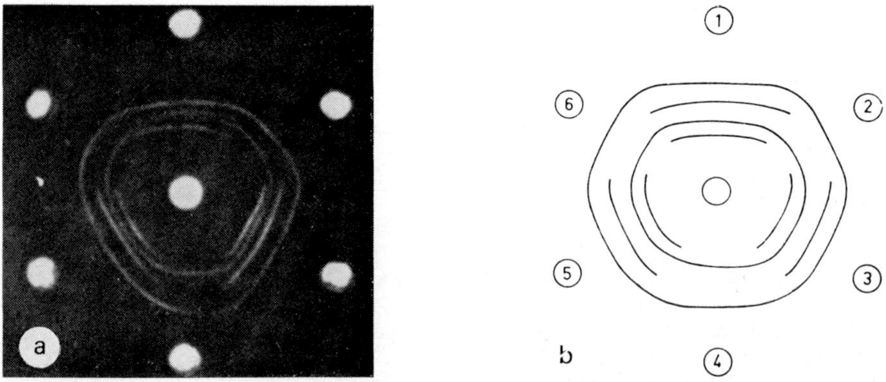

Fig. 12.
a: micro-immunodiffusion pattern obtained from the reaction of crown gall antiserum (central well) versus both strain A_6 extract (peripheral wells 1, 3, 5) and *Agrobacterium radiobacter* extract (peripheral wells 2, 4, 6);
b: *idem*, diagrammatic representation of Fig. 12a.

with bacterial extracts, but not with normal tissue extracts. These antigens are proteins, as they are sensitive to heat treatment (60 min, 100 °C) and digestible by pronase (100 μg pronase B/ml).

The antigens in the strain A_6 extract reacting with antitumor serum are not identical with the endotoxine of the tumor inducing bacterium, as is shown in Fig. 11a. The peripheral precipitin lines in this figure (one of the lines belongs to the TCA extractable endotoxine) representing the reaction with bacterial antiserum, do not coincide with the precipitin lines common to both antisera. In addition, no reaction of the antitumor serum with the TCA extracts of the *Agrobacterium* strains was detected, neither with concentrated culture media. The last result also indicates that the bacterial antigens reacting with antitumor serum are not secreted by the bacteria into the culture media, even after prolonged cultivation (4 days).

The results with *A. radiobacter* extracts again show the close genetic relationship of this species with *A. tumefaciens* as is also evident from the results of the complementarity tests. It was already known that, even on DNA base composition and DNA homology, it is impossible to distinguish between *A. tumefaciens* and *A. radiobacter* (Tinbergen, 1966; Heberlein *et al.*, 1967). But in view of the difference in the reaction of the antigens of *A. tumefaciens*, and *A. radiobacter* with crown gall antiserum, more detailed and quantitative hybridization experiments are needed before an attempt can be made to discriminate between the reactions of crown gall DNA with c-^{3}H-RNA of *A. tumefaciens* and with c-^{3}H-RNA of *A. radiobacter*.

It can be concluded that *A. tumefaciens* DNA is integrated in the genome of crown gall tumor cells and that genetic information of *A. tumefaciens* comes to expression.

Consequently, crown gall formation is considered to be a real genetic transformation of normal cells into tumor cells.

References

Braun, A. C., 1947. Thermal studies on the factors responsible for tumor initiation in crown gall. Am. J. Bot. 34: 234.

Gillespie, D. & S. Spiegelman, 1965. A quantitative assay for DNA-RNA hybrids with DNA immobilized on a membrane. J. molec. Biol. 12: 829.

Heberlein, G. T., J. De Ley & R. Tytgat, 1967. Deoxyribonucleic acid homology and taxonomy of *Agrobacterium*, *Rhizobium*, *Chromobacterium*. J. Bact. 94: 116.

Kirk, J. T. O., 1967. Effect of methylation of cytosine residues on the buoyant density of DNA in caesiumchloride solution. J. molec. Biol. 28: 171.

Milo, G. E. & B. J. S. Srivastava, 1969. RNA-DNA hybridization studies with the crown gall bacteria and the tobacco tumor tissue. Biochem. biophys. Res. Commun. 34: 196.

Nygaard, A. P. & B. D. Hall, 1964. Formation and properties of RNA-DNA complexes. J. molec. Biol. 9: 125.

Quétier, F., T. Huquet & E. Guillé, 1969. Induction of crown gall: partial homology between tumor-cell DNA, bacterial DNA and the G+C-rich DNA of stressed normal cells. Biochem. biophys. Res. Commun. 34: 128.

Schilperoort, R. A., H. Veldstra, S. O. Warnaar, G. Mulder & J. A. Cohen, 1967. Formation of complexes between DNA isolated from tobacco crown gall tumours and RNA complementary to *Agrobacterium tumefaciens* DNA. Biochim. biophys. Acta 145: 523.

Schilperoort, R. A., 1969a. Investigations on plant tumors. Crown gall. On the biochemistry of tumor induction by *Agrobacterium tumefaciens*. Thesis Leiden.

Schilperoort, R. A., 1969b. Onderzoekingen over plantetumoren. Jaarboek Kanker Onderzoek en Kankerbestrijding. J. H. de Bussy, Amsterdam, p. 157.

Schilperoort, R. A., W. H. Meys, G. M. W. Pippel & H. Veldstra, 1969. *Agrobacterium tumefaciens* cross-reacting antigens in sterile crown gall tumors. FEBS Lett. 3: 173.

Srivastava, B. J. S., 1970. DNA-DNA hybridization studies between bacterial DNA, crown gall tumor cell DNA and the normal cell DNA. Life Sci. 9: 889.

Tinbergen, B. J., 1966. Onderzoekingen over plantentumoren. Crown gall. Thesis, Leiden.

Veldstra, H., 1971. Plant tumors and crown gall, an analysis of autonomous growth. These Proceedings: 213.

On the interaction of Agrobacterium tumefaciens with cells of Kalanchoë daigremontiana

R. J. Bogers

Department of Biochemistry, University of Leiden, The Netherlands

Abstract

The interaction of *Agrobacterium tumefaciens* strain B_6 and of *Agrobacterium radiobacter* with the cell walls, after inoculation in a wound on *Kalanchoë daigremontiana*, was studied under the electron microscope.

The attachment to the cell wall observed in the case of *A. tumefaciens* may be an initial stage in tumor induction, whereas the similar behaviour of the non-tumorogenic *A. radiobacter* could explain its capacity to inhibit competitively the tumor initiation by *A. tumefaciens*. *Escherichia coli*, never found associated with the plant cell wall, lacks this inhibitory property. The morphological changes of both *Agrobacterium* species, occurring in the wounded cells, are also observed when the bacteria are grown in wound sap.

In the biochemistry laboratory at Leiden (Schilperoort *et al.*, 1967; Schilperoort, 1969) and later on elsewhere (Quétier *et al.*, 1969; Milo & Srivastava, 1969) evidence has been obtained for the presence of *Agrobacterium tumefaciens* DNA in crown gall tumor cells, this DNA being possibly the tumor inducing principle. This leads of course to an analysis of how this bacterial genetic material is transferred to normal plant cells.

The aim of our experiments was in the first place to obtain information on the interaction between *Agrobacterium tumefaciens* and wounded cells of *Kalanchoë daigremontiana*, the host plant mainly used in our studies on tumor-induction.

As the results showed that *A. tumefaciens* and the non-tumorogenic *A. radiobacter* are remarkably similar, we also studied the interaction between *A. radiobacter* and *Kalanchoë* cells.

In addition, we performed experiments on the influence of *A. radiobacter* on the tumor induction by *A. tumefaciens* (Schilperoort, 1969).

For the experiments, groups of *Kalanchoë* plants about 7 months old were used. For each group plants of approximately equal length and stem diameter were chosen. The plants were punctured in the second and third internodes from one side to the other with a rectangular punch of $1\frac{1}{2} \times 3$ mm. Then conditioning of the cells in the wound area took place for about 28 h at 22–23 °C and at a relative humidity of 60–70 %. Subsequently the holes were filled with 0.03 ml of a bacterial suspension in 0.16 M NaCl with the aid of a tuberculin syringe with a blunt needle.

First, the relation between the number of *A. tumefaciens* cells (strain B_6) in the

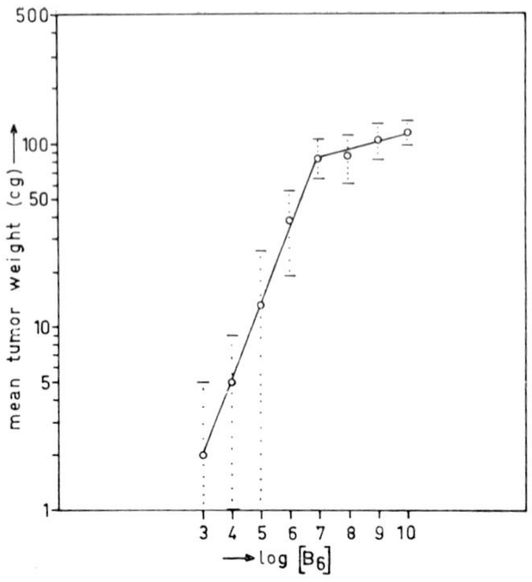

Fig. 1. The relation between the mean tumor weight and the number of inoculated B_6 bacteria.
log [B_6]: logarithm of the number of B_6 cells/ml inoculum.

wound and the size of the tumor was examined. Six weeks after infection the tumors were cut along the plant stem and weighed. The solid tumor and the adventitious roots were weighed together, since we found in all experiments that their separate weights were directly proportional.

Fig. 1 shows that the logarithm of the mean tumor weight was directly proportional to that of the concentration of the B_6 bacteria. It also shows that a maximum tumor weight was obtained with 10^7 bacterial cells per ml inoculum, i.e. 3.10^5 bacteria per wound. Inoculation of more bacteria into the wound did not give significantly larger tumors.

To see whether *A. radiobacter* would have any effect on tumor growth, increasing numbers of bacterial cells, mixed with a fixed number (10^7) B_6 cells per ml inoculum, were injected into the conditioned wounds. As a control, the wounds were identically treated with increasing numbers of *Escherichia coli* cells (strain B). From the results, shown in Fig. 2, it can be concluded that, contrary to *E. coli* cells, cells of *A. radiobacter* inhibit the formation of tumors by *A. tumefaciens* (strain B_6). They lose this capacity when heated at 60 or 100 °C for 1 hour. Fig. 2, suggests that the observed inhibition by *A. radiobacter* is due to a one to one competition with B_6 for some essential and limiting factor, possibly a restricted number of attachment sites for bacteria in the wound.

Recently, Lippincott & Lippincott (1969) came to the same conclusion. However, they found *Agrobacteria*, treated at 60 °C, also able to cause inhibition of tumor induction. This discrepancy may be caused by the quite different infectivity assays.

If the hypothesis of specific attachment sites for the bacteria is correct, addition of a large excess of *A. radiobacter* cells, 24 h after infection with *A. tumefaciens*, should not influence the mean tumor weight. This was confirmed by our experiments (Schilperoort, 1969). Also according to expectation, other experiments showed that inoculation of the plants with different concentrations of *A. radiobacter*, 6 h before infection with B_6, resulted in an inhibition of tumor formation.

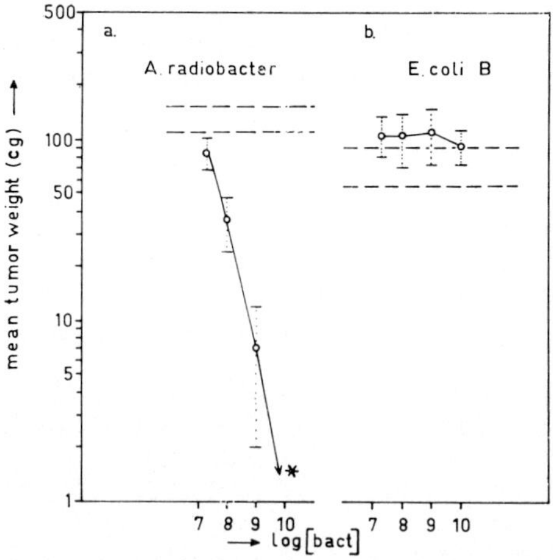

Fig. 2. The effect on the mean tumor weight of increasing numbers of cells of *A. radiobacter* or *E. coli B* mixed with a fixed number of B_6 cells (10^7 bacteria/ml). * The value of the mean tumor weight for 10^{10} *A. radiobacter* cells/ml is 0.3. The confidence interval of the mean tumor weight induced by 10^7 B_6 cells/ml inoculum is indicated by horizontal dotted lines ($\rho = 0.05$).
log [bact]: logarithm of the number of bacteria/ml inoculum.

As it is possible that the inhibition of tumor formation is caused by some toxic product secreted by *A. radiobacter*, the growth interference of both strains on peptone agar plates was investigated. These tests revealed no growth inhibition of one bacterial species by the other.

Deeming it probably that complex formation between *A. tumefaciens* and the plant cells is a prerequisite for transformation of the latter, the electron microscope was used to see whether any attachment of *A. tumefaciens* (strain B_6) to the cell wall of *Kalanchoë daigremontiana* could be detected. Because of the results of the competition experiments the non-tumorogenic *A. radiobacter* were also included. After conditioning, the wound was filled with the bacterial suspension. At different times after inoculation with either B_6 or *A. radiobacter*, samples were taken from the wound region (Fig. 3). They were fixed with glutaraldehyde and osmium tetroxide, and prestained with uranyl acetate, according to Millonig (1961a). Ultrathin sections were made and stained with lead tartrate (Millonig, 1961b).

Until 4 h after inoculation, hardly any bacterial cell could be found in the plant samples. In so short a time it seems that only a very limited number of bacteria has the opportunity to attach themselves to the plant cell wall. Most of them will be rinsed away during fixation or dehydration.

From 10–11 h after infection, however, bacterial cells were observed in damaged cortex and pith cells of the plant, whereas some bacteria were found in the intercellular spaces. Although many bacteria were situated in the plant vessels, they never penetrated very far from the wound surface. In fact, according to Bopp & Leppla (1964) only cortex cells can be transformed into tumor cells.

In samples taken 11 h after inoculation, the bacteria appeared to have kept their normal shape. No difference between *A. tumefaciens* (strain B_6) and *A. radiobacter* could be observed.

After 22 h some bacteria had a rather irregular form, the distance between inner and outer membrane being no longer constant. Sometimes aggregation of nuclear

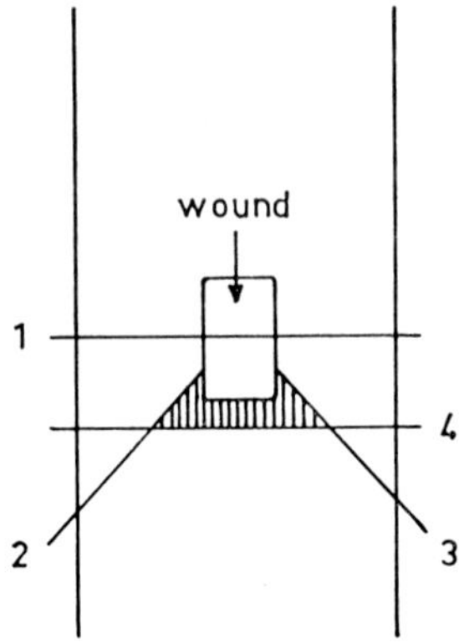

Fig. 3.
a. Excision of the wound area of a *Kalanchoë daigremontiana* plant; side-view of the wound internode.

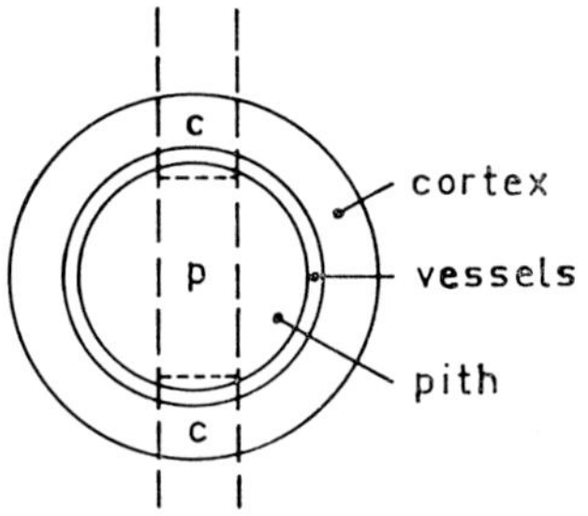

b. Investigated parts of the wounded internode as seen from above, after section 1 (Fig. 3a) has been made.

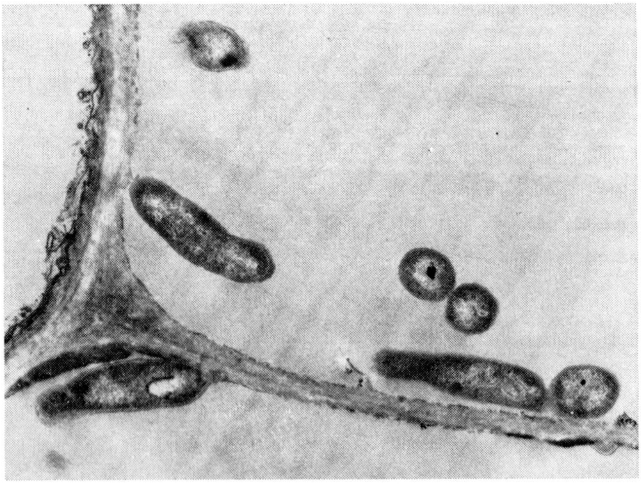

Fig. 4. *Agrobacterium radiobacter* cells 21 h after infection. The bacteria are situated in damaged cells of the cortex of a *Kalanchoë daigremontiana* plant (× 14,650).

material in the bacteria was visible. The same phenomena occurred in bacteria grown in *Kalanchoë* wound sap for the same time (unpublished data). In a normal synthetic medium, however, *Agrobacterium* cells keep their usual form for at least 48 h, even if the pH is as low as 5.1 (as it is in wound sap); 45 h after inoculation not a single bacterium possesses its normal form.

In cortex and pith, both *A. tumefaciens* (strain B_6) and *A. radiobacter* are usually found close to the plant cell wall (Fig. 4–10). But in the cortex cells there is often a kind of thickening of the cell wall in the direction of a bacterium (Fig. 5), generally absent in the pith.

Electron microscopy suggested that among both *A. tumefaciens* B_6 and *A. radiobacter* cells in the cortex many were firmly associated with the plant cell wall. Even if on the photograph a bacterium seemed to be lying at a certain distance of the cell wall it is possible that it is attached to it in the next section, especially when it is flattened on the side turned towards the cell wall (Figs 6a, b).

In the pith the cell walls had a more fluffy appearance than in the cortex. Here two types of bacterial association with the cell wall were observed. The first one is similar to the attachment found for bacteria in the cortex (Figs 9, 10a). The other type showed the bacteria to be covered by and embedded in a membrane of the damaged cell

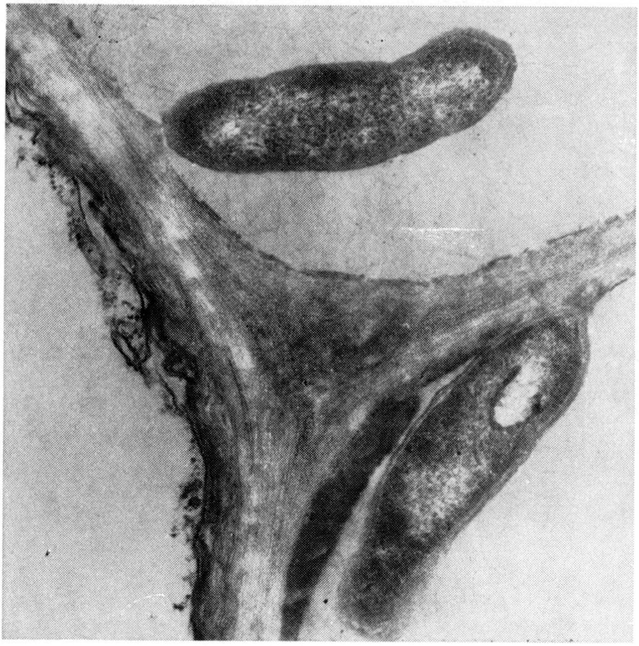

Fig. 5. Close-up of a few *Agrobacterium radiobacter* cells from Fig. 4 (× 30,000). Note the thickening of the plant cell wall at the point of bacterial attachment and the inclusion of DNA-like material by a membrane-like structure.

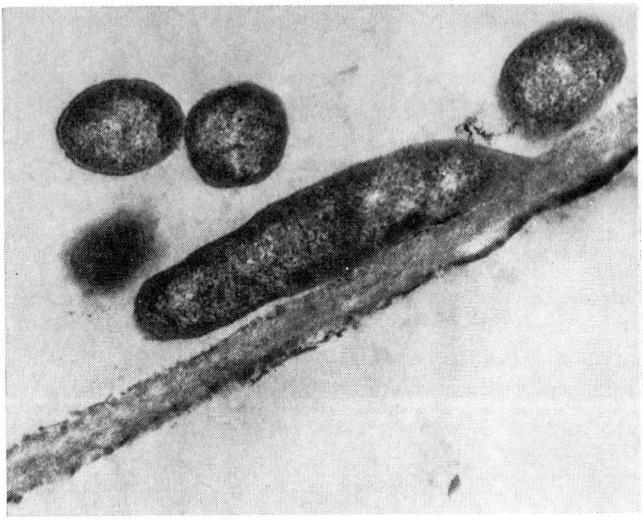

Fig. 6a (b, c, d). A series of sections through a few of the *Agrobacterium radiobacter* cells shown in Fig. 4 (× 30,000). Note the polar cap of one of the bacteria. This cap may be plant cell-wall material (compare thickened cell wall in Fig. 5).

Figs 6c and 6d also demonstrate the fusion of one of the bacteria with the plant cell wall.

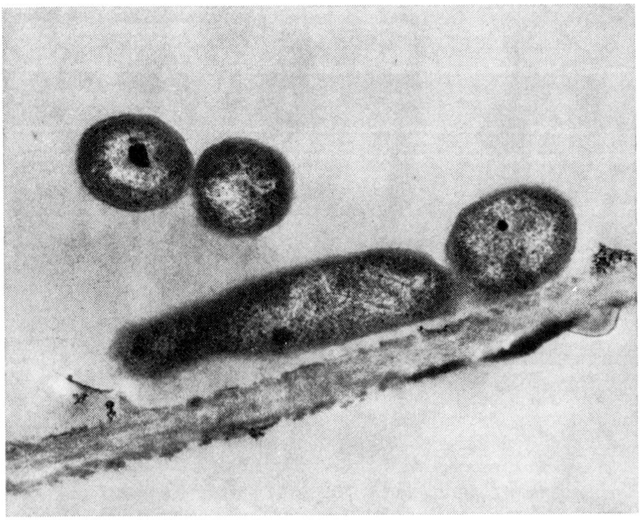

Fig. 6b. See explanation to Fig. 6a.

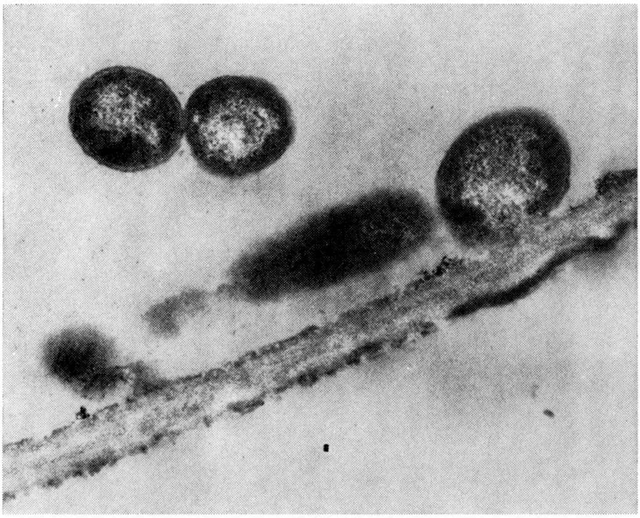

Fig. 6c. See explanation to Fig. 6a.

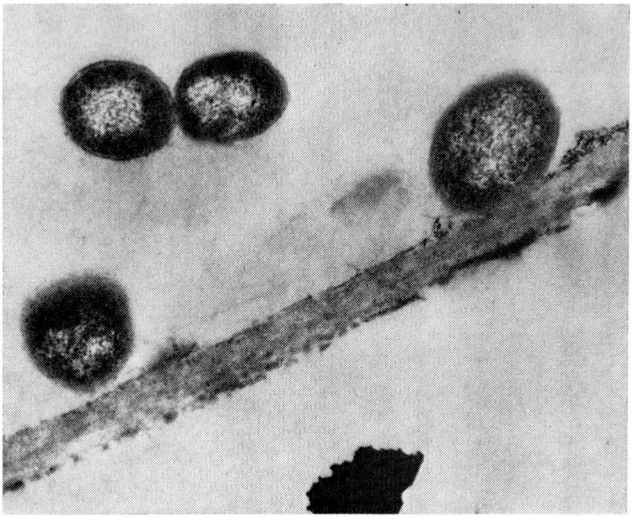

Fig. 6d. See explanation to Fig. 6a.

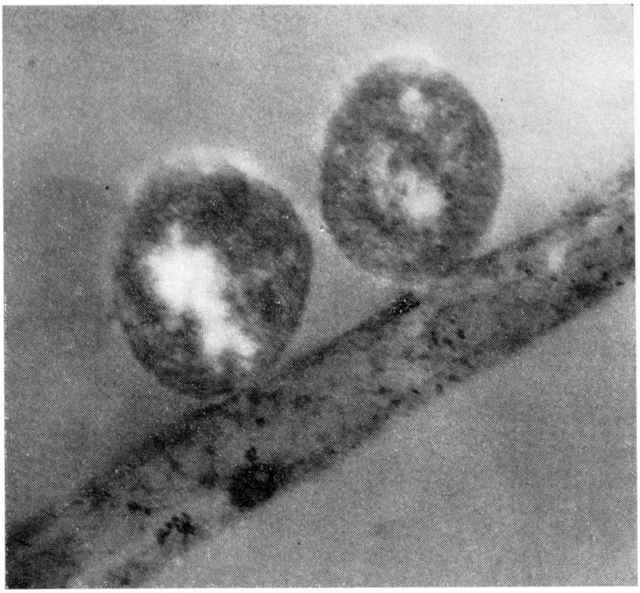

Fig. 7. *Agrobacterium tumefaciens* B_6 cells 22 h after infection in the cortex ($\times$ 57,000).

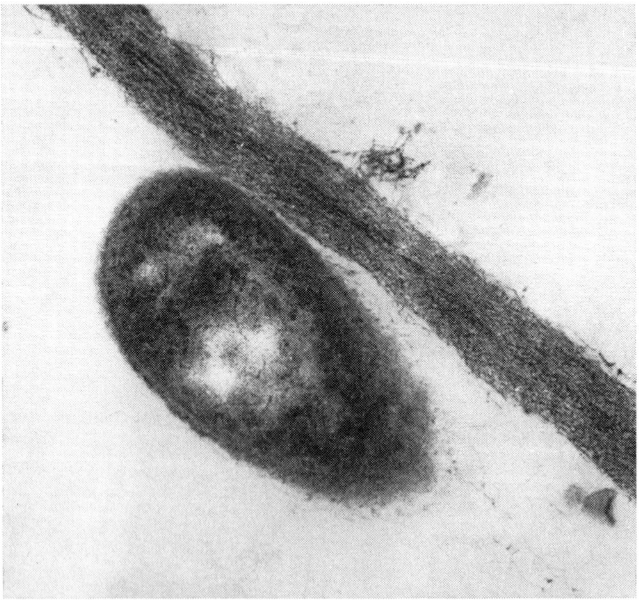

Fig. 8. *Agrobacterium tumefaciens* B_6 cell 10 h after infection in the cortex ($\times$ 52,000).

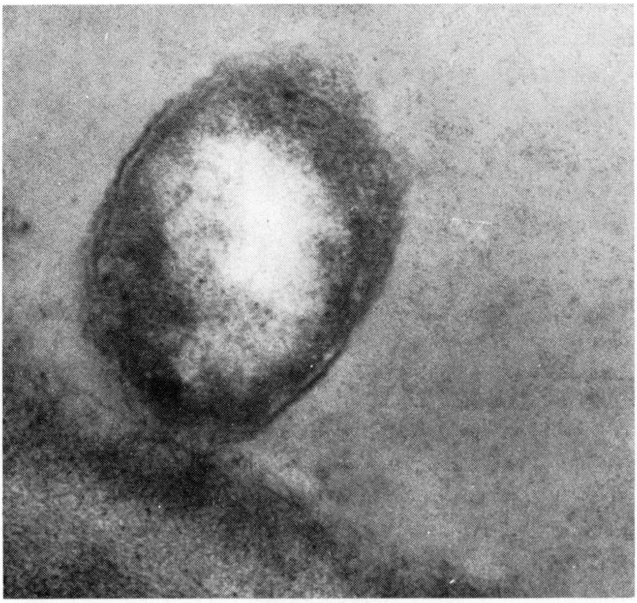

Fig. 9. *Agrobacterium tumefaciens* B_6 cell 10 h after infection in the pith ($\times$ 69,000).

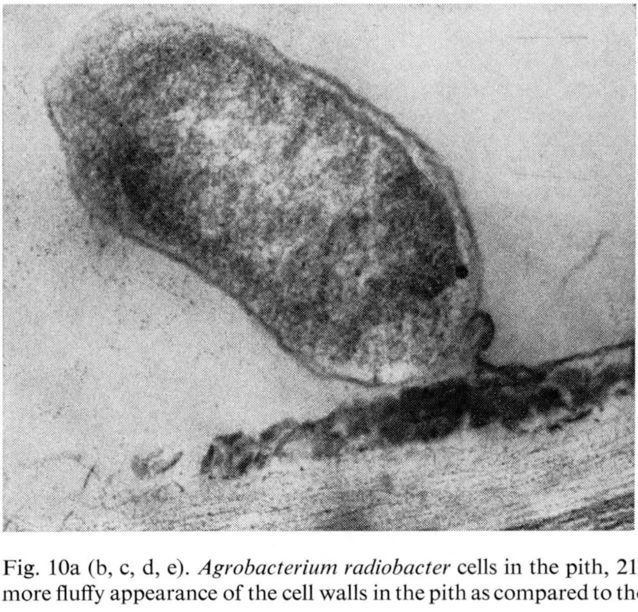

Fig. 10a (b, c, d, e). *Agrobacterium radiobacter* cells in the pith, 21 h after infection. Note the more fluffy appearance of the cell walls in the pith as compared to those in the cortex ($\times$ 69,000).

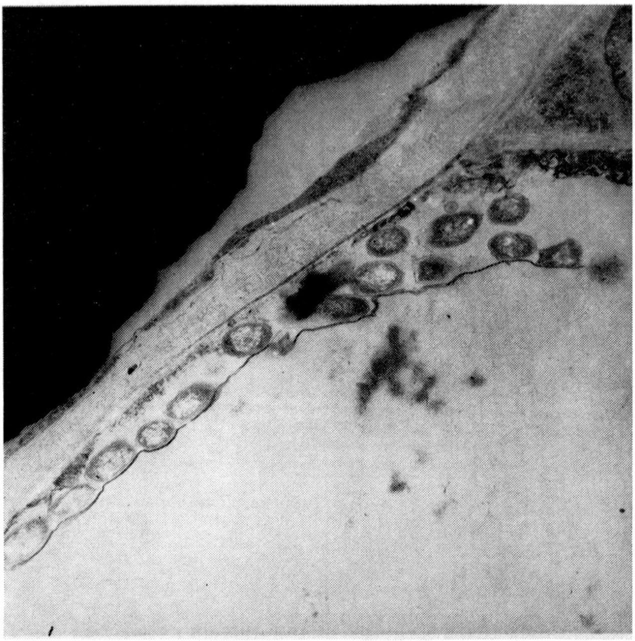
Fig. 10b. See explanation to Fig. 10a.

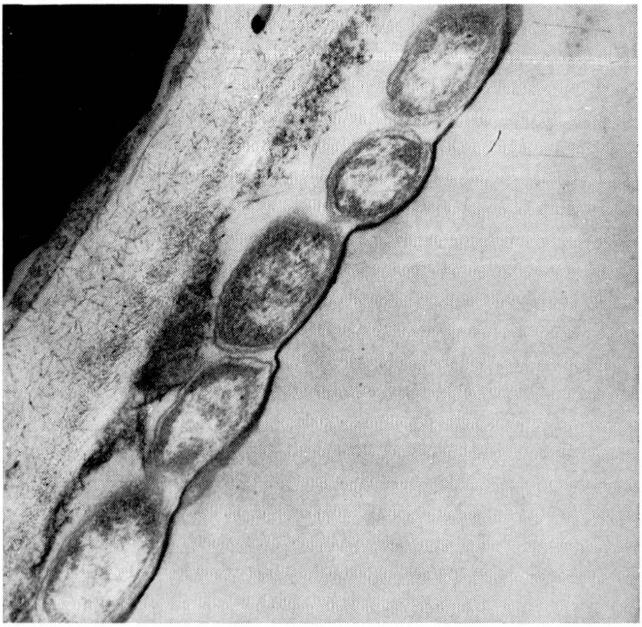
Fig. 10c. See explanation to Fig. 10a.

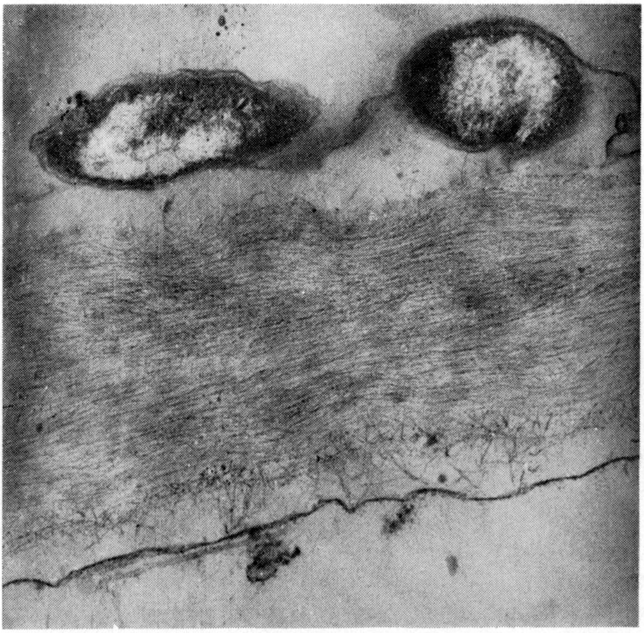

Fig. 10d. See explanation to Fig. 10a.

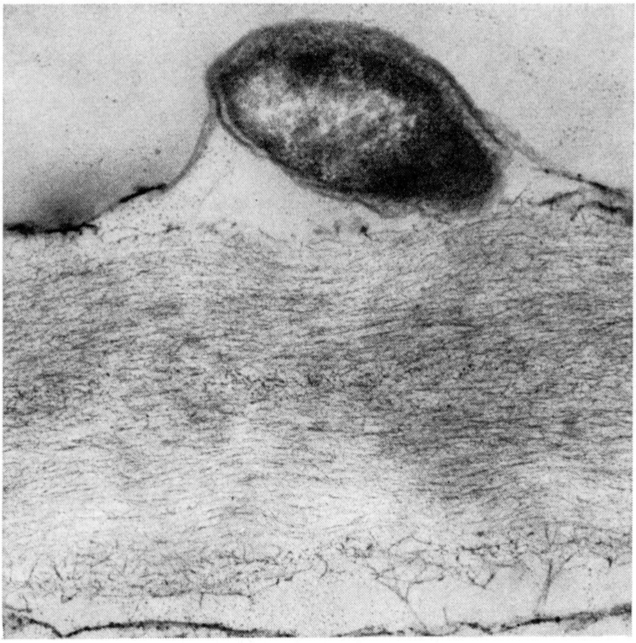

Fig. 10e. See explanation to Fig. 10a.

(Figs 10b, c, d, e). This membrane appeared to be closely associated with the bacterial cells (Fig. 10c). In cortex cells such enveloped bacteria were never observed.

The search for an interaction between *Kalanchoë cells* and *Escherichia coli* B or *Rhizobium leguminosarum* PRE, gave negative results: none of these bacteria could be detected in ultrathin sections of the plant cells. Probably these two organisms cannot attach themselves to the cell wall and are washed away during fixation. As already mentioned, *E. coli* B cannot inhibit the induction of tumors by *A. tumefaciens* (strain B_6). Thus the electron microscopic observations agree with the results of the experiments on the inhibition of tumor induction.

All this can be plausibly explained by supposing that there is a limited number of specific bacterial attachment sites in the conditioned wounds. The attachment of the tumorogenic strains to these sites may be an essential stage in tumor initiation. These sites can also be occupied by cells of the closely related *A. radiobacter*, but not by non-related *E. coli* or *Rhizobium* cells or by heated cells of *Agrobacterium*.

This latter fact indicates that either the viable bacteria are essential for the competition, or that some component on the bacterial surface, essential for bacterial attachment, is very heat-sensitive.

If bacterial attachment is considered the initial stage in tumor induction, the existence of an essential phase in the tumor induction period, called the bacterial adjustment time (Lipetz, 1966) would be explained. Under the circumstances necessary for this adjustment, the bacteria possibly undergo a change in their surface properties essential for the bacterial attachment to occur. If so, this explains why viable cells of *A. radiobacter* are required for inhibition of tumor induction.

References

Bopp, M. & E. Leppla, 1964. Vergleich der Histogenese der Wurzelhalsgallen an Blättern und Sprossachsen von *Kalanchoë daigremontianum*. Planta 61: 36.

Lipetz, J., 1966. Crown gall tumorigenesis. II. Relations between wound healing and the tumorigenic response. Cancer Res. 26: 1597.

Lippincott, B. B. & J. A. Lippincott, 1969. Bacterial attachment to a specific wound site as an essential stage in tumor initiation by *Agrobacterium tumefaciens*. J. Bact. 97: 620.

Millonig, G., 1961a. A modified procedure for lead staining of thin sections. J. appl. Phys. 32: 1637.

Millonig, G., 1961b. A modified procedure for lead staining of thin sections. J. biophys. biochem. Cytol. 11: 736.

Milo, G. E. & B. J. S. Srivastava, 1969. RNA-DNA hybridization studies with the crown gall bacteria and the tobacco tumor tissue. Biochem. biophys. Res. Commun. 34: 196.

Quétier, F., T. Huquet & E. Guillé, 1969. Induction of crown gall: partial homology between tumor-cell DNA, bacterial DNA and the G+C-rich DNA of stressed normal cells. Biochem. biophys. Res. Commun. 34: 128.

Schilperoort, R. A., H. Veldstra, S. O. Warnaar, G. Mulder & J. A. Cohen, 1967. Formation of complexes between DNA isolated from tobacco crown gall tumours and RNA complementary to *Agrobacterium tumefaciens* DNA. Biochim. biophys. Acta 145: 523.

Schilperoort, R. A., 1969. Investigations on plant tumors. Crown gall. On the biochemistry of tumor induction by *Agrobacterium tumefaciens*. Thesis Leiden.

Agrobacterium: intrageneric relationships and evolution

J. De Ley

Laboratorium voor Microbiologie en Microbiële Genetica van de Rijksuniversiteit te Gent, Belgium

Abstract

A preliminary account is given of a study on intrageneric relationships and evolution within the genus *Agrobacterium*. Measurements of DNA:DNA hybridization, thermal stability of DNA hybrids, electrophoretic protein profiles and phenotypic tests were carried out.

All authentic *A. tumefaciens* and *A. radiobacter* strains come under two large and a few small clusters. Phenotypically they are nearly indistinguishable, except for phytopathogenicity. Each cluster has a typical protein band profile. They hybridize for some 50 %. Most mutational changes took place in the cistrons outside the active center of the enzymes. Within each cluster, strains are at least 80 % homologous in their genome. All 3-ketolactose positive strains without exception belong here. All other strains belong mainly in a very dense cluster together with the named *A. rhizogenes* strains. Within this cluster, the strains are genotypically and phenotypically indistinguishable, except for their pathogenic root proliferation induction. Their genomes are for about 20 % similar with those of the other clusters.

Three very small clusters of 1 to 3 strains are being studied.

Introduction

This contribution is a brief report on an extensive study of the genus *Agrobacterium*, effected by the complete team of the laboratory at Gent. We attempt mainly to understand the relationship between the individual strains, and to elucidate the evolutionary development within this genus.

Material

We first collected as many strains as possible, some from culture collections, but most from individual investigators all over the world. In every case we requested all information available, in particular on pathogenicity, host plants and site of isolation.

Methods

We used four completely different techniques.

DNA:DNA hybridization The degree of DNA homology between many sets of two organisms was always determined under the most stringent conditions, to avoid

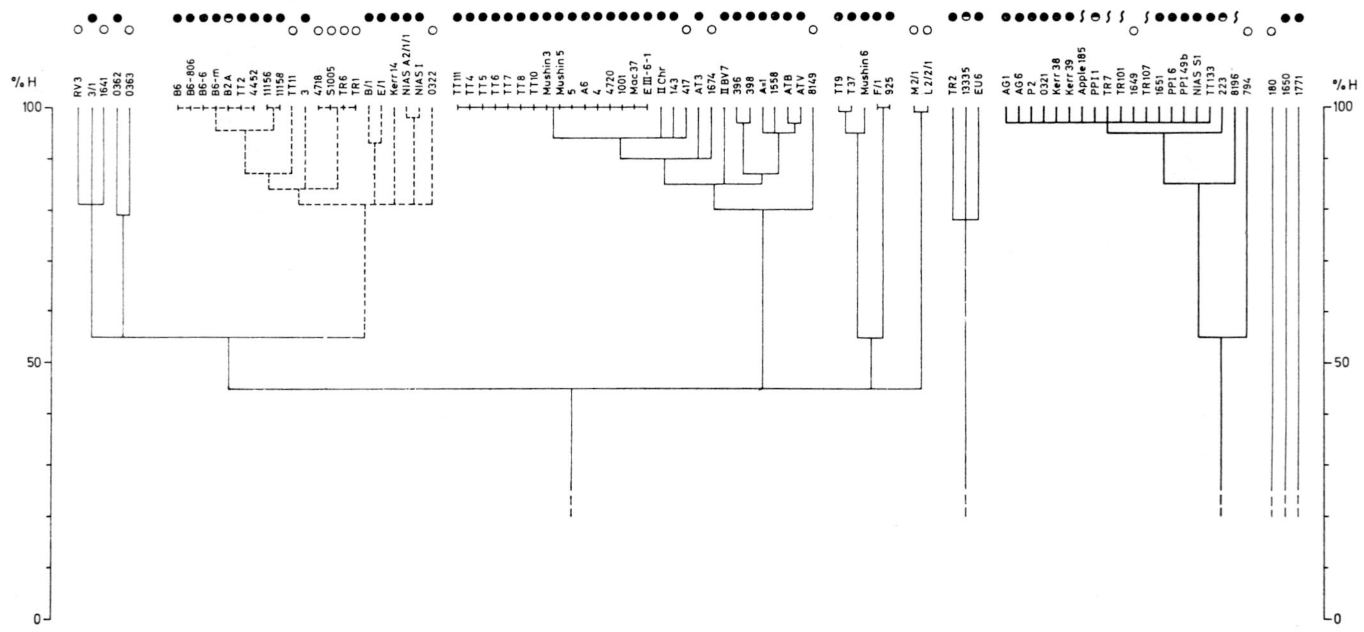

Fig. 1. DNA homology amongst the agrobacteria.
Homology was determined at optimally stringent conditions, with the renaturation rate method and in some cases also with the DNA-filter method in the presence of dimethylsulfoxide.

● crown gall former
∫ root former
◓ doubtfully pathogenic
○ not pathogenic

(De Ley, Reynaerts, Cattoir & Tijtgat, unpublished).

unspecific binding and artefacts, such as the release of DNA from membrane filters at high temperature. We used mainly the renaturation rate method; it allows DNA homology determinations between any set of unlabelled DNA (De Ley *et al.*, 1970). To a limited extent we used also the DNA-membrane filter method in the presence of dimethylsulphoxide (Legault-Démare *et al.*, 1967; De Ley & Tijtgat, 1970) with ^{14}C-labeled reference DNA's in the presence of homologous or heterologous competitive DNA. The results of both methods perfectly agreed.

Thermal stability of DNA:DNA hybrids In DNA:DNA hybrids the degree of matching of nucleotide sequences can vary strongly. With closely related organisms no or very little mutational changes occur and the hybrid is very stable. With increasing evolutionary distances, the number of mutational differences increases in the common part and the thermal stability of the hybrid decreases.

The ^{14}C-labeled hybrids are prepared as usual on membrane filters (see references above) and heated stepwise until complete denaturation. The released label is counted.

Phenotypic features We studied over a hundred features including morphology (colony type, cell size and shape, flagellation, Gram stain), plant pathogenicity (carrot slice, tomato), growth on various amino acids, growth and acid formation on various carbohydrates, sensitivity to various antibiotics, growth on small carbon compounds (succinate, oxalate, etc.) and on various aromatic compounds, gelatine liquefaction, 3-ketolactose test, litmus milk reaction, nitrate and nitrite consumption, absorption of dyes, etc.

Electrophoretic protein band profiles Polyacrylamide gel electrophoresis was carried out on soluble cytoplasmic protein of the strains. The stained gels were photographed; their densities were measured with a densitometer and the graphs compared by a computer method. Clustering was effected by a number of computer programs (unweighed pair group method, different types of factor analyses).

Results

DNA:DNA hybridizations and hybrid stability showed that most *Agrobacterium* strains fell into three large groups (Fig. 1). All authentic 3-ketolactose-positive crown gall producers and *A. radiobacter*, some 160 strains in total, fell under seven clusters. By far the largest cluster was called the TT111 group; it consists of some 90 strains. The second largest, called the B6 group, includes some 60 strains. The remaining 12-odd strains form five small clusters of 2 or 3 strains each. Within each of these groups the genomes are at least 80% homologous. Up to now the work mainly covers the B6 and TT111 groups. The DNA hybrids within each group are nearly as stable as the homologous reference duplex: we estimate that less than 2% mutational changes occurred within the common part. The seven groups hybridize at some 50% homology amongst themselves. Inter-group DNA hybrids are much less stable; we estimate that some 7% mutational changes occurred in the common part (Fig. 2).

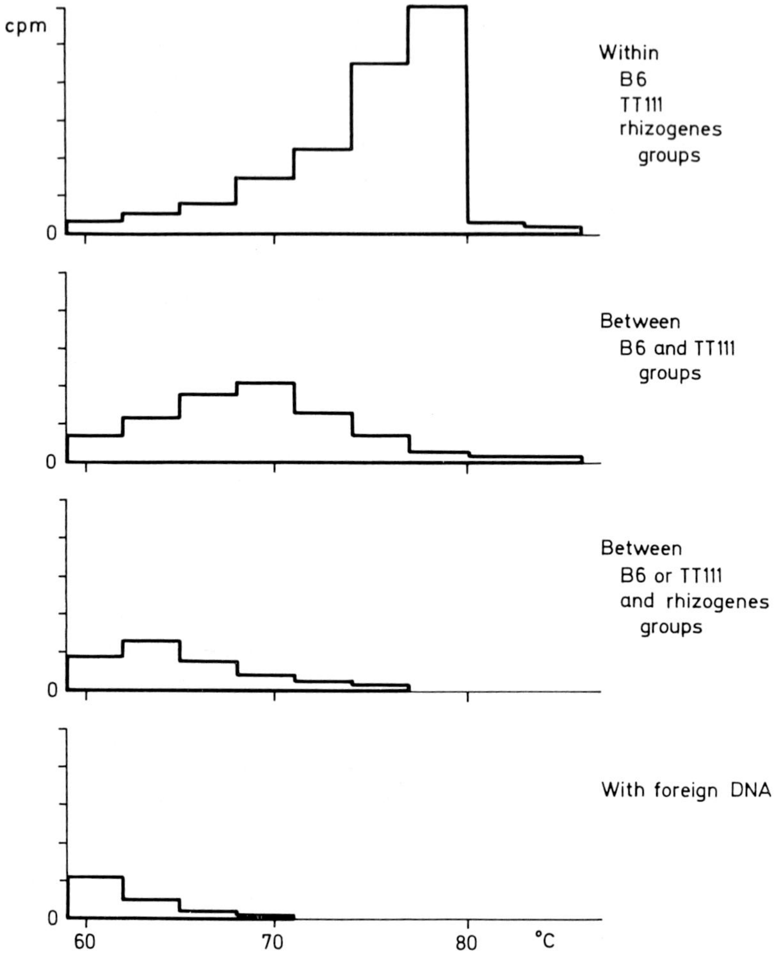

Fig. 2. Thermal stability of DNA:DNA hybrids. The origin of the separate strands is given at the right of each graph. ^{14}C-labelled hybrids were prepared on membrane filters and denatured at stepwise increasing temperatures (De Ley, Tijtgat & De Smedt, unpublished).

These groups contain nearly all named *A. radiobacter* strains. By DNA hybridization and hybrid stability determinations it was not possible to distinguish between *A. radiobacter* and *A. tumefaciens*.

The third large cluster of our collection contains all named *A. rhizogenes* strains, some atypical crown-gall producers, some unnamed *Agrobacterium* strains and some doubtfully pathogenic strains. There are 22 strains, all 3-ketolactose negative. Except for two border cases, their genomes are all for about 97% similar, with little or no mutational changes in the common part. They hybridize for some 20% with the

3-keto-lactose positive organisms and this small common part bears many mutational changes.

There are three very small clusters of one, one and three strains, respectively, which are genetically quite separate and need further study.

About 10% of our collection consists of misnomers, possible contaminants and other mishaps which have to be removed from the genus. They do not hybridize, they are not pathogenic, they are 3-ketolactose negative and phenotypically quite different (Gram positive, spore-formers, polarly flagellate, etc.).

Phenotypic features The fact that the 3-ketolactose positive strains belong to genetically different groups required further understanding. It was greatly surprising to find that they had over a hundred features in common. Usually individual clusters can be differentiated from others by one feature only. E.g. the B6 group grows anaerobically with nitrate as electron acceptor, whereas the TT111 group does not.

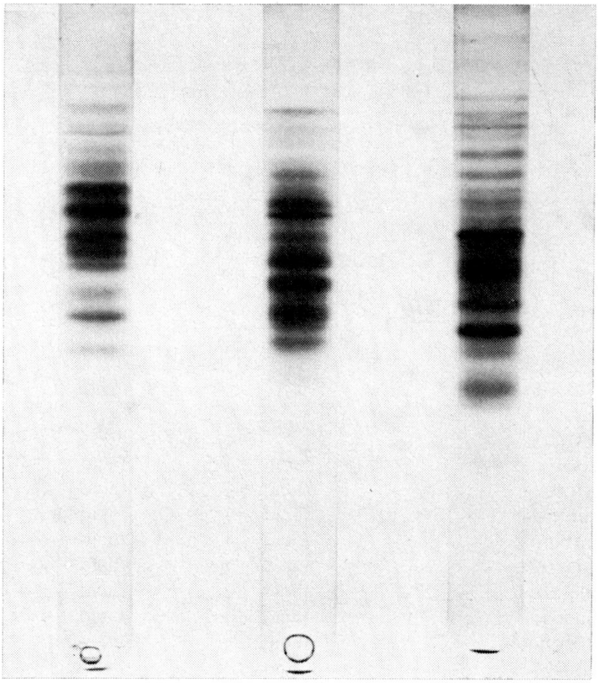

Fig. 3. Electropherograms en polyacrylamide gels of soluble proteins from a representative of each of the tree main *Agrobacterium* clusters. Right and middle: respectively B6 and TT111 cluster in the *A. tumefaciens* - *A. radiobacter* group. At the left: a representative pattern of the *A. rhizogenes* group (Kersters & De Ley, unpublished).

Phenotype of 70 *Agrobacterium* strains (Kersters & De Ley, unpublished data).

Test	B6 group	TT111 group	*A. rhizogenes* group
Common features			
Gram negative rods, peritrichous	+		+
Aerobic	+		+
% GC	60.7 ± 0.9	60.0 ± 0.9	60.8 ± 0.8
H_2S formation from cysteine	+		+
Growth on L-histidine, L-serine, L-alanine, L-arginine, L-proline, L-glutamic acid, L-ornithine	+		+
Growth on succinate	+		+
Sensitive to tetracyclin, novobiocin, neomycin, kanamycin, gentamycin, terramycin aureomycin	+		+
Reduction of nitrite	+		+
Growth on inositol	+		+
maltose	+ (no acid)		+ (no acid)
Liquefaction of gelatin	−		−
Growth in Koser's citrate	−		−
Growth in Simmons' citrate	−		−
Growth on L-valine, L-methionine, L-tryptophane, L-phenylalanine	−		−
Growth on oxalic acid, formate, glyoxylate, benzoate	−		−
Gas formation from nitrate	−		−
Differentiating features			
3-ketolactose production	+		−
Litmus milk reaction	brown, peptonized		pink, acid curd
Growth on L-tyrosine	−		+
Growth on lactate	+		−
Growth on Simmons' citrate + yeast extract	−		+
Brown surface pellicle on ferric ammonium citrate	+		−
Growth on minimal medium of Clark	+		−
Growth on carbohydrates: sorbitol	+++ (acid +)		+ (acid +++)
D-xylose	+++ (acid +)		+ (acid +++)
L-arabinose	+++ (acid +)		+ (acid +++)
D-arabinose	+++ (acid +)		+ (acid +++)
D-galactose	+++ (acid +)		+ (acid +++)
D-mannose	+++ (acid +)		+ (acid +++)
D-glucose	+++ (acid +)		+ (acid +++)
D-gluconate	+++ (acid +)		+, no acid
dulcitol	+++, acid		+, no acid
lactose	++, no acid		+, acid
sucrose	+++, acid		+, no acid
cellobiose	+++, acid		+, no acid
Anaerobic growth on nutrient medium supplemented with 0.5% KNO_3	+	−	−

Phenotype of 70 *Agrobacterium* strains (continued).

Test	B6 group	TT111 group	*A. rhizogenes* group
Nitrate reduction: nitrate produced	+	+	+
complete utilization of nitrate	+	−	−
Variable features			
Growth on L-leucine, L-lysine, L-threonine	+ and −		+ and −
Growth on D-glycerate	+ and −		+
Growth on acetate	+		+ and −
Growth on mandelic acid	+ and −		+ and −
Growth on asparagine	+		+ and −
Oxidase	+		+ and −
Adsorption of aniline blue	+ and −		+ and −
Sensitive to streptomycin, penicillin	−		+ and −
erythromycin	+ and −		+ and −
colimycin	+ and −		+
Growth on glycerol	+ and −		+ and −
Growth on mannitol	+ and −		+
Growth on ethanol	+ and −		−
Phytopathogenicity: crown gall	+ and −		+ and −
root formation	−		+ and −

The only known difference between the named tumefaciens and radiobacter strains resides in their pathogenicity. A taxon separation on this sole criterium may be justified for the practical phytopathologist, but a species separation is biologically not justified. Some 20 features allow a clear separation between the rhizogenes and the tumefaciens-radiobacter groups. A brief survey of these features is given in the table. The phenotypic features of the small groups are being studied.

Electrophoretic protein profiles At this stage one is still faced with the paradox that the organisms of the B6 and the TT111 groups are phenotypically almost identical but genotypically different. This difference had to be detected somewhere, and our hypothesis that it expressed itself, if not in the activity of the enzymes, in their chemical structure, proved to be correct.

All strains of the B6 group display a certain protein band pattern clearly different from that of the TT111 group. The *A. rhizogenes* group is again different (Fig. 3). The small groups also have their typical protein band pattern. The agreement between the electrophoretic data and DNA hybridization seems perfect. Computer methods are being developed as an aid in quick identification of these organisms by their protein profiles. Fig. 4 gives an example of a computerized clustering.

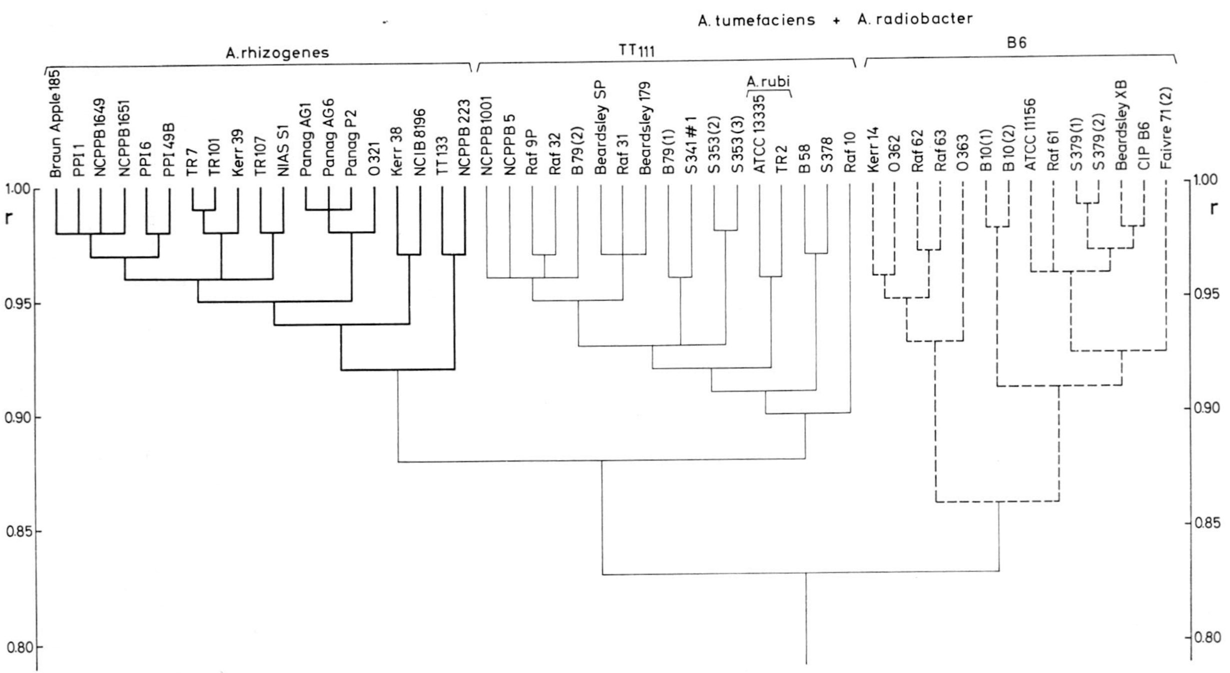

Fig. 4. Dendrogram of correlation between a number of electropherograms, calculated by a computer method for average linkage (De Ley & Kersters, unpublished).

Discussion

We can now present a hypothesis on the evolutionary development within the genus *Agrobacterium*. Let us consider a certain population of ancestral agrobacteria. Through external influences a subpopulation was separated and exposed to different conditions (change in temperature, moisture, host plants, etc.). By mutation and selection, the fittest under the new condition survived. This may have happened several times, and we suppose this to be the origin of the phenotypically different clusters (the 3-ketolactose positive cluster, the *A. rhizogenes* cluster and a few of the small clusters). By mutation, the DNA had been changed drastically between clusters (low hybridization), with many triplets changed (different protein band patterns), severely affecting a number of cistrons (some enzymes different).

Except for the aberrant strains 8196 and 794, the rhizogenes cluster is very homogeneous in DNA hybridization, protein band profiles and phenotypic features. It is obvious that nearly no mutational changes occurred further within this group, in spite of the fact that its members have a world-wide distribution. The cause of this stability is not known, but one can easily imagine a number of explanations. The seven clusters of the tumefaciens-radiobacter group are phenotypically nearly indistinguishable, but their protein profiles are different. Thus mutations happened in the first and second base of the triplets, except around the active center of the enzymes.

For some unknown reason the ancestors of both the B6 and the TT111 group gave rise to large populations. The B6 group appears to occur mainly around the Pacific Ocean (USA, Japan, Australia), the TT111 group around the Atlantic Ocean (USA, Canada, Europe, S. Africa). The 80–100% DNA homology within each group represents the final step in the evolution of a cluster (genetic race) and may be the preparation for the next separation. We think that these organisms are now in equilibrium with their surroundings. Phenotypic changes would be less viable and selected out. The only permissible mutational changes are those at the third base of the triplets. They involve nearly no change in amino acids. Although they hardly affect the phenotype, they decrease DNA homology. These seven clusters deserve to be given one species name, e.g. *A. tumefaciens*. *A. radiobacter* appears to be a very small variation. It would suffice that some time in the future a strain from any one of these clusters would mutate at the active center of a few enzymes, that phenotypically a new species would arise.

References

De Ley, J., H. Cattoir & A. Reynaerts, 1970. The quantitative measurement of DNA hybridization from renaturation rates. Eur. J. Biochem. 12: 133–142.

De Ley, J. & R. Tijtgat, 1970. Evaluation of membrane filter methods for DNA-DNA hybridization. Antonie van Leeuwenhoek 36: 461–474.

Legault-Démare, J., B. Desseaux, T. Heyman, S. Séror & G. Ress, 1967. Studies on hybrid molecules of nucleic acids. Biochem. biophys. Res. Commun. 28: 550–557.

Current taxonomic thinking on the genus Pseudomonas, with emphasis on the plant pathogens

M. N. Schroth and D. C. Hildebrand

Department of Plant Pathology, University of California, Berkeley, California 94720, U.S.A.

Abstract

The genus *Pseudomonas* is divided into two principal groups on the basis of requirements for growth factors. The no-growth factor group contains all of the plant pathogens. The non-poly-beta-hydroxybutyrate accumulation group contains all fluorescent plant pathogens. The arginine dihydrolase negative plant pathogens have been placed into two species. *P. syringae* includes all oxidase negative strains, and *P. cichorii* all oxidase positive strains. In the poly-beta-hydroxybutyrate accumulation group, *P. marginata, P. caryophylli*, and *P. cepacia* are included in the DL arginine utilization section. *P. solanacearum* is the only plant pathogen in the arginine negative section.

Introduction

The titles of this symposium and of this paper are essentially misnomers. It is apparent, when surveying the literature on taxonomy, that there is seldom enough agreement to constitute any real 'current taxonomic thinking'. The diversity in thinking approaches that of the bacteria whose resistance to subjective sorting and cataloguing into definitive taxa is renown. That 'current taxonomic thinking' should always be received with skepticism and considered fuel for experimental challenge is exemplified by some of our taxonomic offerings of the last 10 years. For example, it has been stated that the fluorescent phytopathogenic pseudomonads do not differ substantially from the fluorescent pseudomonads commonly found in water and soil that all should be considered strains of *P. fluorescens*, and that all pseudomonads are oxidase positive. With recent information, the fallacy of these statements is apparent.

The 'current taxonomic thinking' concerning *Pseudomonas* taxonomy is dominated by the school and work of Stanier, Doudoroff, and Palleroni. This is reflected in the proposed 8th edition of Bergey's Manual of Determinative Bacteriology as revised by M. Doudoroff and M. J. Palleroni. The revision contains a key and descriptions of 29 species compared to the 149 listed in the 7th edition. Only those species of which the phenotype has been extensively characterized are included in the key, thereby excluding about 90% of the described species. However, appendices include other nomenspecies which have been assigned to the genus, and certain nomenspecies originally assigned to other genera, but which appear to conform to the present definition of *Pseudomonas*.

General discription of the genus

Pseudomonads are gram negative, single, straight, or curved rods. Motility is by polar flagella, monotrichous or multitrichous. Metabolism is respiratory. Some are facultative chemolithotrophs using H_2 or CO as an energy source. They are strict aerobes except for species which use denitrification as a means of anaerobic respiration. Molecular oxygen is the universal electron acceptor. Some denitrify using nitrate as an alternative acceptor.

The genus is divided into two principal groups on the basis of requirements for growth factors. A key adapted from the various reports characterizing the pseudomonads (Ballard *et al.*, 1970; Billing, 1970a; Gilardi, 1971; Lelliott *et al.*, 1966; Palleroni *et al.*, 1970; Redfearn *et al.*, 1966; Sands *et al.*, 1970; Stanier *et al.*, 1966) is presented at the end of this paper. A key, however, should be considered only as a guide as there are usually exceptions to any one character. The largest group, 'no growth factors required', contains all plant pathogens. This group is further divided by their capacity to accumulate poly-beta-hydroxybutyrate. The non-poly-beta-hydroxybutyrate accumulation section contains all fluorescent plant pathogens, and the saprophytes, *P. aeruginosa*, *P. putida*, *P. fluorescens* (four biotypes), *P. chlororaphis*, *P. aureofaciens*, *P. stutzeri*, *P. mendocina*, and *P. alcaligenes*.

The arginine dihydrolase test separates most fluorescent plant pathogens from the saprophytes with the exception of *P. stutzeri*. In the arginine dihydrolase negative group, *P. syringae* includes all ill-defined oxidase negative fluorescent nomenspecies, such as *P. phaseolicola*, *P. tomato*, and *P. savastanoi*. *P. cichorii* contains all fluorescent oxidase positive strains. *P. stutzeri*, the third species, is also oxidase positive but nonfluorescent.

The arginine dihydrolase positive group includes both plant pathogens and saprophytes and all are oxidase positive. Of these, the plant pathogens *P. marginalis* is included in *P. fluorescens* biotype II. *P. polycolor* is placed in *P. aeruginosa*. *P. fluorescens* biotype II, in our opinion, has not been adequately characterized. On the basis of some of our tests with *P. marginalis*, which is now included in this biotype, it appears that the biotype may represent an assemblage of diverse strains. Further study may result in the biotype being divided into several groups with some given species status.

The poly-beta-hydroxybutyrate accumulation group consists entirely of nonfluorescent organisms and is divided into two sections on the basis of DL-arginine utilization. The arginine positive section includes the plant pathogens *P. marginata*, *P. caryophylli*, and *P. cepacia*. *P. alliicola* is included in *P. marginata*. *P. solanacearum* is the only plant pathogen in the arginine negative section.

The classification system

The present classification system is basically a monothetic system, since taxa possess a single character or a series of unique characters that are sufficient and necessary for inclusion in the defined group. However, most taxa containing plant pathogens were predicated on the basis of a detailed nutritional characterization of strains (Ballard *et al.*, 1970; Lelliot *et al.*, 1966; Misaghi & Grogan, 1970; Sands *et al.*, 1970; Stanier *et al.*, 1966). The number of characters used, over 140, are considerably more than have been used in many past numerical taxonomic studies, and in many respects represent a more random selection of characters. Admittedly, there still is bias because of the emphasis on nutrition. Although one may debate the advantages of a polythetic opposed to a monothetic system of classification, the differences are obscure with *Pseudomonas* taxonomy; the development of a useful monothetic system over a period of time will be based principally on overall similarities between organisms (Sneath, 1962). During repeated revisions the groupings become acceptably phenetic.

It seems fruitless to engage in polemics over the construction of the *Pseudomonas* taxa, since ranking is an arbitrary exercise. Regardless of appealing pontifications used to rationalize taxa, in the final analysis, they are what a competent worker decides to circumscribe. So what's next?

A change in the presently proposed scheme will probably involve the ranking of the fluorescent plant pathogenic pseudomonads. Regardless of whether or not most nomenspecies should be designated as *P. syringae* because of poor resolution, it is an untenable and totally non-utilitarian scheme for a researcher who is interested in the distinctive and diverse pathogenic capabilities of these pathogens. Therefore, it is essential that a decision be made to adopt a system of infra-specific subdivisions such as designating present nomenspecies as pathotypes and races for those with known pathogenic capabilities. The alternative is to adhere to *status quo* and ignore the proposed new taxonomic revision on *Pseudomonas*. Accordingly, we recommend that this international organization immediately consider the appointing of a representative group of taxonomists to devise and recommend a satisfactory nomenclatural system for bacterial plant pathogens in view of the impending 8th edition of Bergey's Manual.

Although all of the arginine dihydrolase-negative oxidase-negative pseudomonads are being lumped into *P. syringae*, it is likely that taxonomists will divide it into separate species as soon as additional data are available. There are indications that considerable physiological differences exist among these pseudomonads. For example, nutritional tests with *P. phaseolicola* have indicated that distinctive characters are associated with specific host-pathogen combinations; the more dissimilar the host range of *P. phaseolicola* strains, the greater their variability to utilize substrates (Schroth *et al.*, 1971). Further suggestion of distinct characters among plant pathogens is revealed by the nature of the symptoms of diseased plants. A plant might be regarded as a complex diagnostic medium, and the series of specific observable events which occur after inoculation as a reflection of the physiological properties of the pathogen. In this sense, the plant is a superior medium for indirectly revealing phenotypic characters that escape our present battery of determinative tests. Many of these tests have been primarily developed for the differentiation of saprophytes and other genera. These unique parasites have not received the attention, with resultant production of data, necessary for making lasting taxonomic decisions. Accordingly, previous numerical taxonomic studies using the available data at that time (Lysenko, 1961; Moffett, 1966; Rhodes, 1961) even failed to clearly differentiate them as a distinct group of organisms separable from fluorescent saprophytes.

Further study should attempt to determine and include tests that reflect the special capabilities of these pathogens. Emphasis, for example, might be focussed on enzyme systems which enable them to degrade and metabolize plant constituents. Attention also should be given to their capacity to produce toxins and other metabolites which alter cellular metabolism of the host. Taxonomic studies on plant pathogens have not included such characters as production of beta-glucosidase, pectic enzymes, cellulases, indoleacetic acid, syringomycin, tabtoxinine, and other substances which would be of considerable value in differentiation of nomenspecies. Furthermore, it has been indicated (Billing, 1970a, b; Hildebrand & Schroth, 1971) that a number of diagnostic tests of determinative value are now available differentiating nomenspecies.

DNA hybridization experiments with the arginine dihydrolase negative fluorescent plant pathogens, such as *P. syringae*, *P. phaseolicola*, *P. tomato*, and *P. cichorii* have revealed varying degrees of homology ranging from 39 to 69% at 69 °C and from 12 to 52% at 80 °C (Ballard, 1969). The significance of the amount of genetic non-homology among the pathotypes to possible future circumscriptions of species is conjectural; it is, however, a further indication that there are probably many undetected characters of differential value.

Studies on genetic homologies among bacteria are of considerable significance to taxonomy and for indicating relationships. For example, DNA-DNA homology studies (Ballard *et al.*, 1970) have confirmed data on nutritional tests (Sands *et al.*, 1970) indicating that the plant pathogen *P. cepacia* is similar to *P. multivorans*, thus resulting in the latter becoming a synonym. Similarly, *P. alliicola* appears to be a synonym of *P. marginata*. DNA-DNA homology studies also indicate that *P. cepacia*, *P. marginata*, and *P. caryophylli* have a genetic relationship with the animal pathogens *P. pseudomallei* and *P. mallei*, thus confirming the previously shown nutritional relationships (Ballard *et al.*, 1970). This has interesting evolutionary implications for animal and plant pathogens: it is noted that many plant pathogens are related to animal pathogens. Obvious examples are the genus *Corynebacterium* which is composed principally of animal and plant pathogens; *Erwinia*, which includes many plant pathogens and the recently reported animal pathogen *E. herbicola* (Gilardi *et al.*, 1970); and the genus *Streptomyces*, which includes several animal pathogens and the plant pathogens *S. scabies* and *S. ipomoeae*. In retrospect, the recent finding that mycoplasmas infect plants as well as animals should have been intuitive on the basis of previous relationships among animal and plant pathogens.

The relationship between DNA homology and phenotypic characterization studies is not clear. For example, how many phenotypic characters would differ using 200–300 randomly selected characters among two strains that showed a 20% difference in DNA homology? Furthermore, the precise significance of DNA homology studies is conjectural at the present since biophysic-technology is improving, and this effects the resultant degree of homology. De Ley (1968), for example, reports that *P. rimafaciens*, a synonym for *P. syringae*, showed 64% homology to *P. fluorescens*-DNA. However, using different techniques, Ballard (1969) has shown the homology of *P. syringae*-DNA with that of several strains of *P. fluorescens* to range only from 15 to 29% at 69 °C. De Ley (1968) also reports DNA homology between *Xanthomonas* and *Pseudo-*

monas spp. at about 50 to 80%. Further work, however, may show a lower degree of homology since Johnson & Ordal (1968) have criticized techniques of DNA homology. They have shown that competition between labelled and unlabelled DNA fragments for binding sites on immobilized DNA is greatly influenced by the incubation temperature, and they suggest that the DNA duplexes formed with *Pseudomonas* and *Xanthomonas* strains at 60 °C are probably, in some cases, non-specific.

From a plant pathologist's viewpoint, an improvement in taxonomy is also consonant with the attaining of a greater understanding of the ecological and pathological properties of plant pathogens. This is especially evident since data on the nutritional and physiological characters required for taxonomic ordering of bacteria are also requisite to understanding the parasite relations between a pathogen and the host, and the process of disease development. Hopefully, the next symposium on 'The Current Concepts of *Pseudomonas* Taxonomy' will provide more of these ellusive data.

Key for Pseudomonas species

I. Growth factors not required
- A. Poly-β-hydroxybutyrate not accumulated
  - 1. Arginine dihydrolase negative
    - a. Oxidase negative . . . . *P. syringae*
    - b. Oxidase positive
      - (1) Fluorescent pigment . . . . *P. cichorii*
      - (2) No fluorescent pigment . . . . *P. stutzeri*
  - 2. Arginine dihydrolase positive
    - a. Growth at 41 °C
      - (1) Growth on 2-ketogluconate . . . . *P. aeruginosa*
      - (2) No growth on 2-ketogluconate
        - (a) Growth on geraniol and L-valine . . . . *P. mendocina*
        - (b) No growth on geraniol and L-valine . . . *P. alcaligenes*
    - b. No growth at 41 °C
      - (1) Hydrolyses gelatin
        - (a) Contains chlororaphin . . . . *P. chlororaphis*
        - (b) Contains phenazine-1-carboxylate . . . . *P. aureofaciens*
        - (c) Not as above . . . . *P. fluorescens*
      - (2) Does not hydrolyse gelatin . . . . *P. putida*
- B. Poly-β-hydroxybutyrate is accumulated
  - 1. DL-arginine used as sole carbon source
    - a. Arginine dihydrolase positive
      - (1) No flagella . . . . *P. mallei*
      - (2) Flagella > 1
        - (a) Utilizes L-rhamnose and saccharate . . . *P. caryophylii*
        - (b) Does not utilize L-rhamnose and saccharate
          - (i) Hydrolyses gelatin . . . . *P. pseudomallei*
          - (ii) Does not hydrolyse gelatin . . . . . *P. pseudoalcaligenes*

b. Arginine dihydrolase negative
  (1) Utilizes tryptamine and meta-hydroxybenzoate . . . *P. cepacia*
  (2) Does not utilize tryptamine and meta-hydroxybenzoate . . . . *P. marginata*

2. DL-arginine not used as sole carbon source
  a. Autotrophic growth with H_2
    (1) Has carotenoid pigments
      (a) Utilizes fructose . . . . *P. flava*
      (b) Does not utilize fructose . . . . *P. palleroni*
    (2) No carotenoid pigments
      (a) Hydrolyses starch . . . . *P. saccharophila*
      (b) Hydrolyses gelatin . . . . *P. facilis*
      (c) Not as above . . . . *P. ruhlandii*
  b. No autotrophic growth with H_2
    (1) Hydrolyses poly-β-hydroxybutyrate
      (a) Utilizes glucose . . . . *P. delafieldii*
      (b) Does not utilize glucose . . . . *P. lemoignei*

References

Ballard, R. W., 1969. The taxonomy of some phytopathogenic *Pseudomonas* species. University of California Thesis.

Ballard, R. J., N. J. Palleroni, M. Doudoroff & R. Y. Stanier, 1970. Taxonomy of the aerobic pseudomonads: *Pseudomonas cepacia*, *P. marginata*, *P. alliicola* and *P. caryophylli*. J. gen. Microbiol. 60: 199–214.

Billing, E., 1970a. Further studies on the phage sensitivity and the determination of phytopathogenic *Pseudomonas* spp. J. appl. Bact. 33: 478–491.

Billing, E., 1970b. *Pseudomonas viridiflava* (Burkholder, 1930; Clara, 1934). J. appl. Bact. 33: 492–500.

De Ley, J., 1968. DNA base composition and hybridization in the taxonomy of phytopathogenic bacteria. A. Rev. Phytopath. 6: 63–90.

Gilardi, G. L., 1971. Characterization of *Pseudomonas* species isolated from clinical specimens. Appl. Microbiol. 21: 414–419.

Gilardi, G. L., E. Bottone & M. Bernbaum, 1970. Unusual fermentative, gram-negative bacilli isolated from clinical specimens. I. Characterization of *Erwinia* strains of the 'lathyriherbicola group'. Appl. Microbiol. 20: 151–155.

Hildebrand, D. C. & M. N. Schroth, 1971. Identification of the fluorescent pseudomonads. Proc. 3rd int. Conf. Plant Pathogenic Bacteria, Wageningen.

Johnson, J. L. & E. J. Ordal, 1968. Deoxyribonucleic acid homology in bacterial taxonomy: effects of incubation temperature on reaction specificity. J. Bact. 95: 893–900.

Lelliott, R. A., E. Billing & A. C. Hayward, 1966. A determinative scheme for the fluorescent plant pathogenic pseudomonads. J. appl. Bact. 29: 470–489.

Lysenko, O., 1961. *Pseudomonas* – an attempt at a general classification. J. gen. Microbiol. 25: 379–408.

Mishagi, I. & R. G. Grogan, 1969. Nutritional and biochemical comparisons of plant pathogenic and saprophytic fluorescent pseudomonads. Phytopathology 10: 1436–1450.

Moffett, M. L., 1966. The new bacterial leaf spot of *Antirrhinum* seedlings caused by a subspecies of *Pseudomonas fluorescens*. Migula, 1895. Qd. J. Agric. Anim. Sci. 23: 121–132.

Palleroni, N. J., M. Doudoroff, R. Y. Stanier, R. Y. Solánes & M. Mandel, 1970. Taxonomy of the aerobic pseudomonads: the properties of the *Pseudomonas stutzeri* group. J. gen. Microbiol. 60: 215–231.

Redfearn, M. S., N. J. Palleroni & R. Y. Stanier, 1966. A comparative study of *Pseudomonas pseudomallei* and *Bacillus mallei*. J. gen. Microbiol. 43: 293–313.

Rhodes, M. E., 1961. The characterization of *Pseudomonas fluorescens* with the aid of an electronic computer. J. gen. Microbiol. 25: 331–345.

Sands, D. C., M. N. Schroth & D. C. Hildebrand, 1970. Taxonomy of phytopathogenic pseudomonads. J. Bact. 101: 9–23.

Schroth, M. N., V. B. Vitanza & D. C. Hildebrand, 1971. Pathogenic and nutritional variation in the halo blight group of fluorescent pseudomonads of bean. Phytopathology (in Press).

Sneath, P. H. A., 1962. The construction of taxonomic groups. *In* Twelfth Symposium of the Society for General Microbiology. Microbial Classification, Vol. 12: 289–332.

Stanier, R. Y., N. J. Palleroni & M. Doudoroff, 1966. The aerobic pseudomonads: a taxonomic study. J. gen. Microbiol. 43: 159–271.

The genus Xanthomonas

R. A. Lelliot

Plant Pathology Laboratory, Hatching Green, Harpenden, Herts., England

Abstract

It is suggested that a classification recognising five species, one with many pathotypes, would be more useful and less misleading than that which at present obtains, and which recognises some 120 species.

Had the 8th edition of Bergey's Manual of Determinative Bateriology been published before this Conference, we would, at this session, have discussed in detail how it treated the more important genera of plant pathogens. But because it is still being prepared, its content cannot be discussed in more than a general way. This situation gives me an opportunity of saying why I support the classification D. W. Dye and I will use in the new edition, and of discussing some of the consequences I think would result from its adoption. In what follows I shall be using the word 'nomenspecies' in the neutral sense of any species whatever the method used to construct it rather than in its usual sense of species delineated by the nomenclatural type; and 'taxospecies' in the sense of species obtained by grouping strains (individuals) that have a high degree of overall similarity, i.e. species obtained phenetically.

There are some one hundred and twenty nomenspecies in the genus *Xanthomonas*, all of which are said to be pathogenic to different plants, or different combinations of plants, and all but four of which are what have often been called 'typical xanthomonads': that is they grow as yellow pigmented, domed, mucoid colonies, 1–3 mm in diameter after 3 days at 27 °C, on nutrient agar containing dextrose.

Dye (1962, 1963, 1966) studied a large number of nomenspecies and isolates of this 'typical' group, but none of the many characters he investigated would differentiate them, nor have preliminary attempts to discern phenetic groups among a limited number of them revealed clusters of isolates at less than high levels of similarity (Colwell & Liston, 1961; Colwell & Mandel, 1964). De Ley and his colleagues (De Ley & Friedman, 1965; De Ley *et al.*, 1966) have shown a high degree of DNA hybridization between twenty-eight of these nomenspecies. Analyzing Dye's data phenetically we have been able to obtain clusters only at very high similarity values and these do not always contain isolates of the same nomenspecies. This evidence supports the conclusion that only by considering their hosts can nomenspecies in this 'typical' group be differentiated from one another. Four other nomenspecies can be readily distinguished by laboratory tests from the 'typical' group and from each other. Dye and I, therefore, will recognise only five taxospecies in the 8th edition of the Manual: the type, *Xanthomonas campestris*, and four other species, *X. albilineans*, *X. axonopodis*, *X. ampelina*, and *X. fragariae*. The remaining nomenspecies, except those that for various reasons we excluded from the genus, will be listed with their hosts as distinguishable from *Xanthomonas campestris* only by plant host reactions.

Although there seems little doubt that the division of the genus into these five taxospecies best expresses the data available, it fails to accommodate pathologists, whose concern is almost wholly with one aspect of bacteria: their ability to cause disease. To supply such particular needs as those of ecologists, geneticists, or pathologists the concept has been developed during recent years of general-purpose and special-purpose classifications. These are sometimes less precisely called natural and artificial. The distinction between these two types of classification, their purposes, and their uses, are discussed with clarity and precision by Heslop-Harrison (1962). Because there is not time to develop fully the arguments for using dual classifications, I would ask you to accept as axiomatic these propositions, most of which are taken almost verbatim from Heslop-Harrison's paper:

1. one of the primary functions of biological systematics is the production of a general classification of living things possessing at once the greatest possible content of information and the maximum convenience in use;
2. a general-purpose classification is not necessarily the most efficient for *all* purposes, and it is not always desirable or indeed feasible to modify it to meet specialized demands;
3. a general-purpose classification cannot be expected to provide the sanction for predicting unique properties of all members of any class; but that for any member correctly referred to a class in the classification, there will be a large number of statements that will probably be true of it, and more probably true of it than if it were a member of another class of equivalent rank;
4. general and special-purpose classifications are not mutually exclusive. They are often, and indeed are usually best, used together. The classification of cultivated plants in economic botany is a well known example. More pertinent is that of the genus *Shigella* where a general-purpose classification of some eight species has been insufficient for epidemiologists who use a special-purpose classification of serotypes, lysotypes and biotypes, grafted to the general-purpose classification.

The classification currently used for the genus *Xanthomonas* is manifestly inadequate and has some absurd consequences: the practical impossibility in most cases of determining the species to which a culture belongs if its host is not known, and the loss of specific identity when an isolate is no longer virulent are examples. But more importantly, if less obviously, it is misleading. The uninitiated assume, naturally enough, that species differ from one another not only in host range but in other, and easily determinable, characters. The descriptions of many species in the genus apparently display an almost frenzied anxiety to find characters, however trivial, to support the claim that what is being described is a new species. On further examination these trivial differences usually disappear into the background of variation within and between the different nomenspecies. Again how many of us have tried to use the differential characters claimed to separate these species, only to have our time wasted and confidence lessened chasing what we later find to be unsubstantial? A more subtle trap awaits those who make general statements about the genus from studies that they think are representative of it. If they accept, as they should be able to, the implication that the 120 or so nomenspecies are of approximately equal taxonomic rank, it is almost inevitable that those they study will not include any outside the *Xanthomonas campestris* group; then, any generalizations they make will be applicable to that group only and not necessarily to the other four taxospecies in the genus. One imagines that Starr & Stephens (1964), who expressed the '... belief that a unique carotenoid occurs universally in the yellow members of the genus *Xanthomonas*' after studying single isolates of nineteen species of the *Xanthomonas campestris* group, were misled in this way. Similarly the valuable study by De Ley *et al.* (1966), in which twenty-eight nomenspecies were shown to possess such a high degree of DNA homology that they could be considered a single genospecies, would have been more valuable if those studied had included species outside the *X. campestris* group.

Fundamentally the classification we use is defective as a method of storing information about the genus. It does not express what we now know: that there are five clusters of strains which differ markedly from one another and which can be easily differentiated, and that one of these clusters, the species *X. campestris*, contains a large number of similar but host-differentiated pathogens. In essence, it is an unsuccessful mixture of a general-purpose and a special-purpose classification. It seems to me that only by separating clearly these two classifications can the systematics of the genus be made both meaningful and useful. The five species would form the general-purpose classification and a division of *Xanthomonas campestris* into pathotypes would constitute the special-purpose classification. Such a pathotype classification would not be subject to the rules of the International Code of Nomenclature of Bacteria and could therefore be made to reflect new information on pathogenic properties more easily than can the classification currently used.

References

Colwell, R. R. & J. Liston, 1961. Taxonomic analysis with the electronic computer of some *Xanthomonas* and *Pseudomonas* species. J. Bact. 82: 913–919.

Colwell, R. R. & M. Mandel, 1964. Adansonian analysis and deoxyribonucleic acid base composition of some Gram-negative bacteria. J. Bact. 87: 1412–1422.

De Ley, J. & S. Frieman, 1965. Similarity of *Xanthomonas* and *Pseudomonas* deoxyribonucleic acid. J. Bact. 89: 1306–1309.

De Ley, J., I. W. Park, R. Tijtgat & J. van Ermengen, 1966. DNA homology and taxonomy of the genera *Pseudomonas* and *Xanthomonas*. J. gen. Microbiol. 42: 43–56.

Dye, D. W., 1962. The inadequacy of the usual determinative tests for the identification of *Xanthomonas* spp. N.Z. J. Sci. 5: 393–416.

Dye, D. W., 1963. Comparative study of the biochemical reactions of additional *Xanthomonas* spp. N.Z. J. Sci. 6: 483–486.

Dye, D. W., 1966. Cultural and biochemical reactions of additional *Xanthomonas* spp. N.Z. J. Sci. 9: 913–919.
Heslop-Harrison, J., 1962. Purposes and procedures in the taxonomic treatment of higher organisms. In: G. C. Ainsworth & P. H. A. Sneath, eds: Microbiological classification. 12th Symp. Soc. gen. Microbiol., Cambridge University Press, pp 14–36.
Starr, M. P. & W. L. Stephens, 1964. Pigmentation and taxonomy of the genus *Xanthomonas*. J. Bact. 87: 293–302.

Identification of soft rot coliform bacteria

D. C. Graham

Department of Agriculture and Fisheries for Scotland, Agricultural Scientific Services, East Craigs, Edinburgh EH12 8NJ, Great Britain

Abstract

The soft rot coliform bacteria, members of the genus *Erwinia*, are usually associated with soft rot diseases of plants. They are now generally considered to form a single species, *E. carotovora*, with several varieties *(atroseptica, chrysanthemi, rhapontici* and *cypripedii)*.

To study the phenotypic characters of all supposed members of this genus, 128 cultures have been examined in 90 tests. Seven tests have shown to be sufficiently consistent to be used for distinguishing between *E. carotovora* and the varieties *atroseptica* and *chrysanthemi*, and these, together with tests for fermentative metabolism of glucose, rotting of potato slices and slide agglutination with antiserum prepared against var. *atroseptica* can be used for routine identification. Mainly because only 3 isolates of var. *rhapontici* and var. *cypripedii* were available, they are not considered, though their reactions in the same tests are given for comparison.

The identity of pectolytic *Erwinia* spp. not belonging to the soft rot group, and certain 'atypical' *Erwinia* spp. is discussed.

Introduction

The soft rot coliform bacteria are gram-negative, peritrichously flagellated rods, having the characters of the Enterobacteriaceae, and which are usually, though not exclusively, associated with soft rot diseases of plants. In the 7th edition of Bergey's Manual these bacteria are all included in the genus *Erwinia*, but this genus contains at least two groups of plant pathogens which differ in pathogenicity: the soft rot coliform bacteria, and those causing necrosis and wilts typified by *E. amylovora*. Waldee (1945) suggested that these two groups should be distinguished at generic level, proposing that the name *Erwinia* be used only for the organisms associated with necrosis and wilts and establishing a new genus, *Pectobacterium*, for the organisms associated with soft rots. From further studies of phenotypic characters, Graham (1964) concluded that separation into these two genera is justified, but this view has not gained general acceptance. There are various reasons for this; for instance, results of DNA base composition analyses have emphasised the close relationship of the two groups, organisms which appear to be intermediate in character have been found, and organisms occur on plants (and in other habitats) which are not plant pathogens but plainly are closely related to them. The latter organisms belong to the species *E. herbicola*.

With regard to the soft rot group, the 7th edition of Bergey's Manual describes five species: *E. carotovora*, *E. atroseptica*, *E. aroideae*, *E. chrysanthemi* and *E. rhapontici*. However, more recent studies have emphasised the close similarity of all five species and, as Graham (1964) and Dye (1969) have concluded, the results indicate that there should be only one species: *E. carotovora*. Within this species, Graham *(op. cit.)* recognised three varieties: var. *aroideae*, var. *atroseptica* and var. *chrysanthemi*, mainly because this concept was useful for the plant pathologist. Since then, however, a wider range of isolates from many plants from various parts of the world has been studied by Dye (1969) and Graham (unpublished). Dye agrees that there should be only one species, but he no longer recognises anaerogenic strains (var. *aroideae*) as a separate taxon and he incorporates them in *E. carotovora*. This is justified because gas production is a variable character and failure to produce gas cannot be correlated with other characters. In his system, Dye includes four varieties of *E. carotovora*, namely var. *atroseptica* (the potato blackleg organism), var. *chrysanthemi*, var. *rhapontici* and var. *cypripedii*. It is likely that this classification will become generally accepted.

Looking for tests suitable for use in identification

About four years ago, I decided to examine a large number of cultures representing all supposed members of the genus *Erwinia*, with a view to perform a numerical taxonomic analysis, though the study was motivated by ecological and epidemiological rather than taxonomic considerations (one main purpose being to try to find tests which could distinguish the potato blackleg organism from all other soft rot coliforms). This study is not yet finished, but work on testing the carotovora organisms has been completed together with a preliminary sorting to pick out characters which could be used for routine identification. In all, 128 cultures were examined in 90 tests. The organisms were isolated from 36 different host plants, as well as one soil and one water sample, in 17 different countries.

The three cypripedii isolates did not liquefy pectate gel or cause soft rot of potato tubers, onion, or cucumber slices and therefore did not appear to be pectolytic. The three rhapontici isolates did not liquefy pectate gel, but slowly rotted potato slices and as distinct from all other cultures, produced a diffusible reddish pigment on several media. Because var. *cypripedii* is not associated with soft rot of plants, and var. *rhapontici* is found only on rhubarb, together with the fact that only three isolates of each variety were available, neither was considered along with *E. carotovora* and vars *atroseptica* and *chrysanthemi*. Some of their properties are given in Table 1.

The data on the remaining 122 isolates were first sorted by picking out those tests in which 95% or more of the organisms gave either positive or negative reaction:

List of tests in which the soft rot coliforms gave 95% or more positive or negative reactions

Positive: motility at 26 °C; fermentative metabolism of glucose; acid from glucose, salicin, xylose, sucrose, arabinose, mannitol, glycerol, mannose, ribose, cellobiose; H_2S production; gelatin liquefaction; pectate gel liquefaction; utilization of citrate, acetate, mucate, galac-

Table 1. Reactions of *Erwinia carotovora* var. *rhapontici* and var. *cypripedii* in the tests listed in Tables 2 and 3.

Test	*E.c.* var. *rhapontici*	*E.c.* var. *cypripedii*
O/F test (glucose)	F	F
Potatoslice rotted	+	—
Acid from lactose	+	— or ×
maltose	+	+
trehalose	+	+
α-methylglucoside	+	—
Indole	—	—
Growth in 5% NaCl	+	+
Lecithinase	—	—
Phosphatase	+	+
Reducing substances from sucrose	+	—
Sensitivity to erythromycin	S	S
Agglutination with atroseptica antiserum	—	—

Notation as in Tables 3 and 4.

turonate; reduction of nitrate; potato slice rotted; catalase; presence of haem; ONPG; sensitivity to chloramphenicol, kanamycin, neomycin, nitrofurantoin, polymyxin, streptomycin, tetracycline.

Negative: Gram at 48 h; acid from erythritol, adonitol, dulcitol; occurrence of symplasmata, biconvex bodies, hydrolysis of starch, arbutin, hippurate; gluconate; oxidase; growth in 10% NaCl; haemolysis; yellow non-diffusible pigment; red non-diffusible pigment; blue pigment on CYC agar; ornithine and lysine decarboxylase; phenylalanine deaminase; urease; DNase; utilization of tartrate; sulphatase, diffusible pink pigment; yellow diffusible pigment on potato; sensitivity to fucidin; methicillin, novobiocin, oleandomycin.

Using the results of the remaining tests, the organisms were sorted into three groups representing *E. carotovora* and vars *atroseptica* and *chysanthemi.* For var. *atroseptica* this was done by comparing reactions of authentic strains (checked for pathogenicity on potato stems) with all other isolates and picking out those which gave very similar reactions. As expected, most of these organisms were isolated from potato; those from other sources were tested for pathogenicity on potato stems of the susceptible cultivar Majestic grown in controlled environment chambers at 65 °F. Those not causing blackleg were classified as *E. carotovora* (details of pathogenicity tests are in Table 2). The var. *chrysanthemi* was selected by comparing the properties of authentic cultures of var. *chrysanthemi* with those of the rest of the organisms; no pathogenicity tests were made. The cultures whose characters did not correspond with those of var. *atroseptica* or var. *chrysanthemi* were regarded as *E. carotovora.*

In all, 30 tests were made, 10 of which gave reasonably consistent positive of negative reactions within each of the three groups (80% or more of the isolates giving positive or negative results). These tests are given in Table 3. The other 20 tests (see page 277) gave variable results within each group (21%–79% of cultures positive).

Table 2. Results of pathogenicity tests on potato stems with organisms having the morphological, physiological and biochemical properties of *Erwinia carotovora* var. *atroseptica*, isolated from sources other than potato.

Code number	Host	Origin	Blackleg production	Remarks
J3	Chinese cabbage	Japan	−	agglutinates with atroseptica antiserum
J9	carrot	Japan	−	no agglutination with atroseptica antiserum
C399	tomato	UK	+	agglutinates with atroseptica antiserum
C403	tomato	UK	+	no agglutination with atroseptica antiserum
C407	cauliflower	UK	+	no agglutination with atroseptica antiserum
GEJ50	water from water-cress bed	UK	+	agglutinates with atroseptica antiserum
G107	Iris rhizome	UK	+	agglutinates with atroseptica antiserum
G103	tomato	UK	+	agglutinates with atroseptica antiserum

All organisms producing blackleg were considered to be var. *atroseptica*.

Table 3. Tests giving 80% or more positive or negative reactions, which differ between *Erwinia carotovora* and its varieties *atroseptica* and *chrysanthemi*.

Test	*E. carotovora* (60 isolates)	var. *atroseptica* (39 isolates)	var. *chrysanthemi* (623 isolates)
Acid from lactose	+ (96)	+ (100)	− or × (100)
maltose	− (80)	+ (87)	− (100)
trehalose	+ (94)	+ (98)	− (100)
α-methylglucoside	− (85)	+ (94)	− (100)
Indole	− (89)	− (100)	+ (100)
Growth in 5% NaCl	+ (100)	+ (94)	− (100)
Lecithinase	− (100)	− (100)	+ (92)
Phosphatase	− (100)	− (100)	+ (100)
Reducing substances from sucrose	− (90)	+ (98)	− (96)
Sensitivity to erythromycin	R (90)	R (94)	S (100)

Notation: lactose + = acid produced in 2 days in 1% lactose + 1% peptone water + bromothymol blue; lactose − or × = either no acid produced in 14 days or acid produced after at least 5 days incubation; maltose, trehalose and α-methylglucoside + = acid produced in 1% peptone + 1% sugar + bromothymol blue within 14 days; indole test + = positive in peptone water after 2 days incubation using Ehrlich's reagent; growth in 5% NaCl peptone water + = positive after 7 days; lecithinase + = opaque zone around colonies on egg yolk agar after 7 days; phosphatase + = pink colour of and around colonies grown for 48 h on 0.05% w/v sodium phenolphthalein diphosphate agar and treated with gaseous ammonia; reducing substances from sucrose + = positive in 4% sucrose peptone water (shake culture) in 48 h tested with Benedict's reagent; sensitivity to erythromycin: R = organism shows no inhibition zone around colonies using Multodisk test discs (potency 50 μg), S = clear inhibition zone visible. Incubation temperature for all tests 25 °C. Figures in brackets are percentages of cultures giving the respective positive and negative reactions.

List of tests giving variable results within E. carotovora *and vars* atroseptica *and* chrysanthemi

Gas from glucose; acid from raffinose, inositol, sorbitol, inulin, on ethanol agar; MR; VP; liquefaction of carboxymethyl cellulose gel; utilization of malonate; hydrolysis of tributyrin, casein; mucoid growth on 5% sucrose agar; growth at 37°C; arginine dihydrolase; KCN; sensitivity to ampicillin, colomycin, penicillin.

On the basis of these tests, the variety *chrysanthemi* includes *E. carotovora* f.sp. *parthenii*, *Pectobacterium parthenii* var. *dianthicola*, *E. carotovora* f.sp. *zeae*, *Pectobacterium carotovorum* var. *graminarum*, *E. dieffenbachiae*, *E. cytolytica* (NCPPB 1385), the organism isolated from pineapple rot in Malaya (NCPPB 551), and an organism received from France as *E. betivora* isolated from soft root of beet. Dye (1969) says that the majority of the strains of var. *chrysanthemi* produce a blue insoluble pigment (indigoindine) on glucose-yeastrel chalk agar, but I have found this to be inconsistent with many isolates, though it is distinctive of this variety when it occurs. Sometimes it is produced in the rotted tissue of potato slices, especially with the cytolytica and betivora isolates, and one of the f.sp. *zeae* isolates.

Routine identification at the East Craigs laboratory

For routine identification of soft root bacteria from plant tissue, material is plated on MacConkey-pectate gel medium (Stewart, 1962), and also on nutrient agar or King medium B when isolations are made from sources other than potato. Colonies are picked and transferred to nutrient agar slopes and the organisms are routinely put through the tests indicated in Table 3, together with those for acid from lactose and trehalose, growth in 5% NaCl reducing substances from sucrose, lecithinase, phosphatase and sensitivity to erythromycin given in Table 3. The three tests mentioned in Table 4 giving less than 90% positive or negative reactions within each group are not used for routine identification.

Agglutination with atroseptica antiserum is not specific for the blackleg organism, but is a very useful and rapid test for its presence in diseased tissue, and also for a quick check on colonies on isolation plates (good reactions can be obtained with

Table 4. Additional identification tests used for soft rot coliform bacteria newly isolated from plant tissue.

	E. carotovora	var. *atroseptica*	var. *chrysanthemi*
O/F test (glucose)	F	F	F
Potato slice rotted	+	+	+
Agglutination with atroseptica antiserum	d	+	−

Notation: F = fermentative metabolism of glucose. Agglutination tests done on slides; d = different isolates give consistently different reactions.

colonies on MacConkey-pectate gel plates). Two isolates with all the morphological, physiological and biochemical characters of var. *atroseptica*, and which caused potato blackleg on inoculation into potato stems, did not give slide agglutination reactions with atroseptica antiserum. One was isolated from a stem rot of tomato, the other from a rotted cauliflower head, both in Eastern England. This is in contrast to about 250 isolates of atroseptica from potato (obtained mainly in Scotland), which all gave good agglutination reactions. It is noteworthy that one culture, identified as var. *chrysanthemi*, was isolated from potato blackleg in Brazil. It is the only case in my experience where this organism has been found associated with blackleg, but so far I have not checked its pathogenicity on potato stems. All chrysanthemi isolates gave slide agglutination reactions with antiserum prepared against NCPPB 1385, whereas no agglutination occurred with atroseptica isolates. At least some carotovora isolates reacted with chrysanthemi antiserum, but the number of tests has as yet been insufficient to see if reactions occur so frequently as to make the test of no value in routine identification.

Pectolytic Erwinia species not included in the carotovora group

E. salicis causes watermark disease of willow, rots potato slices, and forms depressions in pectate gel without actually liquefying it (the rotted potato tissue becomes yellow, the pigment is diffusible).

E. rubrifaciens causes a disease of walnut, produces depressions around colonies on pectate gel, but does not rot potato slices.

E. quercina, the cause of a disease of acorns, gives a slight soft rot of potato slices but is not pectolytic on pectate gel (the effect on potato is the production of a water-soaked area in which the tissue is slightly softened; it is not a typical soft rot).

Erwinia salicis and *E. rubrifaciens* resemble *E. amylovora* in most respects, but *E. quercina* is, on the whole, most like *E. carotovora*. None of these organisms has ever been isolated from other hosts.

Identity of some atypical Erwinia species

E. dissolvens NCPPB 1862, and the organism from Kona coffee cherries described by Frank, Lum & Dela Cruz (1965) as *E. dissolvens*, have the characters of *Klebsiella aerogenes* and do not rot potato slices or liquefy pectate gel. *E. dissolvens* NCPPB 1850 is non-motile but otherwise has the characters of *Enterobacter cloacae*.

E. nimipressuralis NCPPB 440, the cause of 'wet wood' of elm in the USA, was received from Professor Burkholder as an authentic culture. It does not rot potato slices or liquefy pectate gel and has the characters of *Enterobacter cloacae*. An organism from elm in the UK, received from Dr J. Rishbeth, produces depressions around colonies on pectate gel plates but does not rot potato slices. It has the characters of an *Erwinia*, but so far I have not been able to identify it (though it resembles the carotovora group more closely than the amylovora group).

E. carnegieana NCPPB 439, said to be the cause of a disease of the giant cactus, was also received from Professor Burkholder as an authentic culture. It does not liquefy pectate gel or rot potato slices, and I am uncertain on its identity, though it may be an *Enterobacter*. *E. carnegieana* NCPPB 671 and 672, more recently isolated from the giant cactus, are *E. carotovora*.

References

Dye, D. W., 1969. A taxonomic study of the genus *Erwinia*. II. The 'carotovora' group. N.Z. J. Sci. 12: 81–97.

Frank, H. A., N. A. Lum & A. S. Dela Cruz, 1965. Bacteria responsible for mucilage-layer decomposition in Kona coffee cherries. Appl. Microbiol. 13: 201–207.

Graham, D. C., 1964. Taxonomy of the soft rot coliform bacteria. A. Rev. Phytopath. 2: 13–42.

Stewart, F. J., 1962. A selective-diagnostic medium for the isolation of pectinolytic organisms in the Enterobacteriaceae. Nature, Lond. 195: 1023.

Identification of the fluorescent pseudomonads[1]

D. C. Hildebrand and M. N. Schroth

Department of Plant Pathology, University of California, Berkeley, California 94720

Abstract

Various physiological tests have been given that are useful for distinguishing phytopathogenic fluorescent pseudomonad species (pathotypes). Tests include oxidase, arginine dihydrolase, levan formation, pigments, pit formation on polypectate gels of different pH, synthesis and rate of synthesis of β-glucosidase, growth rate, and ability to use β-alanine, trehalose, sorbitol, erythritol, D(−) tartrate, L(+) tartrate, DL-lactate, sucrose, and pelargonate for growth. A key based on these tests has been prepared for identification of these pseudomonads.

Introduction

Identification of the plant pathogenic species of fluorescent pseudomonads has been based primarily on the host attacked and the nature of the elicited symptoms. Technical descriptions of the organisms generally have consisted of a morphological characterization supplemented with reported reactions on litmus milk, gelatin, nitrate, and a few other tests borrowed from medical microbiology. The few definitive tests combined with the variability occurring among strains of an organism has previously prevented any real separation of the organisms on any basis except host. Recently, however, additional tests have indicated that the pathogenic pseudomonads can be separated into taxa on physiological and nutritional characteristics. Furthermore, these taxa generally correspond with the taxa previously described based on the host.

The tests given in the table are those that are now the most useful in distinguishing various species of pseudomonad pathogens. This information has been compiled from various studies on pseudomonads (Billing, 1970; Jessen, 1965; Lelliott *et al.*, 1966; Misaghi & Grogan, 1969; Moustata & Whittenbury, 1970; Sands *et al.*, 1970; Stanier *et al.*, 1966) which have appeared during the past six years. With some exceptions, the tests were selected only if there was general agreement in results among laboratories for a given species. Some tests and organisms which have been studied in our laboratory are added. Among these organisms is strain B-62, isolated from walnut cankers, which

1. This investigation was supported in part by Regional Research Funds-2444 (W-96), USDA Grant (12-14-100-9199-34), and a Public Health Service research carreer development award (AI-34, 942) from the National Institutes of Allergy and Infectious Diseases.

Tests useful for distinguishing species or pathotypes of fluorescent pseudomonad pathogens.

Organism	Group[1]	Oxidase	Arginine dihydrolase	β-alanine	Trehalose	Levan	Pigments	Mannitol	Sorbitol	Erythritol	L(+)Tartrate	D(−)Tartrate	DL-Lactate	Sucrose	Pelargonate	Sodium polypectate pH5	Sodium polypectate pH8.5	β-glucosidase formation	β-glucosidase rate[2]	Growth rate[3]
P. syringae	Ia	−[4]	−	−	−	+	F[5]	+	+	+	−	−	+	+	+	−	−	+	1.6 –57.5	119–220 500
P. pisi	Ia	−	−	−	−	+	vF	+	+	v	−	−	+	+	+	−	−	v	0.0 – 3.9	127–150
P. coronafaciens	Ia	−	−	−	−	+	F	+	+	+	−	−	−	+	+	−	−	+	0.4 – 7.4	172–225
P. tomato	Ia	−	−	−	−	+	F	+	+	−	−	+	−	+	+	+	−	+	1.4 – 1.8	168–200
P. eriobotryae	Ia	−	−	−	−	+	B,F	+	+	−	+	−	−	+	−	+	−	+	0.05– 0.5	250–280
P. lachrymans	Ia	−	−	−	−	+	F	+	v	v	+	−	−	+	+	+	+	+	0.4 – 1.9	175–200
P. tabaci	Ia	−	−	−	−	+	F	+	+	+	+	−	−	+	+	+	+	+	0.3	165
P. angulata	Ia	−	−	−	−	+	F	+	+	+	+	−	−	+	+	+	+	+	2.9	150–175
P. mors-prunorum	Ia	−	−	−	−	+	F	+	−	−	+	−	−	+	+	+	−	−	0.0	300
P. mori/*P. glycinea*/ *P. phaseolicola*	Ia	−	−	−	−	+	F	v	−	−	−	−	−	v	+	+	−	−	0.0	280–370
P. garcae	Ia	−	−	−	−	+	vF[6]	+	+	+	−	−	−	+	+	−	−	+	1.0	175
P. savastanoi	Ib	−	−	−	−	−	F	+	v	−	+	−	−	v	v	v	−	−	0.0	
P. sesami	Ib	−	−	0	−	weak	−	−	−	v	v	−	−	+	0	+	+	−	0.0	300
P. viridiflava	II	−	−	−	−	−	F	+	+	+	−	+	+	−	+	−	+	+	0.5	180
P. cichorii	III	+	−	−	−	−	F	+	−	−	+	−	+	−	+	−	−	+	0.5 – 3.2	165–250
B-62	−	+	−	−	+	0	F,Y	+	v	+	−	−	+	+	+	−	−	−	0.0	220
P. putida	−	+	+	+	−	−	F	v	v	v	v	v	v	v	+	−	−	−	0.0	128
P. marginalis	IVa	+	+	+	+	+	F	+	+	+	−	v	+	+	+	−	+	+	0.03– 5.8	112–152
P. fluorescens	Va	+	+	+	+	+	F	+	v	v	−	v	+	+	+	−	−	−	0	90–110
P. polycolor/ *P. aeruginosa*	Va	+	+	+	−	−	F,P	+	−	−	−	−	+	−	+	−	−	−	0	95

1. Lelliott *et al.*, 1966.
2. Arbitrary enzyme unit based on O.D. change/number of cells.
3. Generation time (minutes) in an inorganic basal medium with glucose as the added carbon source.
4. + = positive reaction, or utilizes growth; − = negative reaction, or cannot utilize for growth; v = variable results; 0 = not tested.
5. B = brown pigment; F = yellow fluorescent pigment; P = phenazine pigment; Y = yellow non-fluorescent pigment.
6. Weak

forms both a diffusable fluorescent pigment and a non-diffusable, yellow pigment. This strain has been included because it shares several characteristic properties with the *Pseudomonas syringae* group and several with the *P. fluorescens* group.

As to the specific tests for determinative purposes, most of the procedures are found in the studies previously referred to (Billing, 1970; Jessen, 1965; Lelliott *et al.*, 1966; Misaghi & Grogan, 1969; Moustata & Whittenbury, 1970; Sands *et al.*, 1970; Stanier *et al.*, 1966) and, consequently, are not given here.

Oxidase test

This test apparently detects the presence of cytochrome *c* in bacteria (Stanier *et al.*, 1966). It is useful for distinguishing the *P. syringae* group of pathogens, which do not possess cytochrome *c* (Sands *et al.*, 1967), from *P. cichorii*, *P. marginalis*, and the saprophytic pseudomonads which do. The N,N-dimethyl-*p*-phenylenediamine reagent is usually preferred to the tetramethyl form, as it is less sensitive to nitrites which may be formed in the culture medium. The composition of the medium may affect test results (Klinge, 1960a, b). We therefore usually grow the organisms on King's B medium (King *et al.*, 1954) since it has given satisfactory results in contrast to other media such as YDCP (Leben *et al.*, 1970) where negative results are sometimes obtained for normally positive organisms such as *P. fluorescens*.

Arginine dihydrolase

Pseudomonas cichorii and one unnamed pseudomonad (B-62) can be separated from the rest of the oxidase positive pseudomonads on the basis of this test. The method described by Thornley (1960) is rapid and reliable.

Levan formation

Levan, formed by levan sucrase (Fuchs, 1959), is produced by most pseudomonads which utilize sucrose (Stanier *et al.*, 1966). Only *P. aeruginosa*, which does not utilize sucrose, *P. putida*, including a few strains which grow on this sugar, and occasionally other sucrose-positive strains do not form levan.

Pigments

Most members of the fluorescent pseudomonad group (the group which does not accumulate poly-β-hydroxybutyrate as an intracellular reserve material) form a greenish fluorescent pigment on King's B medium. Fluorescence should be examined with UV light, however, as several members of the non-fluorescent group such as *P. cepacia* and *P. marginata* form similar pigments on this medium, except that they do not fluoresce. Several other pigments are produced by various fluorescent pseudomonads. Pyocyanine production by *P. polycolor*/*P. aeruginosa* can be detected on King's A medium (King *et al.*, 1954), as can production of chlororaphin (green) by

P. chlororaphis, phenazine-1-carboxylate (orange) by *P. aureofaciens*, and a non-diffusable blue pigment by Biotype IV of *P. fluorescens*. A number of strains of *P. eriobotryae* rapidly turn an organic medium dark brown and appear to be distinguishable on this basis. Non-diffusable yellow carotenoid pigments are formed in large amounts by the unnamed bacterium B-62 which also form a fluorescent pigment. This organism is yellow in culture and could easily be mistaken for a xanthomonad except for the formation of the fluorescent pigment. Spectral tests of the yellow pigment of B-62 indicate that it is different from the xanthomonad pigment.

Pectolytic enzymes

Detection of pectolytic enzymes is by pit formation on polypectate gels. We use a basal medium of 1000 ml distilled water, 1 ml 1.5% alcoholic brom thymol blue, 6 ml 10% $CaCl_2 \cdot 2H_2O$ and 22 g sodium polypectate. This base is adjusted to pH 4.9 to 5.1 with 3.4 to 4.2 ml 1 N HCl (depending upon the source of polypectate) or pH 8.3 to 8.5 with 6.4 to 8.5 ml 1 N NaOH. Add 100 ml of 4% molten sterile agar after autoclaving. Incubate the plates for 6 days or longer after seeding (Hildebrand, 1971). The use of both a high and low pH is important to distinguish among species as three types of activity are observed. Soft-rotting species such as *P. marginalis* cause pitting only at high pH, angular leaf-spotting species such as *P. lachrymans* form pits at both pH levels, and others such as *P. mori* form pits only at the low pH. In a related phenomenon, tests of the strains that Stanier *et al.* (1966) placed in Biotypes A and B of *P. fluorescens* have indicated that with one exception (strain 411 isolated from seawater near a sewage outlet) only the *P. marginalis* strains grow on polygalacturonic acid.

β-Glucosidase

Both the synthesis of β-glucosidase and the rate of synthesis seem to be characteristic for various species. Arbutin, salicin, or esculin are commonly utilized as inducers of the enzyme. Synthesis is usually detected with the peptone basal medium of Hugh & Leifson (1953) with arbutin as the substrate (Hildebrand & Schroth, 1964). Acid production or formation of a brownish color in the medium is an indication that β-glucosidase has been synthesized. Occasionally an organism synthesizes too little of the enzyme to be detected by this method, and the testing of enzyme extracts with *p*-nitrophenol-β-glucoside (Hildebrand & Schroth, 1968) must be used.

Information concerning the relative rate of β-glucosidase synthesis has also been included in the table to indicate how the rate of synthesis of a specific enzyme may be useful in identification. This was tested in a medium consisting of 1.5 g KH_2PO_4, 3.0 g Na_2HPO_4, 1.0 g NH_4Cl, 0.2 g $MgSO_4 \cdot 7H_2O$, 4 g glucose, and 1 liter distilled water, with 0.2% salicin added after preliminary incubation as the inducer.

Although there is considerable variation among the limited number of strains tested, specific patterns tend to emerge. For example, the greatest degree of β-glucosidase synthesis occurred among *P. syringae* strains. Three of 8 strains of this pathogen produced more than 11 enzyme units/10^8 cells, and 4 of 8 produced between 3.3 and 11 eu. In contrast, 3 of 4 strains of *P. marginalis* produced less than 0.2 eu/10^8 cells, and the other formed 5.8 eu. Most of the other β-glucosidase synthesizing strains formed between 0.5 and 2.0 eu/10^8 cells.

Growth rate

As with the rate of β-glucosidase synthesis, different growth rates are characteristic for the various species. This was tested in the basal medium used for the β-glucosidase rate studies. Only one organism *(P. syringae)* showed considerable variation in its growth rate in this medium. Four strains had a doubling time of 119 to 150 min, two strains doubled every 205–220 min, and two doubled every 500 min. The range of doubling times of other species usually was about 30 to 50 min. These growth rate ranges are mostly based on tests of 3 to 5 strains, and consequently it would be expected that these would increase somewhat as more strains are tested.

Growth on various substrates

Most of the recent pseudomonad studies have followed that of Stanier *et al.* (1966) in using the ability of bacteria to utilize various substrates for growth as the major means of characterization. The replicate plate technique (Lederberg & Lederberg, 1952) is usually employed because of its speed and its capacity to test many strains on various substrates. This method has several disadvantages, however, such as the problem of optimum substrate concentration for best growth, and difficulty in making readings of whether growth of a bacterium has occurred. Therefore, errors occur regarding the specific characteristics. This inherent unreliability of specific results probably means that differences among the various taxa are not as clearly defined as they would be if the problems did not exist.

Sufficient agreement has been obtained with several substrates, however, to support their usefulness in identifying pseudomonads. Utilization of several of these substrates separate some of the major groups of pseudomonads, whereas utilization of others serve to distinguish the various species.

β-*Alanine* This substrate serves to separate the oxidase positive from the oxidase negative groups of fluorescent pseudomonads with the exception of B-62 and *P. cichorii.*

Trehalose Growth on this substrate is characteristic of the *P. fluorescens* group of organisms separating it from the *P. putida*, *P. polycolor/P. aeruginosa*, and the *P. syringae* groups.

Mannitol Most pseudomonads utilize mannitol with the exception of *P. sesami.*

Sorbitol *P. polycolor/P. aeruginosa*, *P. cichorii*, *P. sesami*, *P. mors-prunorum*, and the halo blight of bean group cannot utilize this substrate, in contrast to most other pseudomonads.

Erythritol Members of about half of the pseudomonad species can grow on this substrate.

L(+)Tartrate Utilization of this substrate is characteristic of *P. lachrymans*, *P. tabaci*/*P. angulata*, *P. mors-prunorum*, *P. savastanoi*, and *P. cichorii*.

D(−)Tartrate The few pseudomonads which utilize this substrate are *P. tomato*, *P. lachrymans*, *P. eriobotryae*, and *P. viridiflava*.

DL-Lactate Most of the pseudomonads which possess arginine dihydrolase and utilize β-alanine also utilize this substrate. Among those which do not possess arginine dihydrolase, only *P. syringae*, *P. viridiflava*, and *P. pisi* utilize lactate. We have noted that the ability of some strains of these bacteria to utilize lactate is perhaps a result of mutation. This has been observed particularly with *P. pisi*, as heavy streaks of the organism on lactate media often yield only a few isolated colonies.

Sucrose As noted under levan formation, most organisms which form levan will also utilize sucrose.

The various tests must be used with caution. Many of the data are preliminary, as a sufficient number of strains have not been tested to establish the range of variability existing within a species. Some tests are difficult to read, and better methods should be devised to obtain more reliable data. However, the tests can be used to provide an indication as to the identity of a bacterium. We therefore devised a key from the list of tests for the tentative identification of pseudomonads, with the expectation that it will be improved within a short time.

Key for identification of phytopathogenic fluorescent pseudomonads

I. Oxidase positive
  A. Arginine dihydrolase positive, utilize β-alanine
    1. utilizes trehalose
      a. produces pits on polypectate gels (pH 8.5) and utilizes polygalacturonic acid . . . . *P. marginalis*
      b. does not produce pits on polypectate gels or utilize polygalacturonic acid . . . . *P. fluorescens*
    2. does not utilize trehalose
      a. utilizes erythritol . . . . *P. putida*
      b. does not utilize erythritol . . . . *P. aeruginosa*
  B. Arginine dihydrolase negative, does not utilize β-alanine
    1. utilizes trehalose B-62
    2. does not utilize trehalose . . . . *P. cichorii*
II. Oxidase negative
  A. Produces pitting on polypectate gels at pH 8.5 . . . . *P. viridiflava*
  B. Produces pitting on polypectate gels at pH 5 and pH 8.5
    1. utilizes L(+) tartrate . . . . *P. tabaci*/*P. angulata*/*P. lachrymans*
    2. does not utilize L(+) tartrate . . . . *P. sesami*
  C. Produces pitting on polypectate gels at pH 5
    1. utilizes L(+) tartrate
      a. possesses β-glucosidase . . . . *P. eriobotryae*
      b. does not possess β-glucosidase

i. forms levan . . . . *P. mors-prunorum*
ii. does not form levan . . . . *P. savastani*
2. does not utilize L(+) tartrate
a. utilizes D(−) tartrate . . . . *P. phaseolicola/P. mori/P. glycinea*

D. Does not produce pitting on polypectate gels
1. utilizes lactate . . . . *P. syringae/P. pisi*
2. does not utilize lactate
a. possesses β-glucosidase . . . . *P. coronafaciens*
b. does not possess β-glucosidase
i. utilizes L(+) tartrate, does not form levan . . . . . *P. savastanoi*
ii. does not utilize L(+) tartrate, forms levan . . . . . . *P. garcae*

References

Billing, E., 1970. Further studies on the phage sensitivity and the determination of *Pseudomonas* spp. J. appl. Bact. 33: 478–491.

Fuchs, A., 1959. On the synthesis and breakdown of levan by bacteria. Thesis, Delft: Waltman.

Hildebrand, D. C., 1971. Pectate and pectin gels for differentiation of *Pseudomonas* sp. and other plant pathogenic bacteria. Phytopathology (in press).

Hildebrand, D. C. & M. N. Schroth, 1964. β-Glucosidase activity in phytopathogenic bacteria. Appl. Microbiol. 12: 487–491.

Hildebrand, D. C. & M. N. Schroth, 1968. Removal of pseudomonads from plant leaves and measurement of their in vivo β-glucosidase synthesis. Phytopathology 58: 354–358.

Hugh, R. & E. Leifson, 1953. The taxonomic significance of fermentative versus oxidative metabolism of carbohydrates by various gram negative bacteria. J. Bact. 66: 24–26.

Jessen, O., 1965. *Pseudomonas aeruginosa* and other green fluorescent pseudomonads. A taxonomic study. Munksgaard, Copenhagen.

King, E. O., M. K. Ward & D. E. Raney, 1954. Two simple media for the demonstration of pyocyanin and fluorescin. J. Lab. clin. Med. 44: 301–307.

Klinge, K., 1960a. Die Bedeutung der Oxydase-Reaktion für die Identifizierung von *Pseudomonas aeruginosa*, *Pseudomonas fluoresens* and *Pseudomonas putida*. Arch. Hyg. Bakt. 144: 263–276.

Klinge, K., 1960b. Differential techniques and methods of isolation of *Pseudomonas*. J. appl. Bact. 23: 442–462.

Leben, D., M. N. Schroth & D. C. Hildebrand, 1970. Colonization and movement of *Pseudomonas syringae* on healthy bean seedlings. Phytopathology 60: 677–680.

Lederberg, J. & E. M. Lederberg, 1952. Replica plating and indirect selection of bacterial mutants. J. Bact. 63: 399–406.

Lelliott, R. A., E. Billing & A. C. Hayward, 1966. A determinative scheme for the fluorescent plant pathogenic pseudomonads. J. appl. Bact. 29: 470–489.

Misaghi, I. & R. G. Grogan., 1969. Nutritional and biochemical comparisons of plant-pathogenic and saprophytic fluorescent pseudomonads. Phytopathology 59: 1436–1450.

Moustafa (Darweish), F. A. & R. Whittenbury, 1970. A comparison of some phytopathogenic and nonphytopathogenic pseudomonads. Phytopath. Z. 67: 63–72.

Sands, D. C., F. H. Gleason & D. C. Hildebrand, 1967. Cytochromes of *Pseudomonas syringae*. J. Bact. 94: 1785–1786.

Sands, D. C., M. N. Schroth & D. C. Hildebrand, 1970. Taxonomy of Phytopathogenic Pseudomonads. J. Bact. 101: 9–23.

Stanier, R. Y., N. J. Palleroni & M. Doudoroff, 1966. The aerobic pseudomonads: a taxonomic study. J. gen. Microbiol. 43: 159–271.

Thornley, M. J., 1960. The differentiation of *Pseudomonas* from other gramnegative bacteria on the basis of arginine metabolism. J. appl. Bact. 23: 37–52.

Intra-specific variation in Pseudomonas solanacearum

D. C. Harris

National Agricultural Laboratories, Nairobi, Kenya

Abstract

Approximately 100 isolates of *Pseudomonas solanacearum* from various hosts and parts of the world are compared in their ability to utilise numerous carbon compounds as sole carbon source, and in several standard bacteriological tests.

The collection has proved to be bacteriologically heterogeneous and has been divided into three main groups on the basis of several correlated characters. These groups correspond with Hayward's (1964) biochemical types 1, 2 and combined 3 and 4. The second and third groups are quite uniform and differ in ten characters. The first group can be subdivided into a spectrum of types, in which one end resembles the second group and the other end resembles the third group. Isolates from any one economic hosts predominate in only one of these groups.

The significance of the variation in the taxonomy of *P. solanacearum* is discussed.

Introduction

In a bacteriological examination of *Pseudomonas solanacearum* from world-wide sources, Hayward (1964) found that isolates could be arranged in four biochemical types according to their metabolism of three disaccharides and three hexose alcohols. Buddenhagen & Kelman (1964), when summarising the knowledge of intra-specific variation in *P. solanacearum*, divided the species into three pathotype groups or 'races'. Race 1 attacks solanaceous and other hosts but not triploid bananas; race 2 attacks only triploid bananas and *Heliconia* spp.; and race 3 attacks only potato and tomato. Race 3 corresponded with Hayward's biochemical type 2, but the other races contained representatives of the three other biochemical types.

To obtain more information on biochemical variation in this species, about 100 isolates from various hosts and countries were compared in their ability to utilise numerous carbon compounds as source of carbon, using the methods of Stanier *et al.* (1966), and in a range of standard bacteriological tests.

Results

In a preliminary screening, a representative sample of 25 isolates were examined for 117 characters. Isolates were all positive in 33, all negative in 65 and varied in 19. A further 90 isolates were then examined on these differential characters. The reactions of the four biochemical types in these tests are listed in Table 1.

Table 1. Reaction of four biochemical types of *Pseudomonas solanacearum* in 19 differential tests.

Test	Reaction of biochemical type		
	+	−	+ and −
Oxidation of cellobiose, maltose and lactose	2, 3	1, 4	
2-ketogluconate formation	1, 3		2, 4
Gas from nitrate	3, 4	2	1
Hydrolysis of Tween 80	2, 3, 4		1
Tyrosinase	2		1, 3, 4
Utilization of dulcitol, mannitol and sorbitol	3, 4	1, 2	
glycollate	3, 4	1, 2	
lactate	3, 4	2	1
tryptophan	3, 4	2	1
hippurate	2	3, 4	1
malonate		2, 3, 4	1
fructose	2, 3, 4		1
inositol	3, 4	2	1
trehalose	3, 4	2	1
tyrosine	1, 3, 4		2

Using these characters, it is possible to divide the isolates into three main biochemical groups. The first two correspond with Hayward's biochemical types 1 and 2 and the third embraces biochemical types 3 and 4. Isolates of group 2 were quite uniform, showing differences only in the ketogluconate reaction and tyrosine utilisation. The third group was also uniform and was subdivisible only by the oxidation of the disaccharides and the ketogluconate reaction. There are ten correlated characters separating all isolates in groups 2 and 3. Group 1 isolates shared characters of the other groups and exhibited characters not possessed by the other two groups. This category could be divided into subgroups whose combination of biochemical characters varied within a spectrum of types ranging from near group 2 to near group 3 (see Table 2).

Some of these types appeared to be of restricted geographic distribution and/or restricted host range, but many more isolates will have to be studied to confirm this (see Table 3).

Although there was no absolute correlation, the isolates from any one economic host were predominantly from only one of these three groups (see Table 4). A similar relationship is evident in Hayward's (1964) data.

Discussion

The results indicate that there is substantial biochemical variation within the nomen species *Pseudomonas solanacearum* and they confirm that, to a large degree, this cuts across variation in pathogenicity. There are two ways of interpreting this situation.

Table 2. Biochemical subdivisions of biochemical type 1.

Character	Biochemical type 2	Biochemical type 1									Biochemical type 3/4
		A	B	C	D	E	F	G	H	J	
Gas from nitrate	−	−	−	−	−	−	−	−	+	+	+
Tyrosine	+	−	−	+	−	+	−	−	−	−	+/−
Tween 80	+	+	+	+	+	−	+	+	−	+	+
Tryptophan	−	−	−	+	−	+	+	+	+	+	+
Hippurate	+	+	+	+	+	+	+	+	−	−	−
Malonate	−	−	−	−	−	−	−	+	−	−	−
Fructose	+	+	+	+	−	+	+	−	−	+	+
Inositol	−	−	+	+	−	+	+	+	+	+	+
Trehalose	−	+	−	−	+	−	+	+	−	+	+
Lactate	−	−	−	−	−	−	−	−	−	+	+

Table 3. Origins of biochemical type 1 subtypes.

Biochemical subtype	Number of isolates	Host(s)	Location(s)
A	1	plantain	Peru
B	2	plantain	Honduras
C	9	tomato, tobacco	USA, Puerto Rico
D	3	banana	Honduras
E	3	tomato	USA, Puerto Rico
F	2	banana	Philippines, Panama
G	6	tomato, tobacco, eggplant	Trinidad, Guadaloupe, Martinique, High Volta
H	2	potato	Angola, Madagascar
J	3	potato	Mozambique

Table 4. Host distribution of the four biochemical types of *Pseudomonas solanacearum*.

Host	Number of isolates of each biotype from each host			
	biotype 1	biotype 2	biotype 3	biotype 4
Tomato	13	2	5	3
Potato	8	25	2	
Tobacco	6		1	
Eggplant	1		3	2
Pepper			3	2
Ginger				6
Banana/plantain	8		2	

Firstly, it can be postulated that a biochemically variable parent population has split into biochemically variable groups of pathotypes; that is, the races of Buddenhagen *et al.* (1962). Alternatively, *P. solanacearum* consists of a range of biochemical types of various pathogenic potentials. Some, such as biochemical types 2, 1D, 1E and 1J, have a fairly restricted host range while others, such as 1C, 1G and 3, have a wider host range. In the latter interpretation, some types (3, for example), while biochemically uniform, can be subdivided on a basis of pathogenicity.

Clearly, the interpretation of biochemical variation in *P. solanacearum* is still a matter for debate and it epitomises the contrasting approaches of plant pathologist and bacterial taxonomist. For the plant pathologist, pathogenicity characters are paramount, while the taxonomist currently considers that overall similarity is of primary importance in classifying taxa. The races of Buddenhagen *et al.* (1962) are distinguished primarily on pathogenicity characters but some morphological, physiological and biochemical characters correlate with these groupings. Races 1 and 2 can be distinguished on colony morphology and tyrosinase production (French & Sequira, 1970); race 3 has a lower temperature optimum for growth (Buddenhagen & Kelman, 1964); races 1 and 2 differ in their sensitivity to antibiotics (Wardlaw, 1961); and race 2 gives the 'hypersensitivity' reaction in tobacco leaves (Lozano & Sequira, 1968). However, the instability of some of these characters makes them of doubtful taxonomic value.

This is clearly a situation in which the objectivity of Adansonian techniques is essential for intra-specific classification. An intensive study using many isolates and many criteria, interpreted by numerical taxonomic methods will not only result in the most useful intra-specific classification of *P. solanacearum* but will give the greatest information on the affinities of the different strains which occur naturally.

References

Buddenhagen, I. W. & A. Kelman, 1964. Biological and physiological aspects of bacterial wilt caused by *Pseudomonas solanacearum*. A. Rev. Phytopath. 2: 203–230.

Buddenhagen, I. W., L. Sequeira & A. Kelman, 1962. Designation of races in *Pseudomonas solanacearum*. Phytopathology 52: 726.

French, E. R. & L. Sequeira, 1970. Strains of *Pseudomonas solanacearum* from Central and South America; a comparative study. Phytopathology 60: 506–512.

Hayward, A. C., 1964. Characteristics of *Pseudomonas solanacearum*. J. appl. Bact. 27: 265–277.

Lozano, J. C. & L. Sequeira, 1968. Differentiation of races of *Pseudomonas solanacearum* by a hypersensitive reaction of tobacco leaves. Phytopathology 58: 1058.

Stanier, R. Y., N. J. Palleroni & H. Doudoroff, 1966. The aerobic pseudomonads; a taxonomic study. J. gen. Microbiol. 43: 159–271.

Wardlaw, C. W., 1961. Banana diseases. Longmans, Green & Co., London.

The impact of changes in nomenclature on plant quarantaine measures

A. C. Hayward

Department of Microbiology, Medical School, University of Queensland, Herston Road, Herston, Brisbane, Australia

Statements of the need for a stable nomenclature for the plant pathogenic bacteria tend to the platitudinous. The use of different systems hampers the flow of scientific information and is a source of confusion in literature and legislation on plant quarantine, including distribution maps of plant diseases. Several names are used for the same pathogen, and there is a multiplicity of common names for the diseases caused by pathogens of wide host range, such as *Pseudomonas solanacearum* (Kelman, 1953). Among the genera of plant pathogenic bacteria only the species of *Corynebacterium* are now universally recognised and spared from drastic taxonomic revision. In all other genera radical changes have recently been proposed. One can find reference in literature to either *Erwinia chrysanthemi* or to the carnation strain of *Erwinia parthenii* var. *dianthicola*; similarly the potato blackleg bacterium may be referred to as *Pectobacterium carotovorum* var. *atrosepticum* or as *Erwinia atroseptica.* However, it is the genera *Pseudomonas* and *Xanthomonas* which contain most species and it is the concept of species in these genera that is most controversial.

There is much to be said against rash nomenclatural change, from the point of view of plant pathologists and from the doubtful evidence for alternative proposals. It may indeed, with reference to the genus *Xanthomonas*, be '. . . unfortunate that it is very easy to introduce new species names for xanthomonads', but it is, I suggest, fortunate that it is very difficult to dispose of them. The need is for more research on host range, differentiae, etc., and for stability in the meantime. Plant pathologist need to designate as a separate taxon, any organism that is distinct because of differences in host range, symptoms or virulence. Few authors have considered the implications of their proposals for the plant pathologist, one exception being Dye (1969, p. 232) who said *Erwinia stewartii* should be retained as a separate species rather than made synonymous with the saprophyte *Erwinia herbicola.* Any distinct pathological entity must receive a convenient label by which it can be listed in quarantine regulations. There must be, 'precision to host limited taxa.'

With regard to *Xanthomonas* and *Pseudomonas*, the most radical proposal is that of De Ley (1968) that the entire genus *Xanthomonas* should be merged with *Pseudomonas* as a single species *Pseudomonas campestris.* This proposal is based on the following evidence: (1) alleged adaptation of species to new hosts, (2) the results of extensive phenotypic comparisons using the approach of numerical taxonomy, (3) molecular biological data including studies on the %G + C in the DNA and hybridization of single stranded DNA from different sources. Some of this evidence is now questioned; for example, alleged change in host specificity must be re-examined in the light of knowledge on the nature of the hypersensitive reaction (Klement & Goodman, 1967; Klement, 1968). Furthermore, some of the experiments that determined a high degree of homology between *Pseudomonas* and *Xanthomonas* DNA, were made at 'permissive' annealing temperatures, that allow nonspecific associations. It will be of interest whether there is still a high degree of homology between *Xanthomonas* species when experiments are carried out at 'restrictive' annealing temperatures. Stanier (1970) concluded from DNA-DNA hybridization and competition studies at two annealing temperatures, that the genus *Pseudomonas* should be split up into a series of genera, because even at permissive annealing temperatures many of the *Pseudomonas* species compared showed no detectable evidence of being genetically related. Stanier suggests that the methods for phenotypic characterization of species of *Pseudomonas* may not be adequate since differences in degree of hybridization were sometimes not reflected in phenotypic differences.

References

De Ley, J., 1968. DNA base composition and hybridization in the taxonomy of phytopathogenic bacteria. A. Rev. Phytopath. 6: 63–90.

Dye, D. W., 1969. A taxonomic study of the genus *Erwinia*. III. The 'herbicola' group. N.Z. Jl Sci. 12: 223–236.

Kelman, A., 1953. The bacterial wilt caused by *Pseudomonas solanacearum:* a literature review and bibliography. Techn. Bull. N. Carol. agric. Exp. Stn 99: 1–194.

Klement, Z., 1968. Pathogenicity factors in regard to relationships of phytopathogenic bacteria. Phytopathology 58: 1218–1221.

Klement, Z. & R. N. Goodman, 1967. The hypersensitive reaction to infection by bacterial plant pathogens. A. Rev. Phytopath. 5: 17–44.

Stanier, R. Y., 1970. Towards an evolutionary classification of the bacteria. Contribution to Symposium K, 'Numerical versus conventional taxonomy', at the Tenth International Congress of Microbiology, Mexico City, 1970. For 'Recent Progress in Microbiology, Vol. X'. (to be published).

The use of 'special forms' in the classification of phytopathogenic bacteria

Ž. Tesic

Microbiological Institute, Faculty of Agriculture, Zemun, University of Beograd, Yugoslavia

Abstract

In solving the taxonomic problems of phytopathogenic bacteria two different ways are used. Most microbiologists, in determining their general biological properties, apply the usual procedure and rarely apply inoculation tests. The majority of phytopathologists overestimate the acquired phytopathogenic properties of these bacteria by applying cross-inoculation tests and many general properties are overlooked in this way. It is obvious that neither of these one-sided procedures can solve the problem successfully and that only simultaneous investigations of biological properties and phytopathogenic capacities can produce satisfactory results. Therefore, the author classifies these bacteria in larger genetic species with a number of special forms for the 'erwinia' and 'agrobacterium' groups.

We are still far from a satisfactory solution of the nomenclatural and classification problems in phytopathogenic bacteria. In the past, the specificity of these bacteria was emphasized by many phytopathologists in establishing 'special genera' (such as *Erwinia*, *Xanthomonas*, *Phytomonas* and *Phytobacterium)* and higher taxonomic groups. Contemporary studies by various microbiologists, and partly also by phytopathologists have shown, however, that many groups of bacteria include both saprophytic and phytopathogenic bacteria which should be classified in the same genera or should be included in one species (Krasilnikov, 1949; Tesic, 1962). Such bacteria have undergone, by adaptation to new natural habitats (on the host plants), changes in their physiology which, however, are only in rare cases sufficient to establish 'larger' species (Dowson, 1957).

The hypotheses that phytopathogenic bacteria originate from saprophyte ancestors (Tesic, 1962; Gorlenko, 1965) stresses the biological unity between both groups. Consequently, the 'special genera' should finally be abandoned and they should be grouped into lower taxonomic categories within larger species. In 1965 I suggested that in drawing up a more modern bacterial classification, these bacteria should be included in separate sections and series within the usual genera or, in some cases, in a special category within a species.

In solving such taxonomic problems two methods are used. In determining their biological characteristics, most microbiologists (Colwell & Liston, 1961; Martinec & Kocur, 1963, 1964) do not take into account the special properties of the phytopathogenic bacteria in causing plant diseases, so that they rarely base their conclusions on inoculation tests. The result is that these bacteria are classified as smaller units within larger species, which frequently reduces them to a sub-specific level so that their names become synonyms for some similar species. This procedure, in which essential characteristics of these bacteria, though certainly existing in nature, are left out of consideration in grouping is rather one-sided.

Most phytopathologists (Magrou, 1937; Stapp, 1956; Burkholder, 1957) deal with the problem in another way. The overestimate changes incurred in the course of acquiring phytopathogenic properties, as a rule established from cross inoculation with series of plants and, doing so, they overlook many biological properties, so that the actual relationship between closely allied bacteria is nomore considered. In such cases the adage is: 'a new host plant – a new phytopathogenic bacterium'. This results, as Starr (1959) has critically observed, in the description of an unjustified large number of 'species' which actually do not exist as separate units in nature. Obviously, neither of these one-sided procedures can supply an adequate solution of the problem. Only simultaneous investigations on both biological and phytopathogenic properties can produce satisfactory results.

In earlier papers (Tesic, 1962, 1965) I have often emphasized that in solving taxonomic problems among phytopathogenic bacteria, a single procedure should be followed, including both biological and phytopathogenic properties. In this way the properties acquired in the course of adaptation to particular host plants are adequately appraised. As a rule they can be established by cross-inoculation tests on a series of related plants. I support, therefore, those microbiologists and phytopathologists who include phytopathogenic bacteria in larger species as subspecies (or varieties) or special forms; a consistent taxonomic classification cannot be imagined without the introduction of such categories.

The International Code of Nomenclature for Bacteria (Buchanan, 1966) mentions these taxonomic categories (e.g. in Rule 8), though microbiologists and phytopathologists not yet fully agree on their application (Krasilnikov, 1949; Stapp, 1956; Dowson, 1957). We support Starr's view (1959) that subspecies (varieties) should be based on biological differences appearing in cultures within larger biological species, whereas with respect to special forms recommendation 8a of the Code can be strictly applied: 'A special form ('forma specialis') is an intrasubspecific form included in an intrasubspecific subdivision of a species of a parasitic, symbiotic or commensal host or habitat. It is named, preferably, by use of the scientific name of the host written in the genitive' (Buchanan, 1966). Hence special forms would be based upon phytopathogenic properties.

Consequently, 'subspecies' and 'special forms' among phytopathogenic bacteria, far from excluding each other, fit in perfectly and they complement each other: subspecies (varieties) show general biological characteristics, special forms show phytopathogenic properties in special host plants. It is not justified to leave out the latter category, as some microbiologists use to do (Martinec & Kocur, 1963; Burkholder, 1960), in arguing that they frequently have biological characteristics in common, even when attacking various plants. This may justify their classification in subspecies (varieties), but is of no value for their phytopathogenic properties. The second characteristic can only be revealed with cross-inoculation tests, and special forms should be excluded only if differences in behaviour toward various other plants are absent. It is strange that the same microbiologists consider it superfluous to introduce special forms in the classification of these bacteria and that they accept the extreme phytopathologic view in 'special genera'.

No doubt the introduction of special forms and subspecies (varieties) calls for certain alterations in the present names of these bacteria. When it is proved that, biologically, they belong to larger phytopathogenic species, special forms could be named after the scientific name of the host plant written in genitive (Buchanan, 1966). This also involves certain changes when the previous species names do not include the name of the host plant, which is rather frequent. If it is proved, however, that a bacterium belongs to some saprophyte species (owing to its general biological characteristics), then the special form should be named after the scientific name of the host plant written in the genitive, or by adding the suffix '-vora' or '-phaga', or both if needed. Finally, the use of special forms as well as subspecies (varieties) makes it necessary to extend the binomial nomenclature of phytopathogenic bacteria; for the present this cannot be avoided. It should be mentioned, by the way, that the names of commensal bacteria could end on '-cola'; for symbiotic bacteria the suffix '-phila' might be used in the same way.

Some examples from the 'Erwinia-group' of phytopathogenic bacteria may be added to illustrate these proposals (Martinec & Kocur, 1963; Tesic, 1967).

Bacterium amylovorum f. sp. *rosacearum* Tesic, 1967. Basionyms and synonyms: *Micrococcus amylovorus* Burrill, 1882; *Bacillus amylovorus* Trevisan, 1889; *Bacterium amylovorum* Chester, 1897; *Erwinia amylovora* Holland, 1920.

Bacterium amylovorum f. sp. *cucurbitacearum* Tesic, 1967. Basionyms and synonyms: *Bacillus tracheiphilus* Smith, 1895; *Bacterium tracheiphilum* Chester, 1897; *Erwinia tracheiphila* Holland, 1920; *Erwinia amylovora* Martinec et Kocur, 1963.

Bacterium amylovorum f. sp. *salicis* Tesic, 1967. Basionyms and synonyms: *Bacterium salicis* Day, 1924; *Phytomonas salicis* Magrou, 1937; *Erwinia salicis* Chester, 1939; *Erwinia amylovora* var. *salicis* Martinec et Kocur, 1963.

Bacterium carotovorum f. sp. *carotovorum* Tesic, 1967. Basionyms and synonyms: *Bacillus carotovorus* Jones, 1901; *Erwinia carotovora* Holland, 1920; *Bacterium carotovorum* Lehmann & Neumann, 1927; *Pectobacterium carotovorum* Waldee, 1945.

Bacterium carotovorum f. sp. *solani* Tesic, 1967. Basionyms and synonyms: *Bacillus atrosepticus* van Hall, 1901; *Erwinia atroseptica* Bergey, 1934; *Bacillus phytophthorus* Appel 1902; *Erwinia phytophthora* Bergey, 1934; *Bacterium phytophthorum* Burgwitz, 1935; *Erwinia carotovora* Martinec et Kocur, 1963.

Bacterium carotovorum f. sp. *ulmi* Tesic, 1967. Basionyms and synonyms: *Erwinia nimipressuralis* Carter, 1945; *Bacterium nimipressurale* Tesic, 1949; *Erwinia carotovora* Martinec et Kocur, 1963.

Bacterium carotovorum f. sp. *lathyri* Tesic, 1967. Basionyms and synonyms: *Bacillus lathyri* Manns et Taubenhaus, 1913; *Erwinia lathyri* Holland, 1920; *Bacterium lathyri* Burgwitz, 1935; *Erwinia carotovora* Martinec et Kocur, 1963.

The following examples refer to the 'Agrobacteria' group of phytopathogenic bacteria (De Ley *et al.*, 1966):

Bacterium radiobacter f. sp. *tumefaciens* Tesic, 1967. Basionyms and synonyms: *Bacterium tumefaciens* Smith et Townsend, 1907; *Pseudomonas tumefaciens* Stevens, 1913; *Bacillus tumefaciens* Izrailsky, 1926; *Polymonas tumefaciens* Lieske, 1928; *Agrobacterium tumefaciens* Conn, 1942; *Rhizobium tumefaciens* Izrailsky, 1960; *Agrobacterium radiobacter* var. *tumefaciens* De Ley *et al.*, 1966.

Bacterium radiobacter f. sp. *rubi* Tesic, 1967. Basionyms and synonyms: *Pseudomonas rubi* Hildebrand, 1940; *Bacterium rubi* Hildebrand, 1940; *Phytomonas rubi* Hildebrand, 1940; *Agrobacterium rubi* Starr et Weiss, 1943; *Agrobacterium radiobacter* var. *rubi* De Ley *et al.*, 1966.

Bacterium radiobacter f. sp. *gypsophilae* Tesic, 1967. Basionyms and synonyms: *Bacterium gypsophilae* Brown, 1934; *Pseudomonas gypsophilae* Stapp, 1935; *Phytomonas gypsophilae* Magrou, 1937; *Xanthomonas gypsophilae* Magrou et Prévot, 1948; *Agrobacterium radiobacter* var. *gypsophilae* De Ley *et al.*, 1966.

References

Bergey's Manual of determinative bacteriology, 1957. Baltimore.

Buchanan, R., 1966. International code of nomenclature for bacteria. Ames.

Burkholder, W., 1960. Present-day problems pertaining to the nomenclature and taxonomy of the phytopathogenic bacteria. Acad. Rep. Pop. Romine, Bucaresti.

Colwell, R. & J. Liston, 1961. Taxonomic analysis with the electronic computer of some *Xanthomonas* and *Pseudomonas* species. J. Bact. 82: 913–919.

De Ley, J., M. Bernaerts, A. Rassel & J. Guilmont, 1966. Approach to an improved taxonomy of the genus *Agrobacterium*. J. gen. Microbiol. 43: 7–17.

Dowson, W., 1957. Plant diseases due to bacteria. Cambridge.

Gorlenko, M., 1965. Bakterialjnye bolezni rastenij. Moskva.

Izrailsky, V., 1960. Bakterialjnye bolezni rastenij. Moskva.

Krasilnikov, N., 1949. Opredelitelj bakterii i aktinomicetov. Moskva.

Magrou, J., 1937. In: Hauduroy's Dictionnaire des bactéries pathogènes. Paris.

Martinec, T. & M. Kocur, 1964. A taxonomic study of *Erwinia amylovora* (Burrill, 1882) Winslow *et al.*, 1920. IBBNT 14: 5–14.

Stapp, C., 1958. Pflanzenpathogene Bakterien. Berlin.

Starr, M., 1959. Bacteria as plant pathogens. A. Rev. Microbiol. 13: 211–238.

Tesic, Ž., 1962. Problems of the origin and classification of phytopathogenic bacteria. J. agr. Sci., Beograd 15(50): 127–146.

Tesic, Ž., 1967. The special forms in the classification of phytopathogenic bacteria. Plant Protection, Beograd 93/95: 45–53.

A quantitative method for assessing virulence of Erwinia carotovora var. carotovora and E. carotovora var. atroseptica and susceptibility to rotting of potato tuber tissue

M. C. M. Pérombelon

Scottish Horticultural Research Institute, Invergowrie, Dundee, Great Britain

Abstract

A quantal assay is described which enables the quantifying of both virulence of *Erwinia carotovora* on potato tuber tissue and susceptibility to rotting of the latter under various environmental conditions imposed before and after inoculation. Thin tuber tissue discs are inoculated with serial dilutions of the test organism. The subsequent soft rotting, or not, of the discs at each dose level is the basis of the quantal counts which, after transformation into probits, allow the determination of the ED_{50} from a dose-response curve. Not only is it possible to compare virulence of *E. carotovora* isolates and susceptibility of tuber tissue of different ages under various environmental conditions such as aerobic and anaerobic, but an indication can also be obtained of how the organisms produce the response.

Introduction

Extensive sampling over several years has indicated that nearly all tubers in Scotland are probably contaminated by *Erwinia carotovora* Holland var. *atroseptica* (van Hall) Dye and *E. carotovora* Holland var. *carotovora* (Pérombelon, 1969). In this country, bacterial soft rot in bulk stored potato tubers is a constant potential hazard.

One approach to this problem, which we are pursuing, is a study of the susceptibility to rotting of tuber tissue under various environmental conditions imposed before and after inoculation with *E. carotovora* and other pectolytic bacteria. As part of this study we have devised a quantal assay which also provides a model for the study of the more general problem of the relationship of inoculum density to incidence and development of infection. Similar studies have been reported for various host/pathogen combinations involving especially animals (Meynell, 1957) and occasionally plants (Garrett, 1966), but none to our knowledge for a plant/bacterium combination.

Materials and methods

Serial dilutions of the inocula, using a factor of 4, were prepared from nutrient agar slopes 48h after inoculation with the test organism. Population densities of the inocula were determined on nutrient agar by a spot plate viable count procedure, as described elsewhere (Pérombelon & Lowe, 1971).

Unless otherwise stated, discs of tuber tissue (cv Majestic, stored at 5 °C) 1.0 cm diameter, ca 1.5 mm thick, were washed in sterile water and the water potential deficit of the discs brought to ca −1 atm by standing in sterile aerated water at 1–2 °C for 1.5h. Each experiment was carried out with discs obtained from one tuber (ca 650 discs). After drying under uniform conditions the discs were placed in sterile dishes and 10 or 20 replicated discs/dilution of inoculum were inoculated by placing 1 drop (0.02 ml) on each disc. They were then incubated at 20 °C for 5 days either in air at 100% relative humidity or anaerobically under sterile paraffin oil. The effect of suberization on rotting was examined by delaying inoculation of the discs for 18h during which time they were held at 20 °C in air at 100% relative humidity. The response at each dose level (dilution of inoculum) was assessed by counting the number of discs with at least 1 soft rot lesion.

Results

Table 1 gives the results of a typical infectivity titration experiment in which 20 discs/dose levels were serially inoculated with suspensions of *E. carotovora* var. *atroseptica* (isolate 27) and incubated in air and under paraffin oil. Assuming that the bacteria are randomly distributed between the drops and that every viable organism is capable of producing a response (rotting in a disc), as has often been found elsewhere (Meynell, 1958; Wastie, 1962; McKee, 1964), departure from exponentiality of the responses can be tested using Moran's method (Moran, 1954a, b). The calculated M values were 3.749 and 1.975 for responses under aerobic and anaerobic conditions, respectively, while the value for significant deviation at $p \leqslant 0.01$ is $\geqslant 2.326$, suggesting that in this case only the dose response curve for the anaerobic treatment is not different from the predicted curve. In other experiments it was not uncommon, however, for both curves to be non-exponential.

Table 1. Incidence of rotting in discs of potato tuber tissue (cv. Majestic) inoculated dropwise (0.02 ml/disc) with serial dilutions of a suspension of *E. carotovora* var. *atroseptica* (isolate 27) and incubated aerobically (in air) and anaerobically (under paraffin oil).

Viable counts (bacteria/drop)	Aerobic incubation		Anaerobic incubation	
	discs inoculated	discs rotting	discs inoculated	discs rotting
5.82×10^3	20	19		
1.46×10^3	20	18		
3.64×10^2	20	12		
9.1×10^1	20	10	20	20
3.0×10^1	20	3	20	20
7.6	20	1	20	18
1.8	20	0	20	13
0.45	20	0	20	8
0.10			20	2

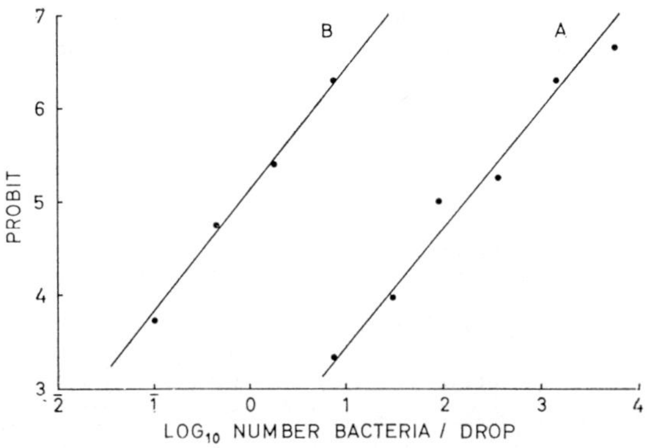

Probit/log dose relationship for potato tuber discs inoculated with *E. carotovora* var. *atroseptica* (isolate 27) and incubated under aerobic (A) and anaerobic (B) conditions.

If the dose response relationship of an infectivity titration using a quantal response is derived from the first term of a Poisson series, variability in response is unavoidable, particularly when using low numbers of replicates (Meynell, 1957). Differences in host resistance further increase the overall variation in the quantal responses. To compensate for this inherent variability and to obtain linearity, it is usual to plot the responses transformed into probits (Fisher & Yates, 1963, Table IX) against the logarithm of the doses. The data of Table 1 are represented graphically in the figure following such a transformation. The parameters of the probit/log dose straight line are its mean (corresponding to a probit of 5 equivalent to 50% responses also known as ED 50) and its slope, the change in probit produced by unit change in log dose (Meynell & Meynell, 1965). Of the various methods available to characterize these parameters, probit analysis is the most efficient and flexible (Finney, 1964).

The regression equations of the dose response data of Table 1 were calculated according to Finney's method using the maximum likelihood solution with one iterative cycle. They gave for the aerobic ($+ O_2$) incubation treatment:

$$y = 2.80 + 1.18\ (\pm\ 0.19)\ x$$

and for the anaerobic ($- O_2$) incubation treatment:

$$y = 5.11 + 1.25\ (\pm\ 0.26)\ x.$$

The χ^2 tests carried out on the two sets of data gave no evidence of heterogeneity of departure from the fitted probit lines ($\chi^2_{(4)+O_2} = 1.59$ and $\chi^2_{(2)-O_2} = 2.63$). A test for parallelism of the regressions gave a $\chi^2_{(1)}$ of 0.06 suggesting similarity of the slopes. The ED_{50} values and their fiducial limits at $p = 0.05$ were found to be 165 (301, 91) and 0.82 (1.44, 0.46) bacteria for the aerobic and anaerobic treatments, respectively.

The computation is however long and tedious, but the ED_{50} and the slope can often be estimated with reasonable accuracy by graphical means. If only ED_{50} and its confidence limits are needed, the moving average method of Thompson (1947) simplified by the use of tables prepared by Weil (1952), is an adequate alternative.

An indication of the nature of rotting potential or virulence of *E. carotovora* and how rotting occurs in tuber tissue may be obtained from a study of dose response relationships. In a preliminary investigation different dose response relationships were examined and their respective ED_{50} values were calculated by means of Thompson's

Table 2. ED_{50} of different dose response combinations obtained by inoculating potato tuber discs with serial dilutions of suspensions of *E. carotovora*, incubated aerobically and anaerobically.

	Isolate	Aerobic incubation ED_{50}	Anaerobic incubation ED_{50}
Experiment 1			
a. discs inoculated at 0 h	27	115.7	2.8
b. discs inoculated at 18 h	27	833.9	30.5
Experiment 2	40	54.1	1.3
	G148	18.8	0.5
	27	178.0	1.0
	C1	508.2	1.5
Experiment 3	362/23	14.1	0.2
	14/B	8.7	0.8
	SR1	293.1	5.2
	714/14	200.5	3.0

Experiment 1: effect of suberization of disc cells on rotting.
Experiments 2 and 3: comparison of rotting potential of 3 isolates of var. *carotovora* (40, G148, 362/23 and 14/B) and 3 of var. *atroseptica* (27, C1, SR1 and 714/14).
Each experiment was carried out on discs obtained from one tuber.

method (Table 2). It is evident that the susceptibility of tuber tissue to rotting by *E. carotovora*, as quantified by the ED_{50} values, depends primarily on the presence or absence of O_2. In the presence of O_2 rotting appears to depend on whether or not the invading organisms can multiply fast enough to reach their critical density (10^6 bacteria/lesion at which a rotting lesion becomes visible) before suberization of the disc cells restricts growth of the bacteria. Similarly the number of sites where growth of the organism can proceed unimpeded is progressively reduced the more extensive suberization is prior to inoculation (Table 2, Exp. 1). Since there is little or no suberization under anaerobic conditions, one bacterium given time (usually 5 days) could give rise to a soft rotting lesion. It is interesting to note that while var. *carotovora* isolates tested appear to rot tuber tissue more easily than those of var. *atroseptica* in air, all of them are equally virulent under anaerobic conditions (Table 2, Expts 2 and 3).

The hypothesis of independent action for bacterial infection (Halvorson, 1935; Meynell & Stocker, 1957) for both aerobic and anaerobic conditions is supported in this example by the following points: (i) Considerations of the similarity in the slopes of the probit/log dose lines for the aerobic and anaerobic treatments (Horsfall, 1956); (ii) the fact the slopes are < 2 (Peto, 1953); (iii) that incubation periods for expression of responses at doses < 1 ED_{50} were found to be equal (Meynell & Meynell, 1958); and (iv) the predominance in rotting discs of one of 2 distinguishable organisms (var. *carotovora* and var. *atroseptica*) when inoculated in a 1:1 ratio at doses < 1 ED_{50} (Meynell, 1957). Therefore, the assumption made earlier that every viable organism is capable of producing a response, provided that a suitable site for growth is also available, is valid.

To conclude we can say that the virulence of *E. carotovora* (equated here with rotting potential on tuber tissue) is perhaps not necessarily an intrinsic characteristic of the organism but a reflection of the availability of suitable sites for growth and infection.

References

Finney, D. J., 1964. Probit analysis. 2nd ed. Cambridge University Press.

Fisher, R. A. & F. Yates, 1963. Statistical tables for biological, agricultural and medical research. 6th ed. Oliver & Boyd, Edinburgh.

Garrett, S. D., 1966. Spores as propagules of disease. In: The fungus spore, p. 309–318. Butterworths, London.

Halvorson, H. O., 1935. The effect of chance on the mortality of experimentally infected animals. J. Bact. 30: 330–331.

Horsfall, J. G., 1956. Principles of fungicidal action. Chronica Botanica, Walthan, Mass.

McKee, R. K., 1964. Observations on infection by *Phytophthora infestans*. Trans. Brit. mycol. Soc. 47: 365–374.

Meynell, G. G., 1957. Inherently low precision of infectivity titrations using a quantal response. Biometrics 13: 149–163.

Meynell, G. G. & E. W. Meynell, 1958. The growth of micro-organisms *in vivo* with particular reference to the relation between dose and latent period. J. Hyg., Camb. 56: 323–346.

Meynell, G. G. & E. W. Meynell, 1965. Theory and practice in experimental bacteriology. Cambridge University Press.

Meynell, G. G. & B. A. D. Stocker, 1957. Some hypotheses on the aetiology of fatal infections in partially resistant hosts and their application to mice challenged with *Salmonella paratyphi-B* or *Salmonella typhimurium* by intraperitoneal infection. J. gen. Microbiol. 16: 38–58.

Moran, P. A. P., 1954a. The dilution assay of viruses. J. Hyg., Camb. 82: 189–193.

Moran, P. A. P., 1959b. The dilution assay of viruses. II. J. Hyg., Camb. 52: 444–446.

Pérombelon, M. C. M., 1969. The Scottish Horticultural Research Institute. Ann. Rep. 1968.

Pérombelon, M. C. M. & R. Lowe, 1971. The effects of bile salts media and age of inocula in quantitative studies of populations of *Erwinia carotovora* var. *carotovora* and *E. carotovora* var. *atroseptica*. J. appl. Bact. (in press).

Peto, S., 1953. A dose-response equation for the invasion of micro-organisms. Biometrics 9: 320–335.

Thompson, W. R., 1947. Use of moving averages and interpolation to estimate median effective dose. Bact. Rev. 11, 115–145.

Wastie, R. L., 1962. Mechanism of action of an infective dose of *Botrytis* spores on bean leaves. Trans. Brit. mycol. Soc. 45: 465–473.

Weil, C. S., 1952. Tables for convenient calculation of median effective dose (LD_{50} or ED_{50}) and instructions in their use. Biometrics 8, 249–263.

Modifying the virulence and host range of weed pathogens

D. C. Sands and A. D. Rovira

Connecticut Agricultural Experiment Station, New Haven, Conn., U.S.A.
CSIRO, Division of Soils, Glen Osmond, South Australia

Bacterial plant pathogens are potentially useful as agents of biological control. For this purpose, they must meet stringent requirements of host specificity, have high virulence and stability, high rates of dissemination, and have minimal ecological side-effects. In the past, a few micro-organisms have been used satisfactorily to control weeds. However, for control of noxious weeds, pathogens that meet all these requirements are simply not available. The usefulness of a given pathogen is most commonly limited by its not being sufficiently host specific. A procedure for irreversibly narrowing the host range of a pathogen is necessary if it is to be modified for biological control of weeds. Experimentally, we have attempted to alter the virulence of *Pseudomonas solanacearum* to tomato and of *P. syringae* to *Echium plantagineum* by use of sublethal doses of the mutagen nitrosoguanidine. This procedure has yielded a range of effects, including auxotrophy, lowering of virulence and, in rare cases, an apparent increase in virulence compared to wildtype. The procedure, to be described, involves a mutation-inoculation cycle, which evaluates methods and specific treatments in narrowing the host range of a pathogen.

This mutation-inoculation cycle involved a repeated sequence of four steps: (1) severe mutagenesis of a multiple host pathogen, (2) inoculation of the mixture of survivors into the desired weed host, (3) isolation of bacteria from infected tissue, and (4) test host range of isolates and repeat cycle. After three such cycles we have mutants of *P. solanacearum* with greatly decreased virulence on tomato and with slightly increased virulence on *Solanum dulcamara*, the weed host. The mutants still infect tomato at lower temperatures and when the light intensity is low. With further cycling and with different mutagens, the prospects remain good for the creation of host-specific mutants.

Inhibition of Pseudomonas syringae by saprophytic bacterial isolates in culture and in infected plant tissues

W. M. Dowler

Plant Science Research Division, U.S. Department of Agriculture, Clemson, South Carolina

Abstract

Peach trees have shown, besides *Pseudomonas syringae*, to be inhabited by a non-pathogenic bacterium of the same size, shape and Gram stain. *In vitro*, the saprophyte grows twice as fast as the pathogen at 25 °C, equally fast at 15 °C. In mixed cultures at 25 °C the saprophyte markedly inhibits the pathogen. Mixed cultures 50:50, when injected into green bean pods, have resulted in a 100-fold increase of the saprophyte compared with tissues inoculated with the saprophyte alone. In the same tissue the pathogen has shown a 16-fold decrease compared with inoculations with the pathogen alone.

Higher temperatures in spring favour the saprophyte which may inhibit *P. syringae*.

Introduction

During fall and in winter, the pathogenic forms of *Pseudomonas syringae*, causing bacterial canker of stone fruits in the southeastern United States, are readily isolated from diseased trees. Under warm spring conditions, the same trees may carry a non-pathogenic fluorescent bacterium similar to this pseudomonas. Petersen & Dowler (1965) and Dowler & Petersen (1966) showed that bacterial canker contributes to the early death of peach trees. However, in spring the presence of the saprophyte makes it difficult to detect the pseudomonad and to prove its association with the disease phenomena. This suggests that saprophytic bacteria might inhibit the activity of virulent pseudomonas strains in diseased tissues.

So research was carried out to determine the effect of the saprophyte on virulent isolates of *Pseudomonas syringae*, both *in vitro* and *in situ*.

Methods and materials

Cultures were grown in liquid media containing 8g nutrient broth (Difco)[1], 5g yeast extract (Difco), 10g dextrose and 1 litre distilled water. Growth was measured turbidi-

1. Mention of a trademark or proprietary product does not constitute a guarantee or warranty of the product by the U.S. Department of Agriculture, and does not imply its approval to the exclusion of other products that may also be suitable.

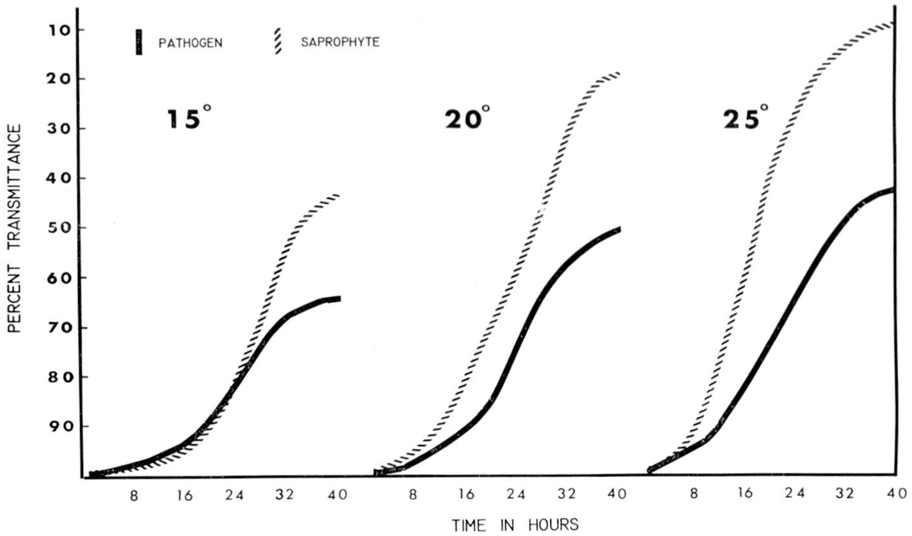

Fig. 1. Comparative growth at 15, 20, and 25 °C of unidentified fluorescent saprophytic bacteria and *Pseudomonas syringae* isolated from peach.

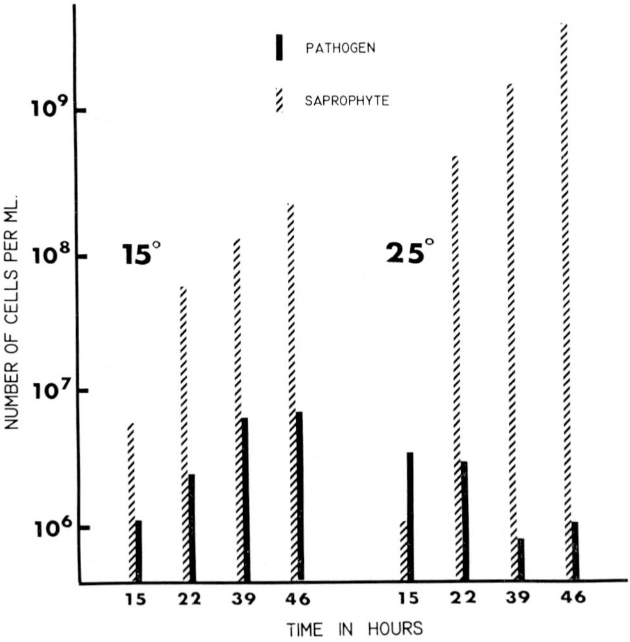

Fig. 2. Cell concentration of saprophytic bacteria and *Pseudomonas syringae* at various time intervals in mixed liquid culture at 15 and 25 °C. Initial inoculum contained equal numbers of each.

metrically at 2 or 3 h intervals by using a Spectronic 20 colorimeter at 590 nm. To determine growth in mixed culture, a virulent strain of *Pseudomonas* resistant to 200 ppm streptomycin and a saprophytic isolate were used. The desired numbers of cells of this strain and the saprophyte were mixed and grown in liquid culture at 15 and 25 °C. Dilution plates were prepared on Difco Pseudomonas F agar (PSF) or on PSF with streptomycin at 200 ppm. This allowed differentiation between the pseudomonad and the saprophyte.

Green bean pods were inoculated with aqueous cell suspensions of *P. syringae*, the saprophyte, or mixtures of both. About 0.1 ml was injected sub-epidermally with a syringe with a 26 gauge needle. This produced a water-soaked area of 6 to 7 mm diameter. Four areas were inoculated per pod. The inoculated pods were placed on moist paper towels in a sterile Pyrex dish, covered with aluminum foil, and incubated 24 or 48 h at 25 °C.

The bacterial populations in the inoculated tissues were determined as follows. The four pieces of inoculated tissue from each of the two pods were excised, weighed, and blended for 30 sec in 100 ml sterile distilled water in a Waring Blendor. The resulting homogenate was serially diluted, and dilution plates were prepared on PSF agar or PSF with 200 ppm streptomycin. Counts of bacterial colonies on the appropriate dilution plates were made after 24 and 48 h at 25 °C.

Peach seedlings grown in a greenhouse were stab-inoculated near the growing tip with a dissecting needle dipped in bacterial suspensions of the same density of pseudomonads, the saprophyte, or mixtures of both. After 48 h, the tips of the inoculated plants were excised and blended in 100 ml sterile distilled water. Suitable dilution plates of PSF and PSF with streptomycin at 200 ppm were prepared to determine the number of bacteria present in the plant tissue.

Results and discussion

Temperature affected the growth of *Pseudomonas* and the saprophyte in liquid cultures in different ways (Fig. 1). At 25 °C the saprophyte multiplied much more rapidly than the pathogen, but at 15 °C there was no difference in growth of the two bacteria within 24 h. The temperature of 25 °C is similar to that in orchards in spring; that of 15 °C is still higher than normal during mid-winter in southeastern peach orchards.

In the mixed liquid culture at 15 °C, the saprophyte grew faster than the pathogen, but the number of cells of the pathogen was still relatively high (Fig. 2). However, at 25 °C the saprophyte strongly inhibited the growth of the pathogen, resulting in a decrease of the original number of cells of *Pseudomonas* (Fig. 2). This indicates the stronger competitive ability of the saprophyte at 25 °C.

Bean pods inoculated with mixtures of the organisms favoured rapid multiplication of the saprophyte, which apparently suppressed the pathogen (Table 1). This was particularly pronounced when the inocula were mixed 50:50. Recovery of saprophytic bacteria from these samples was a hundred times higher than from tissue inoculated with the saprophyte alone, even though only half as many saprophyte cells were

Table 1. Number of fluorescent saprophytic and pathogenic bacteria isolated from inoculated green bean pods after 48 hours at 25 °C.

Type of inoculum	Number of bacteria per mg fresh weight	
	saprophyte	pathogen
Pathogen	0	97,570
Saprophyte	1,202	0
Pathogen + saprophyte (50:50)	155,522	6,172
Saprophyte + pathogen (1:99)	19,130	77,308
Pathogen + saprophyte (1:99)	4,176	119
Control (sterile water)	0	0

Averages of four experiments. The initial injection contained about 10^4 cells per mg tissue.

Table 2. Numbers of fluorescent saprophytic and pathogenic bacterial cells isolated per mg fresh weight from inoculated growing tips of peach seedlings.

Type of inoculum	After 48 h			After 120 h		
	saprophyte	pathogen	ratio	saprophyte	pathogen	ratio
Pathogen	0	6,945			4,400	
Saprophyte	1,110	0		3,450	0	
Saprophyte + pathogen (50:50)	987	453	2:1	150	27	6:1
Saprophyte + pathogen (1:99)	210	945	1:5	45	87	1:2
Pathogen + saprophyte (1:99)	47	253	1:5	150	15	10:1
Control (sterile water)	0	0		0	0	

inoculated initially. This interaction between pathogen and saprophyte shows some of the complexities observed with mixed populations in nature.

Peach seedlings inoculated with mixtures of saprophyte and pathogen appeared to be less damaged than those inoculated with the pathogen alone, but the differences were not statistically significant. In mixtures, saprophyte and pathogen seemed to inhibit each other, resulting in a total number of bacteria in stem tissue that was lower than that with either isolate alone (Table 2).

Apparently the saprophyte can multiply in peach tissue, suggesting the possibility of a biological control of *Pseudomonas*. This is not new: previous reports have already indicated that saprophytic or avirulent bacteria may affect the severity of bacterial plant diseases (Baldwin & Goodman, 1963; Crosse, 1965; Goodman 1965). Since a chemical control of bacterial diseases is often difficult or impossible, this opens the way for biological methods.

It is unknown whether the saprophyte is part of the natural bacterial flora of the tree or has originated as a mutant from the virulent forms.

The effect of temperature on *P. syringae* and a fluorescent saprophytic organism, and their interactions, indicate what may occur under natural conditions in the orchard. At lower temperatures (15 °C) the pathogen grows at a rate comparable to that of the saprophyte, which may explain the relatively frequent isolation of the pathogen during winter. At higher temperatures in spring (25 °C), often reached in our orchards during April and May, growth of the saprophyte is favoured.

References

Baldwin Jr, C. H. & R. N. Goodman, 1963. Prevalence of *Erwinia amylovora* in apple buds as detected by phage typing. Phytopathology 53: 1299–1303.

Crosse, J. E., 1965. Bacterial canker of stone-fruits. VI. Inhibition of leaf-scar infection of cherry by a saprophytic bacterium from the leaf surfaces. Ann. appl. Biol. 56: 149–160.

Dowler, W. M. & D. H. Petersen, 1966. Induction of bacterial canker of peach in the field. Phytopathology 56: 989–990.

Goodman, R. N., 1965. *In vitro* and *in vivo* interactions between components of mixed bacterial cultures isolated from apple buds. Phytopathology 55: 217–221.

Petersen, D. H. & W. M. Dowler, 1965. Bacterial canker of stone fruits in the southeastern States. Pl. Dis. Reptr 49: 701–702.

Differences in motility between virulent and avirulent cells of Pseudomonas solanacearum

A. Kelman and Judith Hruschka

Department of Plant Pathology, University of Wisconsin, Madison, Wisconsin 53706

Abstract

Relative motility of virulent and avirulent cells of the wilt pathogen, *Pseudomonas solanacearum*, was investigated as a possible factor influencing population changes and loss in virulence in non-aerated broth cultures. Cells of virulent isolates grown on standard solid or liquid media were either non-motile or less actively motile than cells of avirulent isolates. Actively motile avirulent variants of *P. solanacearum* appeared at the surface of non-aerated broth cultures after incubation for 48–72 hours. Aerotaxis of motile avirulent variants could be detected within 4–6 hours in glass tubes containing semi-solid motility agar.

Shifts in populations from virulent to avirulent cells in non-aerated broth cultures apparently reflect the preferential growth of motile avirulent cells under conditions in which virulent cells are not actively motile.

Introduction

Rapid loss in virulence in certain culture media characterizes the wilt pathogen, *Pseudomonas solanacearum* (Kelman, 1953). In non-aerated liquid media rapid attenuation is correlated with the increase in numbers of avirulent or weakly virulent variants forming butyrous colonies differing markedly in appearance on tetrazolium agar (TZC) from virulent isolates (Kelman, 1954; Husain & Kelman, 1958). Preliminary studies indicated that virulent cells from TZC plates were non-motile in hanging drops, whereas avirulent types were actively motile. The objectives of this investigation were to determine whether differences in motility between virulent and avirulent cells of *P. solanacearum* could be related to the population shifts in non-aerated broth cultures.

Materials and methods

The isolates mainly studied were the virulent fluidal strain of *P. solanacearum* (25-K60) and an avirulent culture (B1-1) (Husain & Kelman, 1958). Other isolates (26) from various solanaceous hosts and different geographic areas were also examined. Avirulent butyrous colony types were selected from well-isolated colonies on tetrazolium agar (TZC) plates streaked from tryptone yeast-extract glucose (TYD) broth cultures of different isolates after growth periods of 6–8 days (Kelman, 1954).

Cell suspensions (approximately 1×10^6 cells/ml prepared from cultures grown on TZC agar plates or in TYD broth in test tubes for 36–48 hours at 32°C) were examined in hanging drops in van Tieghem cells at a magnification of 430 ×. Relative migration distances of virulent and avirulent cells were determined in semi-solid agar plates. The motility agar medium contained: glucose 0.1 g, tryptone 0.1 g, EDTA 1×10^{-4} M, 10 ml pH 7.0 phosphate buffer, and 3.5 gm Difco agar/l of distilled water (Adler, 1966).

In aerotaxis studies, the semi-solid motility agar medium with cell suspensions containing 50:50 mixtures of avirulent/virulent cells as well as avirulent and virulent alone was pipetted into short sections (5.5 cm long) of sterile glass tubing plugged at one end with hard agar and plasticine. After incubation of one set of tubes in air and another in a nitrogen atmosphere without oxygen, the agar columns were extruded from the tubes. Segments (1 mm) were sliced off in serial fashion from the top down, added to sterile water blanks, and each suspension was sampled to determine ratio of avirulent/virulent cells in mixture series.

Results and discussion

In addition to 25-K60, cells of 26 other virulent isolates grown on TZC agar for 24–48 hours were also non-motile. The cells of butyrous avirulent colony types derived from these virulent cultures were actively motile with the exception of two isolates. If test tube broth cultures initially 100% virulent were carefully sampled at the surface and below the surface, the percentage of avirulent cells at the surface was 63% after 48 hours, whereas percentages below the surface were markedly lower (2%). Electron photomicrographs revealed that flagella were present on only a small percentage of virulent cells (2–5%); normal flagella were present on a high percentage of the avirulent cells (80–90%).

Following a procedure similar to that described by Vaituzis & Doetsch (1969), motility track photographs were taken of cell suspensions of the avirulent B1-1 and the virulent K-60. Distinct tracks made by actively motile cells of avirulent strains were visible as white streaks on the dark background. No streaks were evident in the preparations made with virulent cells from TZC agar.

In motility agar plate experiments percentage increases in diameters of migration zones of avirulent cells were 85 and 233% after 3 and 10 hours, respectively, whereas for the virulent isolate these increases were 2.0 and 45%, respectively.

In the aerotaxis experiments distinct migration patterns were evident following incubation for 4 to 6 hours. The percentage of avirulent cells in segments taken in descending order from top to bottom in the tubes in the nitrogen atmosphere showed no distinct migration pattern. In tubes in air, however, the percentage of avirulent cells in the top segment had increased to 70% and in the third segment had decreased to 32% from the original level of 53%. Distinct migration toward the surface was also evident in the tubes with the avirulent alone. These shifts indicated that positive aerotaxis by the avirulent cells had occurred.

The ostensible loss of virulence in non-aerated broth cultures may reflect the inability of non-motile or less actively motile virulent cells to compete with the highly motile avirulent cells that are capable of aerotaxis, with a resultant shift to a high population of avirulent cells.

References

Adler, J., 1966. Chemotaxis in bacteria. Science 152: 708–716.

Husain, A. & A. Kelman, 1958. Relation of slime production to mechanism of wilting and pathogenicity of *Pseudomonas solanacearum*. Phytopathology 48: 155–165.

Kelman, A., 1953. The bacterial wilt caused by *Pseudomonas solanacearum:* a literature review and bibliography. N. Carol. Agr. Expt Sta. Bull. 99.

Kelman, A., 1954. The relationship of pathogenicity of *Pseudomonas solanacearum* to colony appearance in a tetrazolium medium. Phytopathology 44: 693–695.

Vaituzis, Z. & R. N. Doetsch, 1969. Motility tracks: technique for quantitative study of bacterial movement. J. appl. Microbiol. 17: 584–588.

Factors affecting the virulence of Erwinia carotovora

D. Zutra[1], Y. Henis[2] and Z. Volcani[1]

[1]Division of Plant Pathology, The Volcani Institute of Agricultural Research, Bet Dagan, Israel
[2]Department of Plant Pathology and Microbiology, Faculty of Agriculture, The Hebrew University, Jeruzalem, Israel

Abstract

The interaction between several strains of the soft rot pathogen, *Erwinia carotovora*, and its compatible (pepper fruits) and non-compatible host plants (cucumber seedlings) is studied. The threshold of infection in the pepper fruit appears to be about 1 cell per infection site for virulent strains and $10^4 - 10^7$ cells for less virulent strains. Cucumber seedlings are resistant to an inoculum with as many as $10^6 - 10^7$ cells per infection site.

Application of single amines and amino acids, such as L-alanine, L-aspartic acid, L-glutamic acid, L-asparagine and L-glutamine at a concentration of 2×10^{-3} M, either through the roots or directly to the infection site together with the pathogen, significantly increases the infection rate and reduces the threshold of infection.

Introduction

The pathogenesis of soft rot, caused by *Erwinia carotovora*, is correlated with its ability to synthesize enzymes involved in the degradation of cell-wall polysaccharides. The rate of enzyme production is more rapid with pathogenic than with non-pathogenic bacteria (Lapwood, 1957). According to Lovrekovitch *et al.* (1967) pectinase, produced by *E. carotovora*, induces polyphenoloxidase fermentation in potato tubers, by which phenols are oxidized. In plant tissue this oxidation is inhibited by another enzyme system of the bacterial cells. This inhibition increases with the presence of an increasing number of bacterial cells and depends on the multiplication rate of the pathogen, both in compatible and non-compatible host plants. By determining the number of bacterial cells required to initiate an infection in compatible and non-compatible hosts, the factors affecting the multiplication rate of *E. carotovora* in both plants will be discussed.

Materials and methods

For inoculation, bacterial cultures were grown in nutrient broth in a shaker bath for 18 h at 28 °C. Aliquots of 0.01 ml of tenfold dilution series were used to inoculate pepper fruits (5 replicates for each dilution). Some of the dilutions were also grown in a liquid medium in test tubes. Growth in these test tubes were checked after 5 days on the appearance of symptoms on the fruits, which had been placed under optimum incubation conditions. The most-probable-number (MPN) required for infection was calculated from the parallel sets (Alexander, 1965). The cucumber seedlings were injected, with the same dilution series, into the intercellular spaces of the leaf tissue. The inoculated plants were kept at 25 °C and 90–95% relative humidity. Estimations of the bacterial population in the seedlings were made daily for 15 days after inoculation by the dilution-plate method using a Tergitol 7-TTC medium (Collins, 1967). The number of bacteria was calculated by counting the colonies after 4 days at 30 °C.

In vitro studies to investigate the influence of amino acids on bacterial growth were carried out in the minimum medium M-9 (Silver *et al.*, 1968) containing only 0.04% glucose as carbon source.

Results and discussion

The number of bacteria required for infection was calculated by comparing the number of infected pepper fruits 5 days after inoculation with the growth observed in the liquid media inoculated with the same serial suspensions. The results showed that the threshold of infection in the compatible host was near one cell per infection site for a virulent strain, and some 10^4 to 10^7 cells for less virulent strains (see table). These data agree well with the theory of independent action of infective agents, developed by Meynell & Stocker (1957) for *Salmonella paratyphi* in mice.

In contrast to the pepper fruits, cucumber seedlings were resistant to infection at an inoculum concentration as high as 10^6 to 10^8 cells per ml.

In the inoculated cucumber seedlings exponential growth of the pathogen continued for 2–3 days. After these first three days the number of bacterial cells remained constant for about 15 days (see 'Material and methods').

Addition of certain amino acids to the bacterial suspensions which were used for injection in the leaf tissue, or application of these compounds to the seedling roots prior to the infection of the bacteria in the leaf tissue, significantly increased the infection rate and reduced the threshold of infection in the cucumber seedlings.

A shift towards susceptibility in the cucumber tissue was observed with L-alanine,

Infectivity titration of some strains of *Erwinia carotovora* in compatible and non-compatible host plants.

Strain	Host	Minimum number of cells required for infection
Virulent strain of		
E. carotovora (str. 1)	pepper fruits	10^0–10^1
Less virulent strains of		
E. carotovora (str. 2)	pepper fruits	10^3–10^4
E. carotovora var. *aroidea* (str. 2)	pepper fruits	10^4
E. carotovora var. *atroseptica* (str. 5)	pepper fruits	10^5–10^6
E. carotovora (str. 1)	cucumber seedlings	10^6–10^7

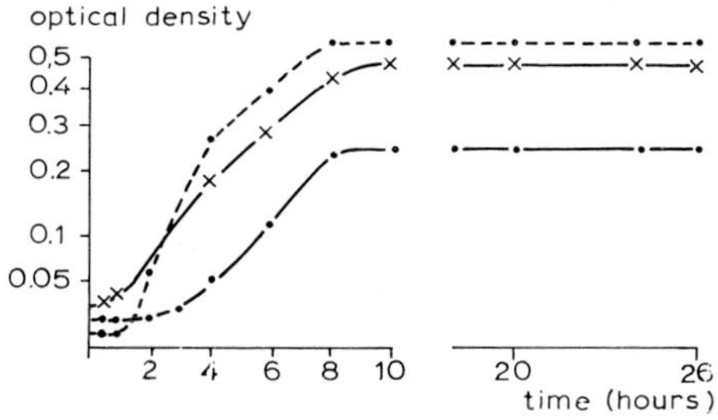

Growth of strain 1 of *E. carotivora* in nutrient browth (·—·), and minimal medium containing NH_4Cl and L-asparagine (○) or L-glutamic acid (×).

L-aspartic acid, L-glutamic acid and their salts at a concentration of 2×10^{-3} M. *In vitro* studies showed that the addition of these amino acids greatly enhanced growth of the bacteria (see figure). This was apparent from the shorter lag period and the increase of the maximum turbidity values. This phenomenon agrees well with Garber's hypothesis (1957) on the nutritional and inhibitory environment in the host as a factor regulating virulence in phytopathogenic bacteria.

Further experiments are being carried out to clarify the role of the host plant in regulating the virulence of *E. carotovora*. We have already found that, e.g., only 5 μg/ml cycloheximide causes a shift towards susceptibility in cucumber seedlings. According to Zucker & El-Zayat (1967), cycloheximide is an inhibitor of the protein synthesis in plant tissue. Accumulation of amino acids in the cucumber leaves may cause the breaking down of natural resistance to this pathogen.

References

Collins, C. H., 1957. Progress in microbiological techniques. Butterworth, London (p. 198).

Garber, E. D. & S. G. Shaeffer, 1957. Free histidine content of turnip varieties and their resistance to histidine requiring mutants of *Erwinia aroidea*. J. gen. Microbiol. 74:392.

Lapwood, D. H., 1957. Studies in the physiology of parasitism. XXIII. On the parasitic vigor of certain bacteria in relation to their capacity to secrete pectolytic enzymes. Ann. Bot. 21: 167–184.

Lovrekovitch, L., H. Lovrekovitch & M. A. Stahman, 1967. Inhibition of phenol oxidation by *Erwinia carotovora* in potato tuber tissue and its significance in disease resistance. Phytopathology 57: 737–742.

Maloy, O. C. & M. Alexander, 1957. The most probable number method for estimating populations of plant pathogenic organisms in the soil. Phytopathology 48: 126–128.

Meynell, G. G. & B. A. D. Stocker, 1957. Some hypotheses on the aetiology of fatal infection in partially resistant hosts and their application to mice challenged with *Salmonella paratyphi-B* or *Salmonella typhimurium* by intraperitoneal infection. J. gen. Microbiol. 74: 392.

Zucker, M. & M. M. El-Zayat, 1967. The effect of cycloheximide on the resistance of potato tuber discs to invasion by a fluorescent *Pseudomonas* sp. Phytopathology 58: 339–344.

A virologist's view on mycoplasma

Jeanne Dijkstra

Laboratory of Virology, Agricultural University, Wageningen, The Netherlands

In 1897 Eriksson introduced the term 'mycoplasma' to designate a certain stage in the development of rust fungi characterized by particles without membrane and which seemed to have a symbiotic relation to the protoplasm of the host cells. Later investigations revealed that in rust fungi such a membrane-less stage did not exist at all, and the name 'mycoplasma' went out of use.

The term emerged again with a completely different meaning when Freundt used it in 1955 to denote an organism with mycelium-like structure but without cell wall, causing pleuro-pneumonia in cattle. Later, when more agents of that type became known, the mycelium-like structure was often absent or it occurred as an artifact due to wrong fixation. Consequently, the name 'mycoplasma' is a misnomer for that group of filterable, highly pleomorphic organisms which lack a rigid cell wall but possess a unit membrane, and often grow and multiply on cell-free culture media.

Some diseases of mammals were originally thought to be caused by viruses, until the mycoplasma character of the causal agent was proven. It seems that similar cases occur among plants. Examples are the yellows-type diseases, characterized by yellowing of the leaves, stunted growth, stimulation of normally dormant buds (causing the witches' broom phenomenon), virescence (greening) and phyllody of flowers, and an abnormally erect growth. These diseases have until recently been thought to be caused by viruses on account of the following characteristics: (1) they were graft-transmissible, (2) they were transmitted by leafhoppers, (3) many were transmitted by dodder (*Cuscuta* species), (4) sometimes the pathogen proved to be filterable, (5) no other pathogens like fungi, bacteria and protozoa were found. However, all efforts to isolate and purify a virus from plants affected by yellows-type diseases were unsuccessful, neither did electron microscopy of ultrathin sections of those plants reveal any virus-like particles.

In 1967 Doi *et al.* reported the presence of mycoplasma-like bodies in the phloem of plants with witches' broom diseases: mulberry, potato, petunia and *Paulownia*. No such bodies were found in healthy plants.

Since that discovery, the number of cases in which mycoplasma-like bodies are associated with yellows-type diseases has risen steadily. In 1967 only 5 were known, in 1969 it was already 24, and it now exceeds 40.

Besides their presence in diseased plants, the mycoplasma-like bodies have also been found in leafhoppers transmitting these diseases and absent in leafhoppers not able to do so (Maramorosch *et al.*, 1968).

Further circumstantial evidence for the mycoplasma character of the above-mentioned pathogens came from chemotherapeutic studies. A number of yellows-type diseases could be arrested by the application of some antibiotics like tetracyclines. Although infection was not prevented, symptom development was delayed and/or suppressed. In addition, vector transmission was often blocked by the tetracyclines.

All such data, however, do not prove that these organisms really cause yellows-type diseases: Koch's postulates have to be satisfied. First of all the agent must be isolated and maintained in pure culture.

In 1969 Hampton *et al.* found in alfalfa, besides the virus causing alfalfa mosaic, a mycoplasma which they called 618 M. In contrast to the mycoplasmas known till that moment it was sap-transmissible and present in parenchyma cells. It could be isolated and grown in a serum-rich liquid medium. The cultured pathogen was introduced into healthy pea plants and it only caused a chlorotic mottle instead of leaf vein necrosis, mild stem necrosis and mild mottle characteristic for the original disease. When these plants were also inoculated with a pure suspension of alfalfa mosaic virus, the original syndrome occurred so that the original disease seemed to have been the result of a mixed infection: a mycoplasma and alfalfa mosaic virus. However, some doubts may arise as to the validity of this conclusion. The main objection is that the authors did not use a pure culture of mycoplasma but a culture obtained by inoculation of several test media with a purified sap suspension. Moreover, the reported size of the 618 M mycoplasma bodies was 15–250 nm, which is very unusual for the mycoplasmas described so far.

More evidence for the mycoplasma etiology of a plant disease was reported in 1970 by Lin *et al.*, by Faivre-Amiot *et al.* and by Chen & Granados.

Lin *et al.* studied white leaf disease of sugarcane which is transmitted by a leafhopper, *Epitettix hiroglyphicus*. Ultrathin sections of diseased leaves showed mycoplasma-like bodies in the sieve tubes. Symptom development was suppressed by tetracyclines. Attempts were made to obtain mycoplasma cultures by cutting buds from diseased stalks into small pieces, washing them in 5% calciumhypochlorite, then placing them in Morton's PPLO media and incubating them at 37°–38°C for 36–72 h. The liquid culture was inoculated to Bacto-PPLO agar plates containing Bacto-PPLO serum fraction and incubated again at 37°–38°C. After 36–48 h colonies resembling fried eggs, typical for mycoplasmas, developed on the agar plates. From 30 buds, two mycoplasma cultures could be obtained. The mycoplasma was further purified and maintained by alternative passages through the liquid and agar media. A suspension of the mycoplasma culture was inoculated into sugarcane by immersing healthy cuttings and buds in the suspension followed by repeated pin-pricking of the tissues. The plant parts were left in the suspension for 18 h at 37°–38°C before they were planted in pots in the glasshouse. Out of 72 inoculated plants in three tests 5 developed white leaf symptoms. Controls remained healthy.

Unfortunately, the authors did not mention whether they used primary cultures or subcultured mycoplasmas in their inoculation trials. Neither did they compare the infectivity of primary cultures with that of subsequent subcultures.

Some investigators have shown the presence of mycoplasmas in the phloem of clover plants affected by clover phyllody which is transmitted by leafhoppers such as *Euscelis plebejus* and *Euscelidius variegatus*. Faivre-Amiot *et al.* (1970) have attempted to isolate and cultivate mycoplasma from clover phyllody-affected white clover plants and periwinkle *(Vinca rosea)*. They washed stem pieces in soap solution, sodium hypochlorite and sterile water, cut them into smaller pieces and transferred these pieces to liquid media such as Difco-PPLO media and solid media. Cultures obtained from both clover and periwinkle revealed the presence of mycoplasma-like bodies as well as bacteria, whereas in the cultures from clover bacterial L forms were also present. Efforts to isolate the mycoplasmas by purifying the suspensions by means of dilution and filtration failed. The cultures consisting of a mixture of bacteria and mycoplasmas could be subcultured two or three times at the most. Healthy periwinkle plants were inoculated by pin-pricking or other mechanical means, whereas clover plants were inoculated by means of leafhoppers which had been injected with the culture suspensions or which had acquired these suspensions by feeding through an artificial membrane. Some of the inoculated periwinkle plants developed yellowing one year after inoculation but they never showed the typical phyllody symptoms. The clover plants, however, developed all symptoms typical of clover phyllody.

It is hard to see how these results fulfill Koch's postulates as the mycoplasma was not obtained in pure culture.

In 1968 Granados *et al.* and Maramorosch *et al.* reported that mycoplasma-like bodies were present in sieve elements of maize plants infected with corn stunt. The bodies were also found in salivary glands, malpighian tubules, intestinal tract, brain and ventral ganglia of the leafhopper-vectors, *Dalbulus* species, infected with this disease. Chen & Granados (1970) made attempts to cultivate the mycoplasma-like agent of corn stunt on several kinds of artificial media. After treatment with alcohol, stem pieces were rinsed in 5% sodium hypochlorite and washed in sterile distilled water. The outer sheath leaves were removed from the stem and the stem pieces were placed in a dish of fresh culture medium. Small portions of the centre of the stem were put in test tubes with the medium or they were first crushed in the medium and a small aliquot of the resulting suspension was transferred into a test tube with the medium. The cultures were maintained at 25 °C. To assay for infectivity, two cultures were combined, centrifuged, the resulting pellet resuspended in fresh medium and injected into healthy adult leafhoppers. In three out of five experiments infectivity was obtained with primary cultures 5, 8, and 14 days old. Electron microscopy of the infective cultures revealed organisms resembling the mycoplasma-like bodies observed in sections of diseased plants and infective insects. The corn stunt agent was not stable in buffer at room temperature for more than 4 h, but in the complex medium it could be maintained for several weeks. The highest concentration of mycoplasma-like bodies was found in 43-day old subcultures. Three out of 30 surviving insects which had been injected with samples of these subcultures transmitted corn stunt to healthy maize seedlings. Although the concentration of mycoplasmas was highest in the 43-day old subcultures, the transmission was only 10% as compared to 20, 43 and 39% in case of primary cultures of 5, 8, and 14 days, respectively. This result is hard to explain.

Summarizing we may say that the results obtained by Lin *et al.* (1970) and by Chen

& Granados (1970) point to mycoplasmas being the causal agents of white leaf of sugarcane and corn stunt, respectively. But only for white leaf of sugarcane has the agent been cultivated on a solid medium, has produced mycoplasma colonies, suspensions of which reproduced the original disease after inoculation into healthy plants.

Isolation and cultivation of mycoplasmas from plants affected by other yellows-type diseases in which mycoplasma-like bodies have been demonstrated, have failed so far.

Unless more becomes known about the requirements of the different mycoplasmas associated with plant diseases, so that cultivation and subsequent characterization and identification will be facilitated, we have to be cautious in concluding that these microorganisms are the causal agents of plant diseases.

Only recently, Ploaie (1971) has reported the presence of virus-like particles associated with mycoplasma in periwinkle affected with clover dwarf. The bacilliform virus-like particles, 85–88 nm long and 31–33 nm wide, occurred free in the phloem cells or were adsorbed to mycoplasma bodies. This suggests that the virus-like particles either infect the mycoplasma or are only carried by the mycoplasma.

In sandal plants affected with spike disease, inclusion bodies typical of virus infection were found in leaf epidermal cells (Dijkstra, 1968). So far there is no evidence of inclusion bodies having been caused by mycoplasmas in any other yellows-type disease.

With these facts in mind we still cannot exclude the possibility that in yellows-type diseases both mycoplasmas and viruses are involved.

References

Chen, T. A. & R. R. Granados, 1970. Plant-pathogenic mycoplasma-like organism: maintenance in vitro and transmission to *Zea mays* L. Science 167: 1633–1636.

Dijkstra, J., 1968. The occurrence of inclusion bodies in leaf epidermis cells of sandal affected with spike disease. Neth. J. Pl. Path. 74: 101–105.

Doi, Y., M. Teranaka, K. Yora & H. Asuyama, 1967. Mycoplasma or PLT group-like microorganisms found in the phloem elements of plants infected with mulberry dwarf, potato witches' broom, aster yellows, or *Paulownia* witches' broom. Ann. phytopath. Soc. Japan 33: 259–266.

Eriksson, J., 1897. Vie latente et plasmatique de certaines Urédinées. C.r. Acad. Sci. 124: 475–477.

Faivre-Amiot, A., J.-P. Moreau, M. T. Cousin & T. Staron, 1970. Essai de mise en culture de l'agent de la phyllodie du trèfle. Ann. Phytopath. 2: 251–258.

Freundt, E. A., 1955. The classification of the pleuropneumonia group of organisms (Borrelomycetales). Int. Bull. bact. Nomencl. Taxon. 5: 67–78.

Granados, R. R., K. Maramorosch & E. Shikata, 1968. Mycoplasma: suspected etiologic agent of corn stunt. Proc. natn. Acad. Sci. U.S.A. 60: 841–844.

Hampton, R. O., J. O. Stevens & T. C. Allen, 1969. Mechanically transmissible mycoplasma from naturally infected peas. Pl. Dis. Rptr 53: 499–503.

Lin, S-c., C-s. Lee & R-j. Chiu, 1970. Isolation and cultivation of, and inoculation with, a mycoplasma causing white leaf disease of sugarcane. Phytopathology 60: 795–797.

Maramorosch, K., E. Shikata & R. R. Granados, 1968. Structures resembling mycoplasma in diseased plants and in insect vectors. Trans. N.Y. Acad. Sci. Ser. II 30: 841–855.

Ploaie, P. G., 1971. Particles resembling viruses associated with mycoplasma-like organism in plants. Revue roum. Biol.-Bot. 16: 3–6.

Mycoplasmas of man and animals

B. E. Andrews

Mycoplasma Reference Laboratory, Central Public Health Laboratory, Colindale Avenue, London NW 9, England

Abstract

A general review of mycoplasmas from the point of view of a medical microbiologist. The taxonomic position of mycoplasmas is stated, followed by brief comments on some important characters relevant to their classification and identification. Problems in defining disease associations of human and animal mycoplasmas are discussed. Reference is made to some practical points concerning techniques for isolation of mycoplasmas that may be relevant to work in the field of plant pathology.

The causal agent of contagious bovine pleuropneumonia was first cultivated in 1898 and subsequently referred to as the *Pleuropneumonia* organism (PPO). Similar organisms isolated from other diseases were called PPLO's (*Pleuropneumonia*-like organisms). The name has now, by international agreement, been changed to mycoplasma and because they are quite distinct from bacteria and viruses the new class Mollicutes and the new order Mycoplasmatales has been created for them. The order contains two families: the Mycoplasmataceae and the Acholeplasmataceae, each with one genus *(Mycoplasma* and *Acholeplasma)*.

The properties of the Mycoplasmatales are intermediate between bacteria and viruses; in the past this has caused some confusion, in particular with viruses. The more important characters of the mycoplasmas, together with brief comments, are given below. It will be seen that several of them resemble those of viruses.

Taking the characters in order, the following remarks can be made.

1. *Growth on cell-free media*

All known mycoplasmas have been cultivated in cell-free media. These media are complex and usually consist of a broth base with serum and yeast extract, but often additions such as boiled blood extract, DNA or DPN give better growth. Most species require a pH of 7–8 for growth but some need pH6. Others only grow in tissue culture type media containing Hank's solution and lactalbumin hydrolysate.

2. *Smallest reproductive units 125–150 nm diameter*

The size range of mycoplasmas is very wide but the smallest reproductive units pass filters of 200 nm pore diameter, indicating a size similar to that of the paramyxoviruses. Unlike bacteria and viruses there is no sharp cut-off in filtration experiments using filters of graded pore size.

3. *No rigid cell wall*
The mycoplasmas do not possess a rigid cell wall, neither do they secrete cell wall substances as may be seen with some bacterial L-forms. Because they lack a cell wall mycoplasmas are seen under high powers of the light microscope as pleomorphic cells which characteristically appear as cylinders, spheres and rings. The mycoplasma cell is bounded by a triple layer membrane about 10 nm thick. This membrane is largely composed of lipoprotein.

4. *Require sterol for growth (except* Acholeplasma *spp.)*
Sterol is required for growth of the Mycoplasmatacae but not by the Acholeplasmatacae, though they are able to make use of it. The sterol is incorporated into the cell membrane and the usual source is the cholesterol of the serum fraction of the medium.

5. *Characteristic colonies with central downgrowth into the medium*
Most mycoplasmas produce characteristic small 0.01–0.1 mm diameter colonies on agar consisting of a dense central area of downgrowth into the agar and a peripheral area of surface growth. Variations occur; some species show predominantly central growth and others may produce only surface growth often with a lacy or vacuolated appearance.

6. *Resistant to penicillin, usually sensitive to tetracyclines*
Because they lack a cell wall, all mycoplasmas are completely resistant to penicillin and other cell wall antibiotics. They are, however, susceptible to most other antibiotics amongst which the tetracyclines and Tylosin are particularly effective. As with bacteria, antibiotic resistance may develop.

7. *Growth inhibited by specific antisera*
The growth of mycoplasmas is inhibited by specific antisera. This effect can be shown in solid and liquid media. On solid media, filter paper discs impregnated with high titre antisera are applied to a heavily seeded plate and after incubation a zone of growth inhibition is seen round the disc containing the homologous antiserum. This test does not give useful quantitative results but serial dilutions of serum in broth cultures containing a suitable substrate (e.g. glucose, arginine or urea) and a pH indicator provides a quantitative test in which inhibition of growth is indicated by absence of the change in pH that would indicate growth.

8. *Species are serologically distinct*
The 50 or so recognised species of mycoplasmas are serologically distinct; indeed routine identification is best done by the paper disc growth-inhibition test supported where there is doubt by the quantitative metabolic-inhibition test. The complement fixation test has been widely used but it tends to show cross-reactions, whereas agglutination and indirect haemagglutination tests tend to be very specific and emphasize differences at the sub-species level.

At present the mycoplasma species are best distinguished by serological methods, although many biochemical tests can be applied to mycoplasmas and these may be of volunteers developed pneumonia after inoculation with a tissue culture propagated strain of *M. pneumoniae*; but after passage of the same strain on agar media only upper respiratory tract infections were induced and these became progressively less severe with an increasing number of agar passages. Nevertheless inoculation with the organism, even when no disease resulted, provoked a serological response. It should be noted that even under conditions of natural infection some persons showing serological evidence of infection may show no signs of illness.

Mycoplasma hominis is another human mycoplasma that has been extensively studied. Here the situation is much less clear; however it seems certain that this organism is frequently present in the uro-genital tract in the absence of any evidence of disease. In some cases of abcesses associated with the genital tract, occasionally in blood cultures from cases of puerperal pyrexia and in other similar conditions, *M. hominis* may be recovered in pure culture, and furthermore the cases often fail to benefit from penicillin but respond to treatment with tetracycline. This sort of evidence is, of course, no proof of pathogenicity, but in the absence of susceptible experimental animals due to the extreme host specificity of *M. hominis*, it may never be possible to formally fulfil Koch's postulates. The concept of *M. hominis* as an opportunistic pathogen, i. e. an organism which does not initiate disease but is able to invade and maintain a disease condition in previously damaged tissues, would appear to fit the facts and it may be found that some of the doubtfully pathogenic animal mycoplasmas should be regarded also as opportunistic pathogens. Apart from the T-strain mycoplasmas, there is no evidence that any of the five other species of mycoplasmas found in man ever cause disease.

Similar problems in defining disease association occur in the field of animal mycoplasmas, but here experimental infection is possible and the use of germ-free animals may prove easier to solve most of them fairly easily. But there is one point that has to be borne in mind, particularly in connection with experiments in germ-free animals. It concerns the role of double infections. I am thinking of instances such as the greatly enhanced pathogenicity of *M. gallisepticum* in chickens, when there is a simultaneous infection with certain viruses or bacteria which also, by themselves, may not cause serious disease.

Finally, I want to mention briefly some points about our experience with the isolation of mycoplasmas. Firstly a range of media is required because, although the majority of species grow well on the standard horse serum (yeast extract broth media), some require additions to the medium, some grow better at pH6, and others can only be grown on tissue culture type media. Sometimes it may even be necessary to make the primary isolation in tissue culture, but if this is done, the common contamination of tissue cultures by mycoplasmas is always a source of concern and difficulties. Sometimes primary isolation is best achieved in broth media, in others on solid media, and transfer from one to the other may be difficult unless a very heavy inoculum is used. It is a very common fault to try to subculture with the sort of dilute inocula that would be satisfactory for bacteria. Bacterial contamination of the specimen being tested for the presence of mycoplasmas can be a serious problem and it is usual to include value as a preliminary to serology. The complex media required for growth of mycoplasma has limited the application of biochemical tests and, although there are some important exceptions, so far they have not proved to be of great value for classification. The exceptions include fermentation of glucose, hydrolysis of arginine or urea, and a test for sterol dependence.

Most mycoplasmas belong either to the glucose positive or arginine positive group. Only a few are positive for both reactions and the so-called T-strains are negative for both but give a positive urea reaction and grow best at pH6.0 (other mycoplasmas prefer pH7.0–8.0). Amongst the glucose fermenting group of mycoplasmas a further subdivision can be made into those requiring sterol (cholesterol) and those able to grow in the absence of sterol (sterol-independent). The latter group has recently been proposed as a separate family within the Mycoplasmatales and named Acholesplasmataceae on the grounds that sterol requirement is a fundamental character of the organism. Other evidence supports this division and in particular it has been found that genome size of these organisms differs from that of the Mycoplasmataceae.

Present biochemical and other information shows considerable heterogenicity amongst the mycoplasmas but there are no very clear indications about how or whether they should be further subdivided at the family or genus levels. Study of DNA base ratios shows a spread of values for the named species between 20 and 40% GC. There is some clustering of % GC values within this range and when biochemical characters are taken into consideration there are suggestions of possible logical groupings that would also be of value for routine identification using biochemical rather than serological methods. There is a need for such subdivision because the present serological methods are both expensive and time-consuming.

As mentioned, there are some 50 distinct species of mycoplasmas. Most of them are relatively specific for a particular animal host, but there are several examples of mycoplasma species that may be found in more than one animal. *Acholeplasma laidlawii*, for example, has been found in cattle, sheep, pigs, rodents, birds and man. On occasions it has also been found in sewage and farmyard compost, but it is not clear whether it can survive and propagate indefinitely in these situations. The growth requirements of *A. laidlawii* are relatively simple and in addition it grows readily at 22 °C (unlike most other mycoplasmas, which usually show poor growth below 30 °C). So *A. laidlawii* may well be an exception to the general finding that mycoplasmas cannot survive naturally outside the body of their animal host, and it is possible that the mycoplasma-like particles seen in plants may prove to be related to the acholeplasmas.

Some *Mycoplasma* species are certainly pathogenic, some are of uncertain status, but about half the recognised species seem to be non-pathogenic, at least under normal conditions. In man, *Mycoplasma pneumoniae* is undoubtedly the cause of some cases of so-called virus or primary atypical pneumonia. There is a lot of clinical and epidemiological evidence for this, but the experimental evidence is not entirely satisfactory because of such problems as loss of virulence during laboratory cultivation of the organism. This is illustrated by the findings of Couch *et al.* who found that 3/27 bacterial inhibitors in the media. Penicillin in large doses is always safe, but few other antibiotics can be recommended. Thallous acetate is often used, but it is important to realise that this substance is only relatively ineffective against mycoplasmas and the concentration that can be used may be fairly critical. It is not the same for all mycoplasmas.

We have found that, after many subcultures in the laboratory, most mycoplasmas will grow to some extent in almost any of the usual atmospheric conditions used in bacteriology. But on primary isolation, conditions for growth may be much more critical, and whilst for most purposes a choice of air or a CO_2/N_2 mixture is satisfactory there are occasions when strict anaerobic conditions or a CO_2/air mixture give markedly better results. The human and animal mycoplasmas are often rather sensitive to physical conditions in the medium and we find it important, for instance, to avoid overdrying of plates and to incubate in closed containers so that the humidity remains high and no further drying occurs during incubation. Suitable osmotic pressure in the medium is also important, most mycoplasmas being rather susceptible to lysis under hypotonic conditions. The main growth of mycoplasmas in the body appears to be associated with cells, either as growth attached to the cell surface or (probably less often) actually inside the cell, and we find that much higher isolation rates are obtained when advantage is taken of this association by including a proportion of the tissue cells in the inoculum used for isolation of the mycoplasmas. Earliest growth is often seen in the vicinity of cells lying on the surface of the medium.

In preparing this general account of mycoplasmas as seen by a medical microbiologist, I have tried to keep in mind the points which seem to be of interest from a comparative point of view in studying and attempting to isolate mycoplasmas from plants. To summarize very briefly: mycoplasmas have properties which distinguish them from both bacteria and viruses – the 50 or so serologically distinct species are relatively host specific and include pathogens, opportunistic pathogens, commensals and possibly saprophytes – they require complex media, and the composition of the medium (together with the physical conditions of incubation) may be very critical for primary isolation.

References

Key references to most of the points covered will be found in the following publications.

Fallon, R. J. & P. Whittlestone, 1969. In: J. R. Norris & D. W. Ribbons, eds: Methods in microbiology. Vol. 3B: 221. London Academic Press.
Hayflick, L., ed., 1967. Biology of the mycoplasma. Ann. N.Y. Acad. Sci. 143: 1–824.
Hayflick, L., ed., 1969. The Mycoplasmatales and the L-phase of bacteria. New York, Appleton-Century-Crofs, and Amsterdam, N. Holland Publishing Co.
Horne, R. W., 1970. The ultrastructure of mycoplasma and mycoplasma-like organisms. Micron 2: 19.

Pectolytic enzymes of Pseudomonas[1]

D. C. Hildebrand

Department of Plant Pathology, University of California, Berkeley, California 94720

Abstract

Information concerning the detection of pectolytic enzymes of pseudomonads, their general distribution, and the specific complement of these enzymes in certain species has been presented. The relation of pectate gel tests to the reported enzyme complements has been discussed, as has their role in disease.

Introduction

The first description of pectolytic enzymes in *Pseudomonas* was published shortly after the description of these enzymes in *Erwinia*. Severini (1913a) in 1913 described two bacteria as causing rots of gladiolus. One of these was a pseudomonad which he named *Pseudomonas gladioli*, a bacterium which is now considered to be either *P. marginata* or *P. cepacia*. The other was an *Erwinia* sp. Subsequently, he (Severini, 1913b) undertook a comparative physiological study of *P. gladioli* and the other bacterium. This investigation revealed that pectolytic enzymes were formed by the two pathogens both in culture and in rotting tissue. Furthermore, the pectolytic enzymes of *P. gladioli* acted faster in rotting tissue than those of the *Erwinia* sp. Unfortunately, *P. gladioli* was ignored, and consequently the promising beginning made by Severini regarding these enzymes in *Pseudomonas* received little consideration. During the next thirty years most of the results reported consisted of negative data. It was mentioned, for example, that pseudomonads, including *P. fluorescens*, *P. putida* and *P. aeruginosa*, did not attack pectin with the production of acid and gas (Coles, 1926).

Occurrence of pectolytic enzymes in pseudomonads

The next significant investigation was a short survey in 1944 by Oxford (1944). He indicated that several plant pathogenic pseudomonads formed pectolytic enzymes

1. This investigation was supported in part by Regional Research Funds-2444 (W-96), USDA Grant (12-14-100-9199-34), and a Public Health Service research career development award (AI-34, 942) from the National Institutes of Allergy and Infectious Diseases.

Table 1. Occurrence of pectolytic enzymes in pseudomonads.

Species	Number of strains tested	Number of strains showing activity[1] pH5[2]	pH8[2]	pH5 and 8[2]	pH6 to 7[2]	Reference
P. aeruginosa	1					Prunier & Kaiser, 1964
	2					Coles, 1926
P. alboprecipitans	1	1				Hildebrand, 1971
P. alliicola	7	7				Hildebrand, 1971
	5					Burkholder & Starr, 1948
P. andropagonis	2	2				Hildebrand, 1971
P. angulata	2			2		Hildebrand, 1971
	1					Sands *et al.*, 1970
	3					Burkholder & Starr, 1948
P. antirrhinii	1	1				Hildebrand, 1971
P. apii	2					Burkholder & Starr, 1948
P. aptata	2					Hildebrand, 1971
P. areris	1					Burkholder & Starr, 1948
P. atrofaciens	1					Hildebrand, 1971
	1					Sands *et al.*, 1970
	2					Burkholder & Starr, 1948
P. barkeri	1					Hildebrand, 1971
	1					Burkholder & Starr, 1948
P. berberdis	2					Burkholder & Starr, 1948
P. betle	1					Hildebrand, 1971
P. cannabina	2	2				Hildebrand, 1971
P. caryophylli	2	2				Hildebrand, 1971
	2					Burkholder & Starr, 1948
	1					Brathwaite & Dickey, 1970
	2					Gehring, 1962
	3	3				Lange & Knösel, 1970
P. cattleyae	1					Hildebrand, 1971
P. cepacia	5	5				Hildebrand, 1971
P. cerasi	1					Hildebrand, 1971
	1					Oxford, 1944
P. cichorii	5					Hildebrand, 1971
	2					Sands *et al.*, 1970
	1					Burkholder & Starr, 1948
P. citriputeale	1					Hildebrand, 1971
P. coronafaciens	3					Hildebrand, 1971
	4					Sands *et al.*, 1970
	2					Burkholder & Starr, 1948
	1					Smith, 1958
P. delphinii	1	1				Hildebrand, 1971
	1					Sands *et al.*, 1970
	2					Burkholder & Starr, 1948
	1					Smith, 1958
P. dysoxylii	1					Hildebrand, 1971
	1					Sands *et al.*, 1970
P. eriobotyrae	4	4				Hildebrand, 1971
P. flectans	1	1				Hildebrand, 1971

Table 1 (continued)

Species	Number of strains tested	Number of strains showing activity[1]				Reference
		pH5[2]	pH8[2]	pH5 and 8[2]	pH6 to 7[2]	
P. fluorescens	3		3			Fuchs, 1965
	6		1[3]			Hildebrand, 1971
	1					Prunier & Kaiser, 1964
	7					Oxford, 1944
	2		2			Huether & McIntyre, 1969
	1					Sabet & Dowson, 1951
P. garcae	2					Hildebrand, 1971
	1					Sands *et al.*, 1970
P. gladioli	1				1	Severini, 1913b
P. glycinea	6	6				Hildebrand, 1971
	4					Sands *et al.*, 1970
	3					Burkholder & Starr, 1948
P. hibiscicola	1					Hildebrand, 1971
P. lachrymans	1					Prunier & Kaiser, 1964
	8				8	Hildebrand, 1971
	1					Sands *et al.*, 1970
	1					Burkholder & Starr, 1948
	3		3			Lange & Knösel, 1970
P. lapsa	1					Burkholder & Starr, 1948
P. maculicola	1					Burkholder & Starr, 1948
	1					Prunier & Kaiser, 1964
P. mangiferae-indicae	1		1			Hildebrand, 1971
P. marginalis	1				1	Knösel, 1967
	5		5			Hildebrand, 1970
	1		1			Ceponis & Friedman, 1959
	3				1	Smith, 1958
	1				1	Oxford, 1944
	2			2		Nasuno & Starr, 1966
	1		1			Lange & Knösel, 1970
	1		1			Lapwood, 1957
	1		1			Murant & Wood, 1957a
P. marginata	7	7				Hildebrand, 1971
	2					Burkholder & Starr, 1948
P. medicaginis	1					Smith, 1958
P. melea	1					Hildebrand, 1971
	1					Burkholder & Starr, 1948
P. mori	7	7				Hildebrand, 1971
	4					Sands *et al.*, 1970
	3					Burkholder & Starr, 1948
P. mors-prunorum	1				1	Sabet & Dowson, 1951
	1					Hildebrand, 1971
	1					Sands *et al.*, 1970
	9					Prunier & Kaiser, 1964
	4					Smith, 1958
	9					Oxford, 1944
P. myxogenes	1				1	Coles, 1926

Table 1 (continued)

Species	Number of strains tested	Number of strains showing activity[1] pH5[2]	pH8[2]	pH5 and 8[2]	pH6 to 7[2]	Reference
P. oryzicola	2				2	Goto, 1926
	1					Hildebrand, 1971
P. panaci	1					Hildebrand, 1971
	1					Sands *et al.*, 1970
P. papaveris	1					Hildebrand, 1971
P. papulans	1					Burkholder & Starr, 1948
P. passiflorae	2					Hildebrand, 1971
	1					Sands *et al.*, 1970
P. pavonacea	3					Coles, 1926
P. phaseolicola	1				1	Sabet & Dowson, 1951
	2	2				Lange & Knösel, 1970
	18	18				Hildebrand, 1971
	9					Sands *et al.*, 1970
	5					Burkholder & Starr, 1948
	3				2	Smith, 1958
	1					Oxford, 1944
P. pisi	6					Hildebrand, 1971
	1					Sands *et al.*, 1970
	2					Burkholder & Starr, 1948
P. polycolor	2					Hildebrand, 1971
P. pomi	1					Hildebrand, 1971
P. primulae	2					Hildebrand, 1971
	1					Burkholder & Starr, 1948
	1					Prunier & Kaiser, 1964
P. prunicola	1					Smith, 1958
	2					Oxford, 1944
	1					Mills, 1949
P. putida	2					Coles, 1926
	1					Hildebrand, 1971
P. ribes	2					Burkholder & Starr, 1948
P. ribicola	3					Burkholder & Starr, 1948
P. rubrilineans	3		3			Hildebrand, 1971
	1					Burkholder & Starr, 1948
P. rubrisubalbicans	2	1				Hildebrand, 1971
P. savastanoi	6	4				Hildebrand, 1971
	34					Sands *et al.*, 1970
	2					Burkholder & Starr, 1948
P. sesami	3	1		2		Hildebrand, 1971
	2					Burkholder & Starr, 1948
P. setariae	1					Hildebrand, 1971
P. solanacearum	1				1	Knösel, 1967
	3					Husain & Kelman, 1958
	4			4		Hildebrand, 1971
	4	4				Lange & Knösel, 1970
	1				1	Winstead & Walker, 1954
P. striafaciens	1					Burkholder & Starr, 1948

Table 1 (concluded)

Species	Number of strains tested	Number of strains showing activity[1]				Reference
		pH5[2]	pH8[2]	pH5 and 8[2]	pH6 to 7[2]	
P. syringae	1		1			Murant & Wood, 1957
	1		1			Lapwood, 1957
	12					Hildebrand, 1971
	10					Sands *et al.*, 1970
	11					Burkholder & Starr, 1948
	17	17[4]				Knösel & Lange, 1968
	13					Prunier *et al.*, 1970
	4					Prunier & Kaiser, 1964
	5				1	Smith, 1958
	11				2	Oxford, 1944
	2				2	Sabet & Dowson, 1951
P. tabaci	1				1	Sabet & Dowson, 1951
	1			1		Hildebrand 1971
	9					Burkholder & Starr, 1948
	1					Smith, 1958
	1				1	Oxford, 1944
P. tomato	4	4				Hildebrand, 1971
	3					Sands *et al.*, 1970
	3					Burkholder & Starr, 1948
P. viburni	1					Burkholder & Starr, 1948
P. viridiflava	1		1			Hildebrand, 1971
	1				1	Prunier & Kaiser, 1964
	1					Sands *et al.*, 1970
P. viridilivida	1					Burkholder & Starr, 1948
P. washingtonia	1					Burkholder & Starr, 1948
P. woodsii	1	1				Hildebrand, 1971
	2					Burkholder & Starr, 1948
Pseudomonas sp.	many				many	Graham, 1958
	many				many	Paton, 1958, 1960
	1				1	Ercoloni, 1966

1. Negative strains not indicated.
2. Approximate pH.
3. May be a strain of *P. marginalis*.
4. Many of these strains are probably *P. mors-prunorum*.

which could be detected using a simple medium. Among the pathogens forming pectolytic enzymes were strains of *P. cerasi*, *P. tabaci*, and *P. marginalis*, one of eight strains of *P. mors-prunorum*, and two of eleven strains of *P. syringae*. No pectolytic enzymes were formed by seven strains of *P. fluorescens* and strains of *P. phaseolicola* and *P. prunicola*.

This study was shortly followed by other surveys. Unfortunately, it soon became clear that the most striking aspect about pectolytic enzyme formation in Pseudomonas was the near total lack of agreement among investigators. Burkholder & Starr (1948) indicated that none of approximately eighty strains of pseudomonads liquified a pectate gel. These included eleven strains of *P. syringae*. In contrast, Sabet & Dowson (1951) found that all strains of pathogens including two strains of *P. syringae* produced slight to moderate liquefaction of a pectate gel. A compulation of reports concerning the occurrence of pectolytic enzymes among pseudomonads (Table 1) indicates that subsequent reports have continued to share this lack of agreement.

Many of the studies employed polypectate gels to detect pectolytic enzymes. Others have used more sophisticated techniques and characterized the enzymes formed by various strains. Consequently, much of the lack of agreement among the various investigations could be due to differences in techniques and conditions of the tests. There are a number of different pectolytic enzymes which vary widely in pH optima (as has been noted for several pseudomonads; see Lange & Knösel, 1970), response to ions, and in the conditions under which they are active. Often enzymes from different sources will behave quite differently (Bateman & Millar, 1966). This diversity in behaviour plus the indicated disagreement among reports led to an examination (Hildebrand, 1971) of what effect different pH levels and the addition of various supplemental substrates had upon the detection of pectolytic enzymes on pectin and polypectate gels. The findings indicated that pseudomonads could be divided into four groups depending upon their action on polypectate gels in relation to pH (Table 2; figure). One group consisted of pseudomonads which consistently produced pitting on gels only at low pH (pH 5.0). Non-fluorescent, soft-rotting pseudomonads such as *P. marginata* and *P. cepacia* and many of the oxidase negative fluorescent pathogens such as *P. mors-prunorum* and *P. phaseolicola* belong to that group. Another group consisted of pseudomonads which produced pitting at a high pH (pH 8.5). That group also contains both fluorescent (*P. marginalis* and *P. viridiflava*) and non-fluorescent *(P. rubrilineans)* pseudomonads. Members of a third group produced pitting at both high and low pH levels. That group is perhaps the most interesting of all, as it consists of all of the pseudomonad pathogens tested which are described as causing angular leaf spot symptoms. The only other organism in this group is *P. solanacearum*, and this pathogen differs from other pathogens which produce pits at both pH levels by forming much larger pits on the gels. A fourth group of pseudomonads did not consistently produce pitting. Major members of that group were *P. cichorii*, *P. coronafaciens*, and *P. syringae*.

In the same study, addition of supplemental substrates such as glucose, succinate sodium asparaginate, and β-alanine altered the pectolytic activity of various organisms. Most of the changes could be ascribed to changes in pH which occurred upon the metabolism of the supplemental substrate. Thus, growth of *P. cepacia* on yeast extract, sodium asparaginate, or β-alanine raised the pH of the gel to a point where pit formation normally does not occur. In contrast, growth of these and several other pseudomonads on glucose often resulted in the lowering of pH with the resultant formation of pits. Whether the change in pH affects enzyme activity or enzyme

Table 2. Pit formation of pseudomonads on polypectate gels in relation to pH (Hildebrand, 1971).

Pits formed			Usually no pits formed
pH 5	pH 8.5	pH 5 and 8.5	
P. alboprecipitans	*P. mangiferae-indicae*	*P. angulata*	*P. aptata*
P. alliicola	*P. marginalis*	*P. lachrymans*	*P. atrofaciens*
P. andropagonis	*P. rubrilineans*	*P. sesami*	*P. barkeri*
P. antiirrhinii	*P. viridiflava*	*P. solanacearum*	*P. betle*
P. cannabina		*P. tabaci*	*P. cattleyae*
P. caryophylli			*P. cerasi*
P. cepacia			*P. cichorii*
P. delphinii			*P. citriputeale*
P. eriobotryae			*P. coronafaciens*
P. flectans			*P. dysoxylii*
P. glycinea			*P. fluorescens*
P. marginata			*P. garcae*
P. mori			*P. hibisciola*
P. mors-prunorum			*P. mellea*
P. phaseolicola			*P. oryzicola*
P. rubrisubalbicans			*P. panaci*
P. savastanoi			*P. papaveris*
P. tomato			*P. passiflorae*
P. woodsii			*P. pisi*
			P. polycolor
			P. pomi
			P. primulae
			P. putida
			P. setariae
			P. syringae

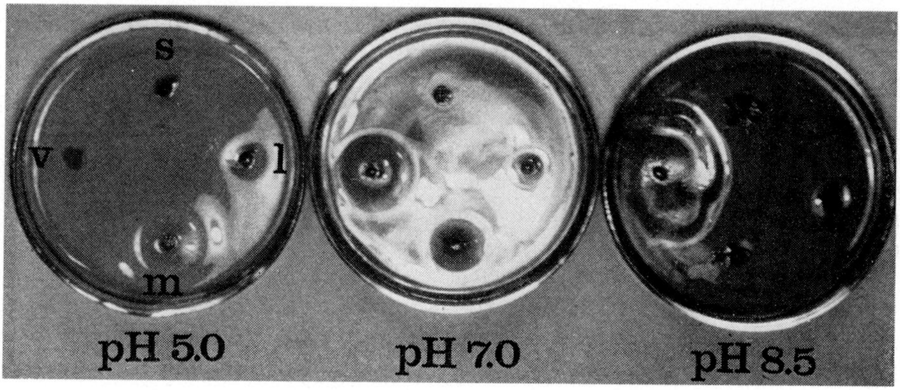

Pitting of polypectate gels in relation to pH by 4 pseudomonads of the *P. syringae* group. Note pitting by *P. mori* (m) at pH 5.0 and 7.0, by *P. viridiflava* (v) at pH 7.0 and 8.5, by *P. lacrymans* (l) at all pH levels, and its absence by *P. syringae* (5) at any pH level.

Table 3. Characterization of pectolytic enzymes of pseudomonads.

Species	Enzyme								Reference
	PME	PG			pectate lyase			pectin lyase, not specified	
		exo	endo	not specified	exo	endo	not specified		
P. caryophylli	–		tr			–			Lange & Knösel, 1970
	–			–			–		Brathwaite & Dickey, 1970
P. fluorescens[1]						+			Fuchs, 1965
	tr	+	+				+		Huether & McIntyre, 1968
							+		Zucker & Hankin, 1970
P. lachrymans	–		–			+			Lange & Knösel, 1970
P. marginalis	+								Ceponis & Friedman, 1959
	–			+			+		Knösel, 1967
	–		–			+			Lange & Knösel, 1970
							+		Lapwood, 1957
							+		Murant & Wood, 1957
	tr		+			+		–	Nasuno & Starr, 1966
	–						+[2]		Smith, 1958
P. phaseolicola	–		+			–			Lange & Knösel, 1970
	–			+[3]					Smith, 1958
P. prunicola	+								Mills, 1949
P. solanacearum	+			+				–	Buddenhagen & Kelman, 1964
	+			+					Husain & Kelman, 1958
	+			+			+		Knösel, 1967
	+		+			–			Lange & Knösel, 1970
	+			+					Winstead & Walker, 1954
P. syringae	–			+[4]			–		Knösel & Lange, 1968
	–			+[5]					Smith, 1958

1. Possibly *P. marginalis* – 2. One of three strains – 3. Two of three strains – 4. A number of the strains tested were probably *P. mors-prunorum* – 5. One of five strains.
tr = trace.

synthesis was not established. Presumably catabolite repression was not involved, as presence of any of the supplemental substrates *per se* had no effect on pitting unless changes in pH occurred.

Strains which formed pits on polypectate gels usually formed pits on pectin gels of a similar pH, although the degree of pitting was less. The major exception was one strain of *P. marginalis* which formed large pits in the pH 5 pectin gel, although none were formed in the polypectate gel of this pH. Also, the action of pseudomonads on pectin versus polypectate gels appears to differ from that of erwinias, as several erwinias which rapidly formed large pits on polypectate gels failed to or produced only slight pitting on pectin gels.

Characterization of pectic enzymes of pseudomonads

Generally, there are three types of pectic enzymes to be considered. Two of these are degradative enzymes which cleave the α-1,4 glycosidic bond between adjacent uronic acid monomers. These are distinguished from each other by their mechanism of cleavage. One group, the glycosidases, effects the splitting through a hydrolytic action, whereas the other group, the lyases, effects the cleavage through a trans-eliminative mechanism. These glycosidases and lyases furthermore are characterized as to their preferred substrate (either pectin or pectinic acid) and to the position in the pectic chain at which cleavage occurs (endo, or random; exo, or terminal). Thus, for example, the glycosidase endo-polymethylgalacturonase (endo-PMG) degrades pectin hydrolytically in a random manner, whereas exo-polygalacturonase (exo-PG) cleaves pectic acid hydrolytically in a terminal manner. Likewise, exo-pectin lyase cleaves pectin terminally in a trans-eliminative manner, and endo-pectate lyase cleaves pectic acid randomly. The third pectic enzyme is pectin methylesterase (PME). This enzyme is also a hydrolyase; but rather than cleaving the pectin chain, it removes the methoxy groups of pectin or pectinic acids to yield pectic acid. Presence of this enzyme may be important in stimulating the activity of PG or pectate lyase (Bateman & Millar, 1966), since pectic acid is the preferred substrate of these latter enzymes.

The pseudomonads whose pectolytic enzymes have been most often characterized are *P. solanacearum* and members of the *P. marginalis*/*P. fluorescens* group. The pectic enzymes complement of several other pseudomonads has also been investigated, and a listing of which enzyme occurrence in the various organisms has been compiled (Table 3). Several reports have not been included for various reasons. The report of Preiss & Ashwell (1963) is not included, since the culture with which they were working was a mixture of two organisms (Nasuno, 1966). Lapwood (1957), and later Murant & Wood (1957), reported on a pectolytic enzyme with a high optimum pH in a strain of *P. syringae*. However, it is likely that the strain with which they were working was misidentified, as no other reports have appeared concerning a high pH enzyme in this species. Furthermore, a number of other strains of *P. syringae* were tested which could not be used in his study because they did not form pectolytic enzymes.

The pectolytic enzymes identified thus far for pseudomonads are pectate lyase, PG, and PME. PMG and pectin lyase have not been detected. The major enzyme for most of the oxidase negative fluorescent strains is PG, whereas the primary enzyme for the oxidase positive groups is pectate lyase. PME, if present, is formed only in trace amounts by these organisms. *P. solanacearum*, on the other hand, forms considerable PME as well as PG.

Relation of pectolytic enzyme complement to action on polypectate gels

PG activity usually can be distinguished quite readily from pectate lyase on the basis at which pH the splitting of pectic acid occurs. Pseudomonad PG has a pH optimum of 5.0 or lower, whereas the optimum for pectate lyase is 8.0 or higher. Normally, each enzyme is virtually inactive at the optimum pH of the other. The use of polypectate gels of pH 5.0 and 8.5 (Hildebrand, 1971) was designed to take advantage of these differences in pH optima and to enable an assessment of the enzyme which was present in a given organism.

Comparison of the data for polypectate pitting (Table 1) and the reported enzymes (Table 3) indicates that for some pseudomonads there is good agreement between polypectate gel pitting and the reported enzyme complement, and that some general assessments may be made. Thus, it appears that the pitting at pH 5 by many of the oxidase negative fluorescent pseudomonads is due to PG. Likewise, the pitting at pH 8.5 by *P. viridiflava* and *P. marginalis*/*P. fluorescens* appears to be due to pectate lyase. PG also occurs in *P. marginalis*/*P. fluorescens* (Huether & McIntyre, 1970; Knösel, 1967; Nasuno & Starr, 1966), although amounts apparently are too small to be detected on the gels.

However, pitting of the gels in other cases cannot be related to the reported enzyme complement. Thus, *P. caryophylli* has been reported to produce no pectolytic enzymes (Brathwaite & Dickey, 1970) or trace amounts of PG (Lange & Knösel, 1970); yet large pits are formed in gels of low pH, and one strain forms some pits at pH 8.5. Likewise, *P. lachrymans* is reported to produce pectate lyase with a pH optimum of pH 8 (Lange & Knösel, 1970), yet this organism forms similar sized pits at all pH levels. Also, the disease it causes is not a soft-rotting type, which formation of lyase would indicate (see next section).

There are several factors which could account for these differences. In some cases the amount of enzyme formed may be too little to be detected by the gels. Also, Ca^{++} is a constituent of the gels, and this ion is inhibitory to endo-PG activity (Bateman & Millar, 1966). Although some of the enzymes are constitutive (Nasuno & Starr, 1966), others may need to be induced. The conditions required for induction are not always the presence of suitable substrate. *P. fluorescens* previously cultured on a glucose containing medium required 10 to 20 generations of growth on an inducing medium before appreciable pectate lyase synthesis occurs, whereas *E. carotovora* shows no lag in induction (Zucker & Hankin, 1970). Inducers vary considerably in effectiveness, and it is not always the best substrate for an enzyme which is the most efficient inducer. A mixture of pectin and potato tissue was a much better inducer of pectate lyase than was sodium polypectate for cells which had been pre-induced by transferring them several times on a medium containing pectin plus potato tissue as the carbon source (Zucker & Hankin, 1970). These observations indicate that the conditions of growth, the conditions of induction, and inducer used, all affect pectolytic enzyme formation with the result that some may remain undetected or the amounts which are possible to be formed are misjudged.

Pectolytic enzymes and disease

The role of pectolytic enzymes in disease can only be evaluated in terms of the normal symptoms produced and the normal progression of the disease process. This is perhaps most easily accomplished with the organisms causing soft rot. Friedman & Ceponis (1959) first established a relationship between the ability to cause disease and formation of pectolytic enzymes in Pseudomonas. They produced an avirulent mutant of *P. marginalis* through UV irradiation and observed that it was unable to form pectolytic enzymes. Similar evidence of the apparent importance of pectolytic enzymes in disease causation has been noted by us. Five strains of *P. marginalis* were examined for pectolytic enzyme formation during the polypectate gel tests (Hildebrand, 1971). Although all formed large pits according to our rating scale, the size of the pits produced varied considerably among the various strains. Tests of the ability of these five strains to rot lettuce revealed that those strains which produced the largest pits also were the most vigorous rotters. Since the pits are formed at pH 8.5 on the polypectate gels, this would indicate that pectate lyase is probably the major enzyme responsible for tissue breakdown. Similarly, *P. viridiflava* also produces pits at this pH, indicating a pectate lyase, and it can produce a soft rot (Lelliott *et al.*, 1966).

A bacterium identified as *P. fluorescens* (*P. marginalis* [?]) is associated with the pinkeye disease of potato tubers (Huether & McIntyre, 1969). Their study of the pectolytic enzymes of this organism indicated that the rate and amount of maceration of potato tuber tissue corresponded most closely with the lyase activity of enzyme preparations, and suggested to them that the lyase was the most important enzyme concerned with the breakdown of tuber tissue.

A virulent strain of *P. solanacearum* formed higher levels of PME and lower levels of PG than by a weakly virulent isolate (Husain & Kelman, 1958). An avirulent strain also formed PME and PG, but the levels were not given. The conclusions from this study are that pectolytic enzymes were not involved directly in wilting, but could cause tissue disorganization which could facilitate the spread of the organism.

In conclusion, with the possible exception of the fluorescent soft-rotting strains, the role of pectolytic enzymes in disease has not been established, although many of the pseudomonad pathogens form these enzymes, and the types which are formed are characteristic for different species. However, there are certain indications such as the pattern of pit formation by the angular leaf-spotting pathogens which suggest that these enzymes are important in the disease process and that this should be a fruitful area of study.

References

Bateman, D. F. & R. L. Millar, 1966. Pectic enzymes in tissue degradation. A. Rev. Phytopathol. 4: 119–146.

Brathwaite, C. W. D. & R. S. Dickey, 1970. Synergism between *Pseudomonas caryophylli* and a species of *Corynebacterium*. Phytopathology 60: 1046–1051.

Buddenhagen, I. & A. Kelman, 1964. Biological and physiological aspects of bacterial wilt caused by *Pseudomonas solanacearum*. A. Rev. Phytopath. 2: 203–230.

Burkholder, W. H. & M. P. Starr, 1948. The generic and specific characters of phytopathogenic species of *Pseudomonas* and *Xanthomonas*. Phytopathology 38: 494–502.

Ceponis, M. J. & B. A. Friedman, 1959. Pectolytic enzymes of *Pseudomonas marginalis* and their affects on lettuce. Phytopathology 49: 141–144.

Coles, H. W., 1926. The digestion of pectin and methylated glucoses by various organisms. Pl. Physiol. 1: 379–385.

Ercolani, G. L., 1966. A pectolytic pseudomonad as an agent of stem necrosis of young tomato plants. Phytopath. Medit. 5: 126–127.
Friedman, B. A. & M. J. Ceponis, 1959. Effect of ultraviolet light on pectolytic enzyme production and pathogenicity of *Pseudomonas*. Science 129: 720–721.
Fuchs, A., 1965. The *trans*-eliminative breakdown of Na-polygalacturonate by *Pseudomonas fluorescens*. Antonie van Leeuwenhoek 31: 323–340.
Gehring, F., 1962. Untersuchungen über den Infektionsverlauf einer durch *Pectobacterium parthenii* (Starr) Hellmers var. *dianthicola* Hellmers verursachten Nelkenbakteriose sowie über enzymatische Eigenschaften dieses Bakteriums im Vergleich mit *Pseudomonas caryophylli* (Burkholder) Starr et Burkholder und einige typischen Nassfäuleerregern. Phytopath. Z. 43: 383–407.
Goto, M., 1965. A comparative study of the sheath rot bacteria of rice. Ann. phytopath. Soc. Japan 30: 42–45.
Graham, D. C., 1958. Occurrence of soft rot bacteria in Scottish soils. Nature, Lond. 181: 61.
Hildebrand, D. C., 1971. Pectate and pectin gels for differentiation of *Pseudomonas* spp. and other bacterial plant pathogens. Phytopathology (In press).
Huether, J. P. & G. A. McIntyre, 1969. Pectic enzyme production by two strains of *Pseudomonas fluorescens* associated with the pinkeye disease of potato tubers. Am. Potato J. 46: 414–423.
Husain, A. & A. Kelman, 1958. The role of pectic and cellulolytic enzymes in pathogenesis by *P. solanacearum*. Phytopathology 48: 377–386.
Knösel, D., 1967. Untersuchungen über Vorkommen und Aktivität von pektolytischen und cellulolytischen Enzymen bei phytopathogenen Bakterien. Z. PflKrankh. PflPath. PflSchutz 74: 150–155.
Knösel, D. & E. Lange, 1968. Einige physiologische Eigenschaften von *Pseudomonas syringae* van Hall geprüft an verschiedenen Isolierungen. Z. PflKrank. PflPath. PflSchutz 75: 467–472.
Lange, E. & D. Knösel, 1970. Zur Bedeutung pektolytischer, cellulolytischer und proteolytischer Enzyme für die Virulenz phytopathogener Bakterien. Phytopath. Z. 69: 315–329.
Lapwood, D. H., 1957. Studies in the physiology of parasitism. XXIII. On the parasitic vigour of certain bacteria in relation to their capacity to secrete pectolytic enzymes. Ann. Bot. 21: 167–184.
Lelliott, R. A., E. Billing & A. C. Hayward, 1966. A determinative scheme for the fluorescent plant pathogenic pseudomonads. J. appl. Bact. 29: 470–489.
Mills, G. B., 1949. A biochemical study of *Pseudomonas prunicola* Wormald. 1. Pectin esterase. Biochem. J. 44: 302–305.
Murant, A. F. & R. K. S. Wood, 1957. Factors affecting the pathogenicity of bacteria to potato tubers. I. Ann. appl. Biol. 45: 635–649.
Nasuno, S., 1966. Pectic enzymes of phytopathogenic bacteria. Doctoral Thesis, University of California, Davis, California.
Nasuno, S. & M. P. Starr, 1966. Pectic enzymes of *Pseudomonas marginalis*. Phytopathology 56: 1414–1415.
Oxford, A. E., 1944. Production of a soluble pectinase in a simple medium by certain plant-pathogenic bacteria belonging to the genus *Pseudomonas*. Nature, Lond. 154: 271–272.
Paton, A. M., 1958. Pectin decomposing strains of *Pseudomonas*. Nature, Lond. 181: 61–62.
Paton, A. M., 1960. The role of *Pseudomonas* in plant disease. J. appl. Bact. 23: 526–532.
Preiss, J. & G. Ashwell, 1963. Polygalacturonic acid metabolism in bacteria. I. Enzymatic formation of 4-deoxy-L-*threo*-5-hexoseulose uronic acid. J. biol. Chem. 238: 1571–1576.
Prunier, J. P. & P. Kaiser, 1964. Study of the pectinolytic activity in phytopathogenic and saprophytic bacteria of plants. I. Research on pectinolytic enzymes. Annl. Épiphyt. 15: 205–219.
Prunier, J. P., J. Lursetti & L. Gardan, 1970. Study on the bacterial disease of fruit trees. II. Characterization of a non fluorescent *Pseudomonas* agent of a new bacterial disease of peach tree. Ann. Phytopath. 2: 181–197.

Sabet, K. A. & W. J. Dowson, 1951. Action of phytopathogenic bacteria on pectate gel. Nature, Lond. 168: 605.

Sands, D. C., M. N. Schroth & D. C. Hildebrand, 1970. Taxonomy of phytopathogenic pseudomonads. J. Bact. 101: 9–23.

Severini, G., 1913a. Una bacteriosi dell'*Ixia maculata* e del *Gladiolus colvilli*. Annali Bot. 11: 413–424.

Severini, G., 1913b. Inttorno alle attività enzimatiche di due bacteri patogeni per le piante. Annali Bot. 11: 441–452.

Smith, W. K., 1958. A survey of the production of pectic enzymes by plant pathogenic and other bacteria. J. gen. Microbiol. 18: 33.

Winstead, N. N. & J. C. Walker, 1954. Production of vascular browning by metabolites from several pathogens. Phytopathology 44: 153–158.

The influence of pectolytic enzymes on bacterial infection of plant tissue

D. Knösel and E. Lange

Department of Microbiology and Phytopathology, University of Hohenheim, Stuttgart-Hohenheim, Federal Republic of Germany

Abstract

To assess the role of extracellular pectolytic enzymes in dissolution of maceration of living plant tissue by phytopathogenic bacteria, relations among enzymes activities, virulence of bacterial strains, and resistance or susceptibility of leaf tissue were studied.

Introduction

Phytopathogenic species from the genera *Erwinia*, *Pseudomonas*, and *Xanthomonas* were tested for production of certain pectolytic enzymes, and infections were carried out to show correlations between virulence and enzyme activity.

The organisms were isolated in our laboratory or were obtained from colleages and type collections (NCPPB, ATCC).

Methods and materials

For enzyme induction the cultures were grown in a basal medium with citrus pectin added. After 24–48 h of cultivation on a shaker at 100 rev/min, the cultures were centrifuged and the supernatant was concentrated, dialyzed against demineralized water, and stored in a freezer.

To determine pectolytic activity, a viscosimetric assay was used (Bell *et al.*, 1955), with sodium polypectate as substrate. Relative activity was expressed as the reciprocal of time (h) required for a 50% increase of the flow-rate of the reaction mixture multiplied by 100.

Breakdown products were detected by chromatography and by spectrophotometry (Albersheim *et al.*, 1960; Koller & Neukom, 1964; Knösel & Garber, 1968).

To determine pectin methylesterase activity, culture filtrates or concentrated and dialyzed water extracts from leaf tissue were incubated with a reaction mixture adjusted to pH 7.0 and continuously titrated with sodium hydroxide. Relative PME activity was expressed as milliequivalents of released methoxy groups per ml filtrate or extract multiplied by 100.

To test virulence, an infiltration technique was used (Lange & Knösel, 1971); *Physalis floridana* and *Phaseolus vulgaris* served as test plants; they were kept in the

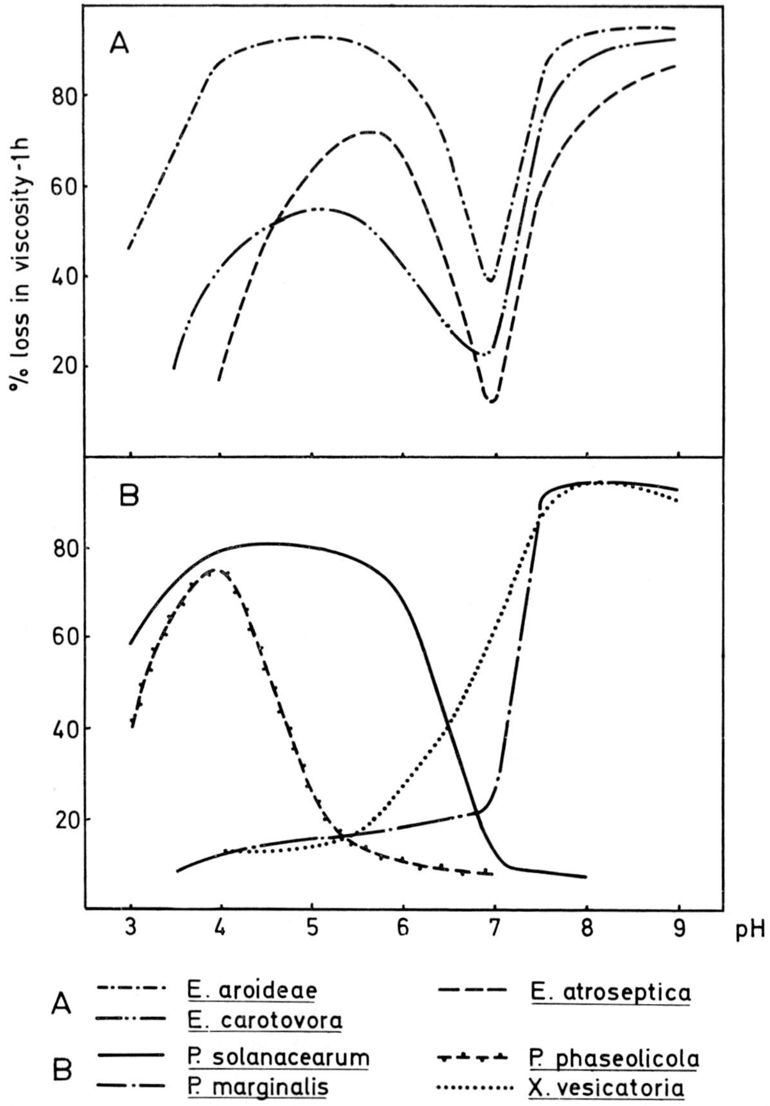

Fig. 1. Influence of pH of reaction mixture on pectolytic activity of culture filtrates of *Erwinia* species and *Pseudomonas* or *Xanthomonas* species.

greenhouse at 26 °C and at high air humidity. Plants of *Vicia faba* were inoculated by injecting cell suspensions into the main vein of the leaves.

Results and discussion

Many culture filtrates contained pectate degrading enzymes. Most of the *Erwinia* strains were active both in the acid and alkaline range. The optima were around pH 5 and between pH 8 and 9; at the neutral point the activity was always very weak (Fig. 1A). With *Pseudomonas* and *Xanthomonas* species degradation was obtained either under acid or under alkaline conditions. The optima were at about pH 4 or pH 8, respectively (Fig. 1B).

After incubation for 24 hours or longer, the reaction mixtures contained breakdown products of pectate. Under acid conditions, galacturonic acid and oligomers could be demonstrated; under alkaline conditions monomer and oligomer unsaturated compounds were detected. The latter showed an absorption maximum at 235 nm (Fig. 2) that suggests the presence of C-4,5-unsaturated galacturonyls.

It was concluded that, under acid conditions, pectate is degraded hydrolytically by endo-polygalacturonases, under alkaline conditions by endo-pectic acid transeliminases.

In infection experiments *Physalis floridana* and *Phaseolus vulgaris* were found

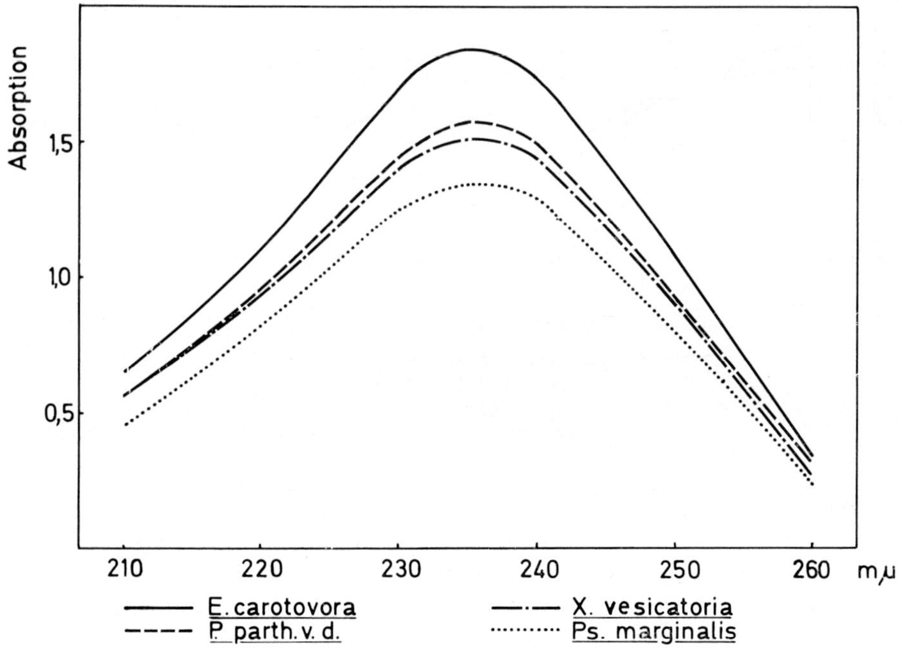

Fig. 2. Absorption spectra of C-4,5-unsaturated galacturonyls in reaction mixtures; endo-pectic acid transeliminases reacting with polygalacturonic acid at pH 8.5, 2 h, 30 °C.

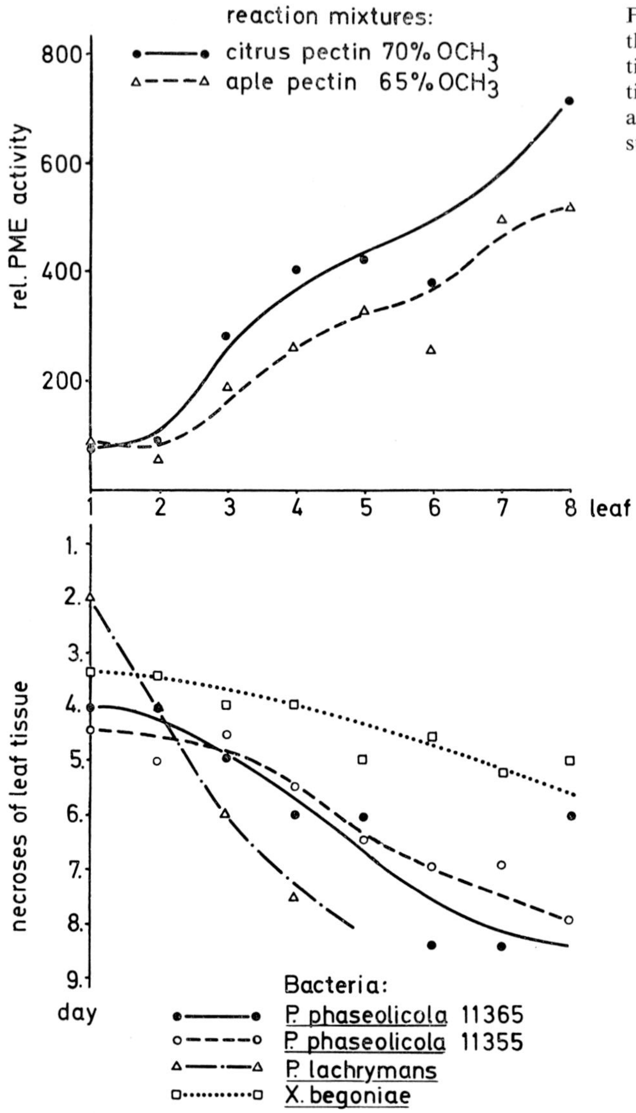

Fig. 3. Relative pectin methylesterase activity of leaf tissue from *Vicia faba* and time of reaction of leaves after injection of bacterial suspensions.

suitable for testing the virulence of species of *Erwinia* and *Pseudomonas* or *Xanthomonas*, respectively. High numbers of bacterial cells, infiltrated into the intercellular spaces of the leaves often induced heavy reactions of the tissue; graduations in virulence could not be detected. With suspensions of 10^4 to 10^6 cells/ml, various symptoms like chlorotic spots, lesions, rotting and necrosis were observed.

There were marked differences in enzyme activity and virulence in several species and strains (Tables 1 and 2). *Erwinia aroideae* gave a high relative PG activity, while

Table 1. Relative pectolytic activity of culture filtrates and reaction of leaf tissue induced by cells of *Erwinia* species.

Species and strains	PME	Endo-PG	Endo-PSTE	Leaf reaction[1]
E. amylovora	21	—	—	+++
E. aroideae	—	740	185	+++
E. atroseptica 435	34	41	53	+
E. atroseptica 549	—	25	104	++
E. carotovora 569	142	10	5000	+
E. carotovora 1125	152	67	5500	+++
E. chrysanthemi	—	—	—	+
E. cypripedii	—	70	11500	+++
P. parthenii BBA	—	10	—	—
P. parthenii H 1	67	—	80	+++

1. Test plant *Physalis floridana;* — = no reaction, + = chlorotic spots, ++ = rotting or necrotic lesions, +++ = complete rotting or necrosis.

Table 2. Relative pectolytic activity of culture filtrates and reaction of leaf tissue induced by cells of *Pseudomonas* and *Xanthomonas* species.

Species and strains	PME	Endo-PG	Endo-PSTE	Leaf reaction[1]
P. caryophylli 609	—	10	—	—
P. caryophylli 11441	—	10	—	+
P. lachrymans G 1	—	—	33	++
P. lachrymans G 2	—	—	52	+++
P. marginalis	—	—	625	+
P. phaseolicola 11365	—	12	—	+
P. phaseolicola 11355	—	43	—	+++
P. solanacearum 1029	108	10	—	++
P. solanacearum 1400	432	63	—	+++
X. begoniae Lo III	—	10	—	—
X. begoniae Lo I	13	10	—	+++
X. pelargonii	—	10	—	+++
X. vesicatoria	92	15	5500	+++

1. Test plant *Phaseolus vulgaris;* — = no reaction, + = chlorotic spots, ++ = necrotic lesions, +++ = complete necrosis.

the activity in *E. carotovora*, *E. cypripedii*, *Pseudomonas solanacearum* and others was intermediate. Whereas several strains of *Erwinia* showed extremely high relative PSTE activity, for most of the species of *Pseudomonas* and *Xanthomonas* it was low or absent.

Sometimes a positive correlation between pectolytic activity and virulence was suggested. Corresponding to the results with strains of *E. carotovora*, *P. phaseolicola* and *P. solanacearum*, PG activity apparently had more influence on virulence than PSTE activity. PME activity occurred only with some organisms.

The effect of pectin methylesterases on maceration of plant tissue is still under discussion. According to Smith (1958), PME activity is involved in the process of rotting, whereas Bateman & Millar (1966) proposed that pectins of cell walls are demethylated by PME activity of the tissue to form insoluble complexes with calcium and perhaps other multivalent ions, rendering resistance to hydrolysis by polygalacturonases from pathogens.

Infection tests on *Vicia faba* indicated that the stage of tissue development distinctly influenced the maceration by virulent bacteria. Using plants in the 8th-leaf stage, an increase of resistance against injected bacteria was observed, ascending from the first to the eighth leaf. There were marked differences in the PME activity of these leaf stages (Fig. 3). The higher resistance of the tissue to bacterial pathogens was attributed to the obviously higher PME activity in the leaves.

Endo-polygalacturonases (poly-α-1,4-D-galacturonidglycanohydrolases) and endo-pectic acid transeliminases (poly-α-1,4-D-galacturonidglycanolyases) degraded pectin, although by different mechanisms and at different pH optima: pectin methylesterases were active at acid pH values only, consequently in the range for endo-PG. Sometimes bacterial virulence was positively correlated with endo-PG activity and, on the other hand, susceptibility of plant tissue was inversely related to PME activity.

References

Albersheim, P., H. Neukom & H. Deuel, 1960. Über die Bildung von ungesättigten Abbauprodukten durch ein pektinabbauendes Enzym. Helv. chim. Acta 43: 1422–1426.

Bateman, D. F. & T. L. Millar, 1966. Pectic enzymes in tissue degradation. Ann. Rev. Phytopath. 4: 119–146.

Bell, T. A., J. L. Etchells & I. D. Jones, 1955. A method for testing cucumber saltbrine for softening activity. U.S. Dept. Agric. Publ. ARS-72-5: 1–15.

Knösel, D. & E. D. Garber, 1968. Separation of pectolytic and cellulolytic enzymes in culture filtrates of phytopathogenic bacterial species by starch gel zone electrophoresis. Phytopath. Z. 61: 292–298.

Koller, A. & H. Neukom, 1964. Detection of oligogalacturonic acids by thin-layer chromatography. Biochim. Biophys. Acta 83: 366–367.

Lange, E. & D. Knösel, 1970. Zur Bedeutung pektolytischer, cellulolytischer und proteolytischer Enzyme für die Virulenz phytopathogener Bakterien. Phytopath. Z. 69: 315–329.

Smith, W. K., 1958. Chromatographic examination of the products of digestion of pectic materials by culture solution of plant pathogenic and other bacteria. J. gen. Microbiol. 13: 211–238.

Pectolytic and cellulolytic enzymes produced by Xanthomonas pelargonii[1]

A. Garibaldi

Istituto di Patologia vegetale, Università degli Studi di Torino, Italia

This study was carried out to characterize the pectolytic and cellulolytic enzymes produced in healthy and in infected pelargonium plants by *Xanthomonas pelargonii*. Isolate 1013, used in all experiments, was obtained from a diseased plant in Northern Italy and selected from seven isolates for its high enzyme production.

To obtain diseased plant material, geranium plants (belonging to a susceptible commercial variety 'hortorum') were inoculated in accordance with the method described by Lemattre (1965) and incubated at 25–26 °C until severe symptoms appeared.

The methods described by Garibaldi & Bateman (1970) were used for production, fractionation, assay, estimation of molecular weight of the enzymes, and protein determination.

In culture fluids the pectolytic and cellulolytic activities of the Xanthomonas gradually increased after the cells had entered the logarithmic phase of growth; they reached a maximum after 48–72 h (Table 1). Beyond this point the activity of the enzymes (ratio of viscosimetric units to mg protein) decreased slightly.

Growth on polygalacturonic acid (PGA) and sodium polypectate (NaPP) produced greater quantities of pectic enzymes than growth on pectin. Very little pectic enzyme was detected in culture fluids with glucose as the sole carbon source. Glucose also greatly reduced pectic enzyme production when added to a medium containing sodium polypectate. The optimum pH for pectolytic activity, as determined by the reducing group, with thiobarbituric acid (TBA) and by viscosity loss methods, was near pH 9.0 (Fig. 1). In the reaction mixture, a product formed that had maximum light absorption at 230 nm and reacted with TBA to give a red chromogen absorbing light with a maximum at 548 nm. This indicated that the pectolytic enzyme of the Xanthomonas was a *trans*-eliminase. Furthermore, a 50 % reduction in the viscosity of a sodium polypectate solution was accompanied by a 1.5 % degradation of the substrate (Macmillan *et al.*, 1964). This indicates that the *trans*-eliminase degraded the substrate in an 'endo' manner (endo-PGTE).

The endo-PGTE degraded PGA more rapidly than esterified (68 %) citrus pectin (Fig. 2), showing that the polygalacturonic acid *trans*-eliminase of *Xanthomonas pelargonii* is specific for PGA rather than for pectin. No trace of hydrolytic polygalacturonase was detected in any of the tested culture fluids.

1. This work was supported by a grant from the National Council of Research of Italy.

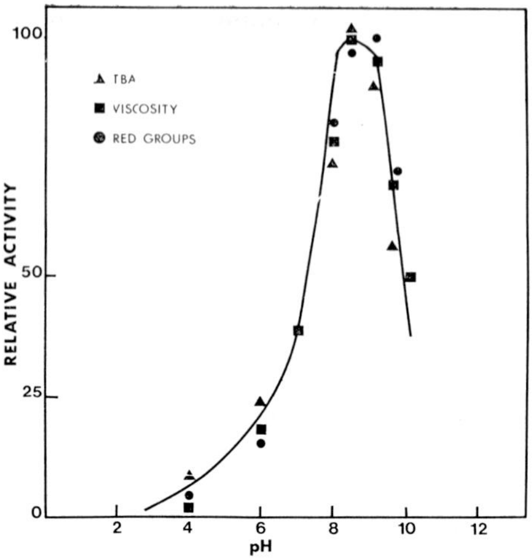

Fig. 1. Effect of pH on PGTE activity in culture filtrates of *Xanthomonas pelargonii*. Activity determined by reducing group, TBA, and viscosity loss methods, using sodium polypectate as the substrate. Buffer system (0.05 M): citrate (pH 4–5), phosphate (pH 6–7), tris-HCl (pH 7–10).

Table 1. Extracellular PGTE and Cx activities in culture filtrates of different age.

Age of culture (in hours)	Activity units $\times$ 10^{-2}/mg protein	
	PGTE	Cx
24	0.54	1.07
48	2.72	7.57
72	9.90	7.29
96	8.95	6.85

PGTE and Cx activities determined by viscosity loss method and expressed as 1000/t where t is the time (in min) for 50% viscosity loss of the substrate at 30 °C.

The hydrolysis of polygalacturonic acid by PGTE was followed over a period of 24 h. The results showed that in the initial stages of the reaction higher oligalacturonides predominated, but that the amount of unsaturated digalacturonic acid progressively increased and became dominants after 24 h.

After purification with ammonium sulphate precipitation, and chromatographical analysis on DEAE-cellulose, polygalacturonic acid-*trans*-eliminase behaved as a single component with pI near pH 9.25 (Fig. 3).

Based on gel filtration in Sephadex G-75, the estimated molecular weight of the *X. pelargonii* PGTE was about 32,000 (Fig. 4).

In extracts of uninoculated healthy plants no polygalacturonic acid *trans*-eliminase was found. Extracts of infected plants, 5–20 days after inoculation, showed little or no

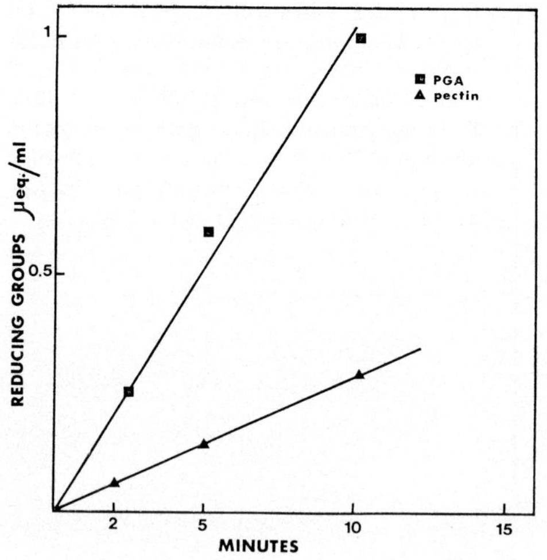

Fig. 2. Rate of hydrolysis of polygalacturonic acid and pectin by partially purified polygalacturonic acid *trans*-eliminase from *Xanthomonas pelargonii*. Reaction mixtures contained enzyme, 0.5% PGA or pectin, 0.05 M *tris*-HCl buffer pH 9.0 and 2×10^{-4} $CaCl_2$. Enzyme activity estimated by reducing group method.

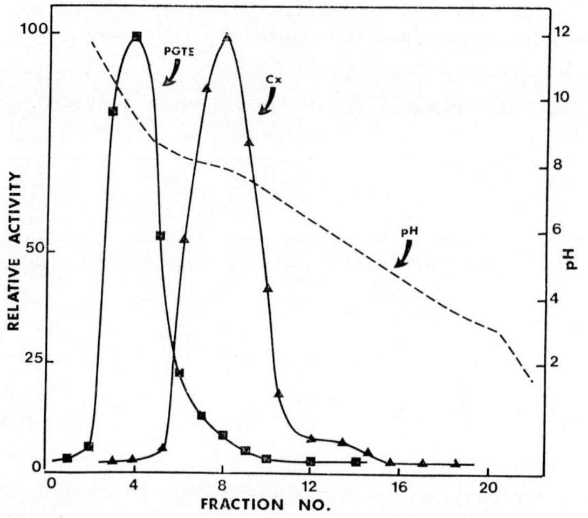

Fig. 3. Electrofocusing of partly purified PGTE and Cx preparations from culture filtrates of *Xanthomonas pelargonii*. PGTE and Cx activities measured by viscosity loss method; 10 ml of PGTE and Cx peak fractions from DEAE Sephadex A-50 were used. Four ml fractions were collected.

PGTE activity (Table 2). The characteristics of PGTE in infected tissue could not be determined because its activity was insufficient.

The highest cellulase formation was observed in the presence of sodium polypectate and pectin, but high concentration also occurred with glucose as a substrate. This clearly shows that in *Xanthomonas pelargonii* cellulase is a constitutive enzyme. This organism was unable to grow when carboxymethyl cellulose (CMC) was the sole carbon source.

The Cx produced by the Xanthomonas was active over a broad pH range, with an optimum near pH 5.5, as determined by the reducing group and viscosity loss methods (Fig. 5).

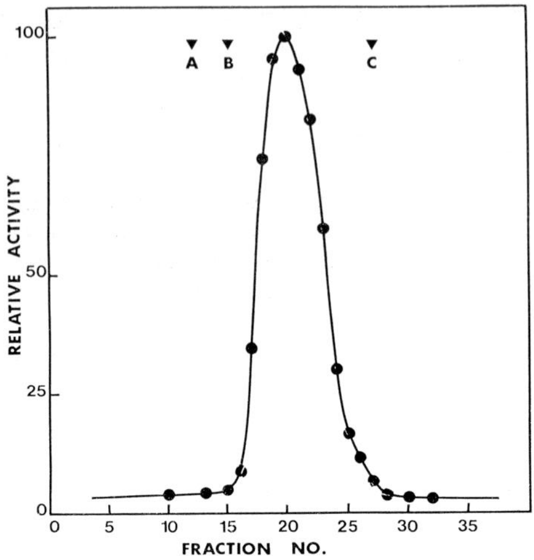

Fig. 4. Gel filtration patterns of PGTE from *Xanthomonas pelargonii* culture filtrates on Sephadex G-75. Markers A, B and C represent elution volumes of Blue Dextran (mol wt 2,000,000), bovin serum albumine (mol wt 67,000) and Cytochrome C (mol wt 12,400), respectively. Enzyme activity estimated by TBA method.

Table 2. Polygalacturonic acid *trans*-eliminase (PGTE) and cellulase (Cx) activities in extracts of healthy and *Xanthomonas pelargonii*-infected geranium stem tissue.

	Days after inoculation	Relative activity in units/mg protein	
		PGTE	Cx
Experiment I			
inoculated	8	0.0	12.1
inoculated	20	3.2	25.2
uninoculated		0.0	0.0
Experiment II			
inoculated	5	1.1	18.5
inoculated	18	5.2	50.8
uninoculated		0.0	0.0

PGTE and Cx activities determined by viscosity loss method using sodium polypectate at pH 9.0 and CMC at pH 5.0.

After purification with ammonium sulphate precipitation, and DEAE-Sephadex chromatography, electrofocusing of the Cx fraction showed only one peak with pI near pH 8.1 (Fig. 3). When the Cx fraction from the DEAE-Sephadex A-50 column was subjected to gel filtration on Sephadex G-75, the enzyme was retained too long by the column and its elution volume gave an abnormally low molecular weight. This phenomenon was already observed with a Cx produced by *Sclerotium rolfsii* (Bateman, 1969) and by *Erwinia chrysanthemi* (Garibaldi & Bateman, 1970).

The products from the action of cellulase on carboxymethyl cellulose were separated

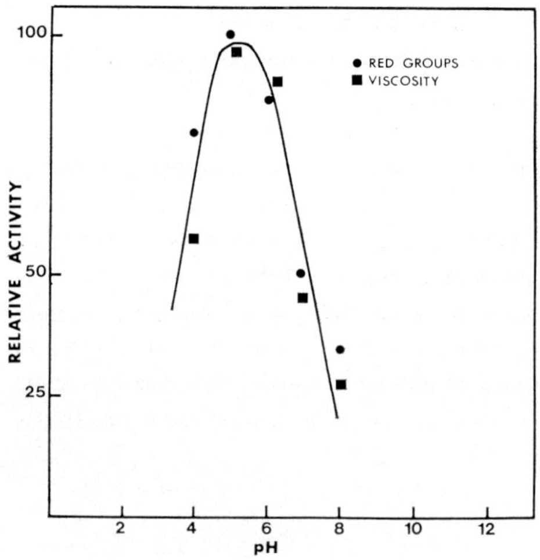

Fig. 5. The effect of pH upon the Cx activity from culture filtrates of *Xanthomonas pelargonii*. Activity determined by viscosity loss and reducing group methods. Same buffers as indicated in Fig. 1.

by ascending chromatography on Whatman no. 1 paper for 15 h with isopropanol in water (4:1, v/v) as mobile phase and developed with aniline reagent (Smith, 1960). After incubation of Cx with CMC at 30 °C for 24 h, the chromatograms revealed the release of glucose, cellobiose and two other products. Probably other intermediate products were also liberated, but they could not be detected by the methods used. The chromatographic analysis of the reaction mixtures in which cellobiose (1 mg/ml) was used as a substrate revealed a slow release of glucose after 24 h incubation at 30 °C.

Extracts of healthy geranium plants did not show Cx activity, but in extracts of infected geranium plants the Cx activity was high, as became apparent 4 to 5 days after inoculation (Table 2). The Cx in extracts of severely infected plants 20 days after inoculation was similar to that produced by Xanthomonas in culture at optimum pH after precipitation with ammonium sulphate by DEAE-Sephadex A-50 chromatography. The only observed difference was that tissue extracts of diseased plants completely degraded both CMC and cellobiose to glucose, whereas in the early phases of the reaction culture filtrates solely degraded CMC to cellobiose. Extracts from healthy plants degraded cellobiose, but no CMC, to glucose. This seems to be due to the presence, in infected as well as in healthy geranium plants, of a β-glucosidase capable of degrading to cellulose the molecules of cellobiose resulting from the CMC breakdown.

These results show that *Xanthomonas pelargonii* possesses the enzymatic mechanisms necessary to degrade pectin and soluble cellulose derivatives. The PGTE from its culture filtrates seems to be very similar to the PGTE produced by *Xanthomonas campestris* (Nasuno & Starr, 1967).

The cellulase produced by this organism is a constitutive enzyme, as has been reported for the Cx of several other bacteria (Hammerstrom *et al.*, 1954; Garibaldi & Bateman, 1970). The role of PGTE and Cx in bacterial stem-rot of geranium is not yet clear. Although the culture filtrates of *Xanthomonas pelargonii* contain a large amount of PGTE; in the extracts of infected plants the PGTE activity was very low. So it seems that pectic enzymes do not play an important role in the pathogenesis of this species, or that some host metabolites inhibit PGTE in the infected plants.

On the other hand it was easy to detect Cx in tissue extracts of infected plants a few days after inoculation. Large amounts of cellulase have also been found in carnations infected by *Erwinia chrysanthemi* (Garibaldi & Bateman, *op. cit.*), and this enzyme seems to play an important pathogenic role in the bacterial wilt of tomato caused by *Pseudomonas solanacearum* (Kelman & Cowling, 1965).

The importance of Cx in the geranium stem-rot caused by *Xanthomonas pelargonii* is supported by the disappearance of the birefringent properties of cell walls, observed in polarized light, in infected geranium stems (Lemattre, 1966). This may be interpreted as evidence for the destruction of crystalline cellulose in the host by cellulase.

References

Bateman, D. F., 1969. Some characteristics of the cellulase system produced by *Sclerotium rolfsei* Sacc. Phytopathology 59: 57–62.

Garibaldi, A. & D. F. Bateman, 1970. Association of pectolytic and cellulolytic enzymes with bacterial slow wilt of carnation caused by *Erwinia chrysanthemi* Burkh., McFad. et Dim. Phytopath. mediterr. 9: 136–144.

Hammerstrom, R. A., K. D. Claus, J. M. Coghlan & R. H. McBee, 1955. The constitutive nature of bacterial cellulase. Arch. Biochem. Biophys. 56: 123–129.

Kelman, A. & E. B. Cowling, 1965. Cellulase of *Pseudomonas solanacearum* in relation to pathogenesis. Phytopathology 55: 148–155.

Lemattre, M., 1965. La maladie bactérienne du pelargonium au *Xanthomonas pelargonii* (Brown) Stam et Burkh. Annls Epiphyt. 16: 5–35.

Lemattre, M., 1966. Etude histologique et histochimique des réactions déclenchées par *Xanthomonas pelargonii* dans les tissues de *Pelargonium zonale*. Annls Epiphyt. 17: 215–228.

Lowry, O. H., N. J. Rosebrough, A. L. Farr & R. J. Randall, 1951. Protein measurement with the Folin phenol reagent. J. biol. Chem. 193: 265–275.

Macmillan, J. D., H. J. Phaff & R. H. Vaughn, 1964. The pattern of action of an exopolygalacturonic acid *trans*-eliminase from *Clostridium multifermentans*. Biochemistry 3: 572–578.

Nasumo, S. & M. P. Starr, 1967. Polygalacturonic acid *trans*-eliminase of *Xanthomonas campestris*. Bioch. J. 104: 178–185.

Comparative biochemistry of the toxic glycopeptides produced by some plant pathogenic Corynebacteria[1]

G. A. Strobel

Montana State University, Bozeman, Montana 59715

Abstract

Toxic glycopeptides have been purified from *Corynebacterium sepedonicum* and *C. insidiosum*. Both toxins have been physically and chemically characterized. The toxin from *C. sepedonicum* has a molecular weight of 21,400, that from *C. insidiosum* of 5×10^6. Although these compounds greatly differ in physical properties, they are similar in that they both contain about the same sugar residues, virtually the same abundance of a keto-deoxy sugar acid and a peptide. At low concentrations both produce complete wilting of plant cuttings and the available evidence suggests that each is present in the host plant infected by the respective pathogen.

Introduction

Phytopathogenic species of the genus *Corynebacterium* generally produce a wilt in their host plants, according to Hodgson *et al.* (1949) due to polysaccharides. This led Spencer & Gorin (1961) to examine *C. sepedonicum* and *C. insidiosum*, two wilt pathogens, for their ability to produce such toxic polysaccharides. They demonstrated that crude preparations of culture fluids of both organisms were capable of inducing wilting in plant cuttings and their preparations indeed contained physiologically active polysaccharides. Subsequent to their work we have examined in more detail the purification, properties, and mode of action of these substances in cultures of the two pathogens. This report compares some physical, chemical, and biological properties of two toxic bacterial glycopeptides. The aim of this paper is to formulate some concepts on the role of glycopeptides in disease production by plant pathogenic bacteria.

Methods

The toxic glycopeptide of *Corynebacterium sepedonicum* was purified from the culture fluid of this organism by Strobel (1967). The techniques involved ion exchange chromatography to remove unwanted cations and anions, followed by flash evapora-

1. The research presented in this report was supported in part by NSF Grant GB 12956 and by Public Health Service Career Development Award 1-k4-GM-42, 475-02 from the National Institute of General Medical Sciences. It is Montana Agriculture Experiment Station Paper 263, Journal Series.

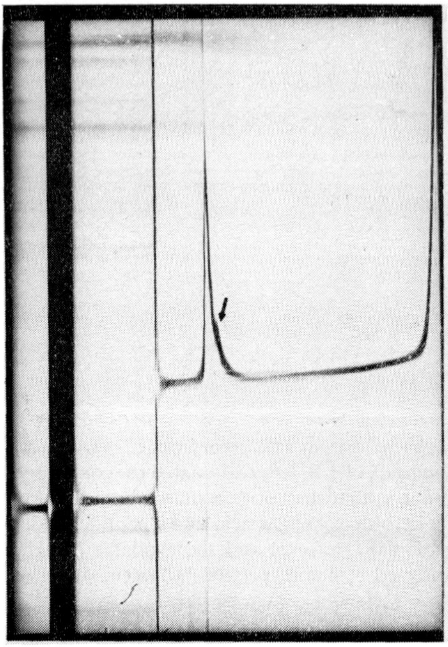

Fig. 1. Ultracentrifuge pattern of the purified glycopeptide toxin of *Corynebacterium insidiosum*. A 2.6 mg/ml of an aqueous toxin solution was centrifuged at 64,000 rpm with a bar angle of 50° for 2 h in a Beckman Model E ultracentrifuge. Sedimentation is left to right.

tion and acetone precipitation. The precipitate was taken up in a small amount of water, centrifuged to remove insolubles, and the supernatant liquid was subjected to chromatography on a column of Sephadex G-200. The final preparation was dialyzed to remove any sugars that might have overlapped the toxin peak. The preparation was shown to be homogeneous by several criteria, including gel electrophoresis, paper electrophoresis, ultracentrifuging, and chromatography on another column of Sephadex G-200 as indicated by Strobel (1970).

The toxic glycopeptide of *Corynebacterium insidiosum* was purified by Ries & Strobel (1970) utilizing acetone precipitation, anion and cation exchange chromatography, followed by two precipitations with ammonium sulphate. Using these procedures, Ries (1971) showed that the specific biological activity of the preparation increased with each purification step. The final preparation showed to be homogeneous by disc gel electrophoresis, paper electrophoresis and column chromatography on Sepharose 2B. However, analytical ultracentrifuging revealed that the preparation contained a small percentage of a slightly faster moving band (Fig. 1). For all practical purposes the preparation was considered homogeneous.

Results

Physical properties Ries (1971) determined the molecular weight (M.W.) of the *C. insidiosum* toxin to be 5.1×10^6 daltons by the light scattering technique utilizing a Zimm plot. This sharply contrasts with the M.W. of the toxin of *C. sepedonicum*

Table 1. Physical properties of two toxic glycopeptides in *Corynebacterium.*

Property of toxin	Corynebacterium sepedonicum	Corynebacterium insidiosum
Estimated molecular weight	21,400	5,000,000
Sedimentation coefficient	0.76 S	1.53 S
Optical rotation $[\alpha]$ 30°	+32°	−166°
Intrinsic viscosity	0.125 dl/g	0.230 dl/g

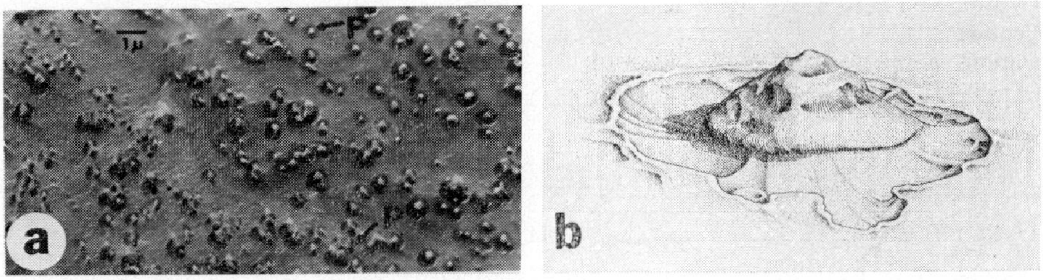

Fig. 2a. (a) Electron micrograph of freeze-etched replicas of the toxin of *Corynebacterium insidiosum*. A 0.1% solution of toxin was freeze-etched and examined by electron microscopy. Particles (P) observed and measured were 80–90 mμ in diameter. Particles (P′) appeared to be aggregates of smaller particles (P).

(b) An artist's conception of the particle (P) as it appears protruding from the surface of the suspending medium.

which Strobel reported to be 21,400 (1967). The M.W. of this toxin was determined from an average value of measurements made by gel filtration, analytical ultracentrifuging, and membrane osmometry. Other physical properties of the two compounds are given in Table 1.

Dr W. Hess of Brigham Young University made freeze etch replicas of the glycopeptide toxin of *C. insidiosum*. This technique permits particles to be seen directly in the suspending medium and virtually prevents artifact formation. Fig. 2 shows the electron micrograph of the replica with an artist's conception of the actual shape of the molecule. The larger particles (P′) are conceivably the ones accounting for the faster moving peak (shoulder in Fig. 1). The average particle diameter (P) was 80–90 nm.

Being much smaller, the molecules of the *C. sepedonicum* toxin cannot be seen with the electron microscope, but the pure toxin occurs as small crystals (Strobel, 1967) and powder diagrams can be obtained by X-ray diffraction techniques (Strobel, 1970).

Chemical properties The toxins from both sources were hydrolysed in dilute acid and the sugar residues subjected to reduction by $NaBH_4$, followed by acetylation and gas chromatography of the alditol acetates (Albersheim, 1967). The peaks were identified and counted using authentic alditol acetates as standards. The results (Table 2) show that both toxins contain the same sugar residues, but in greatly differing

Table 2. Chemical properties of two toxic glycopeptides in *Corynebacterium*, in %.

Compound	*C. sepedonicum*	*C. insidiosum*
Sugars: L-fucose	0.3	38.7
mannose	33.4	3.8
glucose	18.8	20.0
galactose	4.5	20.6
rhamnose	1.1	trace
arabinose	0.7	0
ribose	3.8	0
Organic acid keto-deoxy sugar acid	9.9	8.8
Peptide	5.0	2.6
Cations: copper	0	0.1

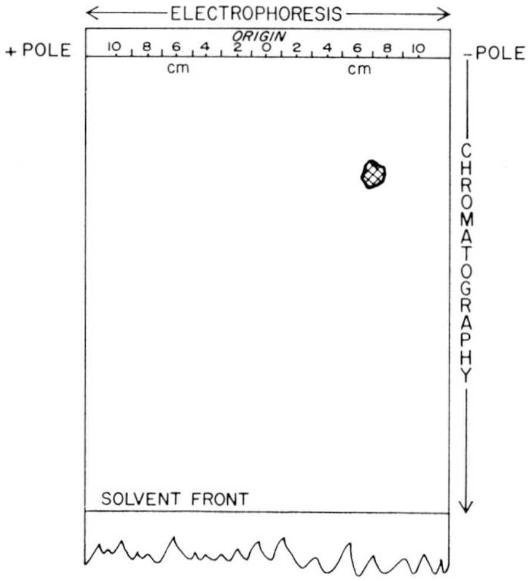

Fig. 3. Reproduction of a chromatogram showing the electrophoretic mobility in an acetic-formic acid system, pH 2.0, followed by chromatography in the second dimension in n-butanol-acetic acid-H_2O 4:1:5 v/v of the peptide fraction of the *Corynebacterium insidiosum* toxin.

amounts. The 2-keto-3-deoxygluconic acid was isolated and identified from the acid hydrolysate of the toxin of *C. sepedonicum* (Strobel, 1970). Likewise a keto-deoxy sugar acid was found in the toxin of *C. insidiosum* by Ries (1971).

About 5% of the weight of the toxin from *C. sepedonicum* was peptide (Table 2). During treatment of the toxin in 0.5 N NaOH for 196 h at 5°C, a β-elimination reaction occurred and the peptide fragment was released. Electrophoresis, followed by chromatography of the peptide fraction, revealed the presence of five peptides, two of which predominated (Fig. 3). Peptide 1, the most abundant, contained equal molar amounts of ala, gly, ser, thr, glu, and aspartic acid with alanine as the N-terminal amino acid. Peptide 2 contained equal molar amounts of ala, gly, ser, and thr with glycine as the N-terminal amino acid. It seemed as if each toxin molecule contained

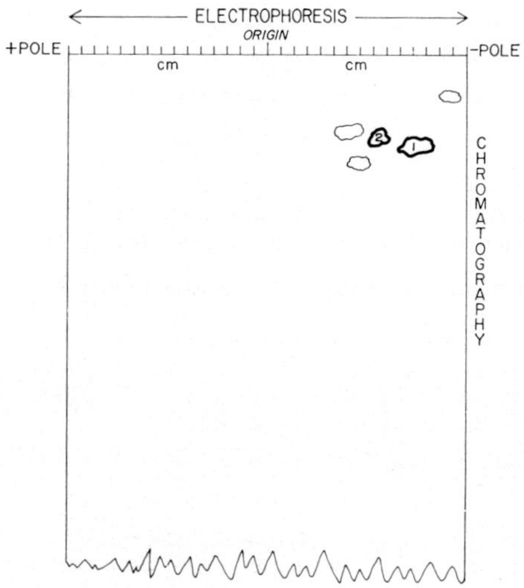

Fig. 4. Reproduction of a chromatogram treated identically to that in Fig. 3 but with the peptide fraction of the *Corynebacterium sepedonicum* toxin. Peptide 1 predominated in its color intensity with ninhydrin followed by peptide 2.

only one peptide, but that variation occurred as to which peptide was attached. These results were consistent with the earlier observation of Strobel (1970), who demonstrated that alanine and glycine are N-terminal amino acids of the toxin.

On the other hand, Ries (1971) showed that the peptide fraction accounted for 2.5% of the weight of the toxin of *C. insidiosum*. Again, using mild base treatment on the toxin, Ries recovered the peptide fraction and subjected it to electrophoresis and chromatography (Fig. 4) and showed that only one peptide was present. Furthermore, there were 56–76 single peptide chains per mole of toxin. The toxin of *C. insidiosum* is bright blue. It binds 75 moles of copper/mole of toxin. Presumably, the peptide plays a major role in binding the copper ions since pronase digestion of the toxin greatly reduces the intensity of the colour.

Structural properties Recently, much work has been done on the structural features of the toxin of *Corynebacterium sepedonicum*. To determine the sugar residue to which the peptide is attached, the toxin was first totally reduced with $NaBH_4$, followed by treatment in base to eliminate the peptide, and ultimately it exposed one free reducing group on each molecule. The toxin containing the free reducing group was treated with $NaBT_4$, effectively labelling the sugar residue to which the peptide was originally attached. After hydrolysis, reduction and acetylation followed by gas chromatography of the alditol acetates, the only sugar labelled with tritium was mannose.

The amino acid in the peptide to which the mannose attached was determined by mild base treatment in the presence of $NaBH_4$. During β-elimination, a dehydro-amino acid was formed which was subsequently reduced to another amino acid by $NaBH_4$. In this experiment the reduction in the amount of threonine was about 50%, with a concomitant production of α-aminobutyric acid. This suggested that the peptide

Table 3. Linkage present in the toxic glycopeptide of *Corynebacterium sepedonicum.*

Sugar and linkage	Mole %	Sugar and linkage	Mole %	Sugar and linkage	Mole %
Terminal mannose	26.0	1 → 3 mannose	7.4	1 → 6 galactose	1.8
1 → 6 glucose	18.0	1,4,6 glucose	5.0	1,2,3, galactose	1.0
1 → 2 mannose	11.1	1 → 4 glactose	4.6	1,3,6, galactose	2.0
1,2,6 glucose	ca 7.0 } 18.0	terminal rhamnose	2.2	1 → 3 galactose	1.8
1,2,6 mannose	11.0–18.0 } 18.0				

peptide–threonine → mannose–toxin
Total % terminal sugars = 28.2
Total % branch sugars = 28.2

had a glycoside link through the -OH of threonine to mannose. Final confirmation of this came with the isolation of a peptide sugar complex from pronase treatment and mild acid hydrolysis of the toxin followed by chromatography. The complex contained equal molar amounts of mannose, threonine, glycine, alanine, serine, and aspartic acid.

In co-operation with Dr K. Talmadge and Dr P. Albersheim of the University of Colorado, the nature of sugar linkages present in the toxin of *C. sepedonicum* was determined, using the techniques of Bjorndal *et al.* (1970). Good agreement was obtained between the proportion of methylated sugars (Table 3) and the sugar analysis reported in Table 2. The methylation results suggested that the toxin was a highly branched heteropolymer, as originally proposed on the basis of NMR data (Strobel, 1967). The majority of branching occurred at 1 → 2, 1 → 6 linked mannose. The major internally linked sugars were (1 → 6) linked glucose, (1 → 2) linked mannose, (1 → 3) linked mannose, and (1 → 4) linked galactose (in that order). Over 90% of the terminal sugars consisted of mannose which comprised 26% of the total residues. The total terminal residues (28.2%) equalled those of the total branched sugars (28.2%), as expected (Table 3). Ribose was not found as a methylated derivative. The toxin seemed to be associated with RNA, since it had some absorption at 260 nm (Strobel, 1970) and staining properties resembling those of RNA (Pearson, 1971). This was not too surprising, as Petrova (1970) recently showed that oligoribonucleotides are capable of forming complexes with starch.

Peptide and sugar linkage determinations have not yet been carried out on the toxin of *C. insidiosum.*

Toxins in host plants In 1970, Strobel isolated the phytotoxic glycopeptide from potato plants infected with *C. sepedonicum.* The crude juices of the plants were subjected to the same purification steps as the culture filtrates of the organism. The final step in purification, Sephadex column chromatography, was repeated twice, and the material collected from the column was compared with the purified toxin obtained from cultures of the organism. Elution from Sephadex columns, disc gel electrophoresis, IR spectra, serological analysis, residue analysis, and X-ray diffraction patterns of the substances from the two sources were identical in virtually all respects. Furthermore, there was enough toxin in the infected plant tissues to produce wilt in potato test plants of a comparable fresh weight.

The only evidence for the presence of the toxin of *C. insidiosum* in infected alfalfa was given by Spencer & Gorin (1961). They demonstrated L-fucose in acid hydrolysates of extracts of infected alfalfa but they were unable to recover this unusual sugar from extracts of healthy plants.

Biological activity The toxins from both sources produce wilting in 6–8 day old tomato plant cuttings in the nanomolar range (10^{-9} moles). Evidence for membrane damage as being the primary factor involved in wilt production was presented by Strobel & Hess (1968) for the toxin of *C. sepedonicum.* Johnson & Strobel (1970) showed that the toxin of *C. sepedonicum* possessed an active site, or sites, as the carboxyl group of 2-keto-3-deoxygluconic acid. Ries (1971) demonstrated that high M.W. substances bracketing the *C. insidiosum* toxin in M.W. cause leaf wilt, but not stem wilt as measured by a wiltometer (Fig. 5). Dye movement and electron microscopic studies by Rai & Strobel (1969) on the toxins of *C. michiganense* in tomate cuttings also pointed to membrane damage induced by the toxin.

In a practical sense, the most important biological property of the glycopeptides is their antigenicity. Antisera prepared to the crude toxin of *C. sepedonicum* have been utilized in the diagnosis of potato ring rot by Strobel & Rai (1968). More recently, Shepard (at Montana State University) has prepared relatively high titred antisera to the toxin of *C. sepedonicum* using whole cells as the inject antigen. This makes

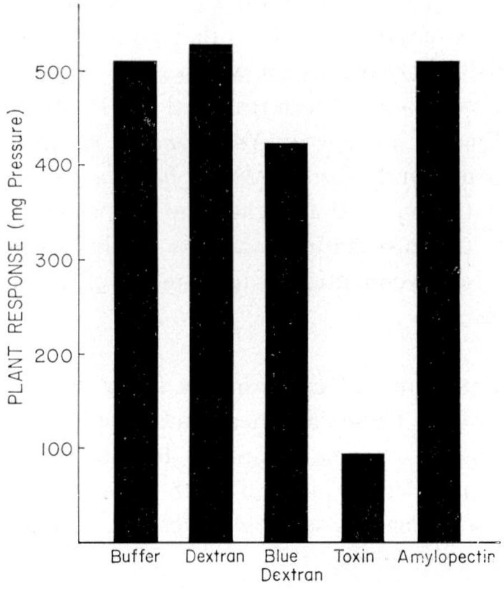

Fig. 5. A histogram showing the ability of variously sized compounds to cause wilt in tomato cuttings. Solutions 0.25% of dextran (2.5 × 10^5 daltons), blue dextran (2 × 10^6 daltons), toxin (5 × 10^6 daltons) and amylopectin (70 × 10^6 daltons) were checked for their ability to produce wilt in tomato cuttings as measured by a wilt-o-meter (see Johnson & Strobel, 1970).

possible tests by double diffusion gel techniques. Further studies are being conducted to determine the specificity of the serum.

Conclusions

The toxins of *Corynebacterium sepedonicum* and *C. insidiosum* vary greatly in their physical properties, but both are glycopeptide in nature as they contain virtually the same sugar residues, a small percentage of peptide and about 10% of a keto-deoxy sugar acid (Tables 1 and 2).

Studies on the sugar composition of the toxin of *C. sepedonicum* have revealed some differences in the relative abundance of glucose and mannose as reported earlier by Strobel (1970). This may be attributed to some contamination of the original final preparation by unwanted sugars, to the method of hydrolysis, the methods of analysis, or to strain differences in the culture of *C. sepedonicum*. The data in Table 2 are very reproducible and are the most realistic for this toxin.

Structural studies on the toxin of *C. sepedonicum* have presented some interesting questions regarding toxin biosynthesis. The fact that each branch on each chain is terminated by mannose or rhamnose suggests that biosynthesis of the toxin molecule is a directed, rather than a random event of transglycosylation (Table 3). Perhaps RNA directs synthesis and this is the reason for its association with the toxin.

The mode of action of the toxin of *C. sepedonicum* has been studied extensively. It seems that cellular membranes are affected by this substance. This idea was expressed by Rai & Strobel (1969) as 'the mode of action' of the toxins of *C. michiganense*. Some studies by Ries (1971) point to a mechanism that, in addition to or different from plugging, may be responsible for the wilt inducing property of the *C. insidiosum* toxin (Fig. 5). Further studies on this are in progress.

Although not discussed in this report, it should be mentioned that the toxins of *C. michiganense* have been isolated and partially characterized by Rai & Strobel (1969) as being glycopeptides. This tends to strengthen the view that glycopeptides are common to plant pathogenic bacteria. Other organisms for which slime or gums or polysaccharide-like materials have been reported have been reviewed by Goodman *et al.* (1967). Among others they include *Erwinia amylovora*, *Xanthomonas phaseoli*, *Pseudomonas solanacearum*, *Erwinia carotovora* and *Agrobacterium tumefaciens*. It would be interesting to know the details of the physical and chemical properties of these substances and to compare them with the information available on the toxins of *Corynebacterium* spp. Perhaps they are also glycopeptides with a mode of action similar to that of the toxins of *C. sepedonicum*.

The results discussed above stress the importance of collecting as many data as possible on any newly discovered toxic compound. These data then can be used for (1) comparison with other toxins, (2) alterations in organic chemicals for active site determinations, (3) studies on pathways with labelled toxins, and (4) the examination of disease control measures involving the toxin, such as serological testing, or mass screening for disease resistance in a breeding programme.

References

Albersheim, P., D. J. Nevins, P. D. English & A. Karr, 1967. A method for the analysis of sugars in plant cell-wall polysaccharides by gas-liquid chromatography. Carb. Res. 5: 340–345.

Bjorndal, H., C. G. Hellerquist, B. Lindberg & S. Svensson, 1970. Gas-liquid chromatography and mass spectrometry in methylation analysis of polysaccharides. Angew. Chem., Internat. Edit. 9: 610–619.

Goodman, R. N., Z. Kiraly & M. Zaitlin, 1967. The biochemistry and physiology of infections. Plant Disease. Van Nostrand Co., New York.

Hodgson, R., W. H. Peterson & A. J. Riker, 1949. The toxicity of polysaccharides and other large molecules to tomato cuttings. Phytopathology 39: 47–62.

Johnson, T. B. & G. A. Strobel, 1970. The active site on the phytotoxin of *Corynebacterium sepedonicum*. Pl. Physiol. 45: 761–764.

Pearson, D., 1971. The biological activity of phytotoxic extracts from *Corynebacterium sepedonicum*. M.S. thesis University of Nebraska.

Petrova, A. N., 1970. Complexes of polysaccharides with oligoribonucleotides and their biological action. Biochemistry, N.Y. 35: 861–866.

Rai, P. V. & G. A. Strobel, 1969a. Phytotoxic glycopeptides produced by *Corynebacterium michiganense*. I. Methods of preparation. Phytopathology 59: 47–52.

Rai, P. V. & G. A. Strobel, 1969b. Phytotoxic glycopeptides produced by *Corynebacterium michiganense*. II. Biological properties. Phytopathology 59: 53–57.

Ries, S. M., 1971. Chemical and biological properties of a phytotoxic glycopeptide from *Corynebacterium insidiosum*. Ph.D. thesis Montana State University, Bozeman.

Ries, S. M. & G. A. Strobel, 1970. Properties of a high molecular weight toxin produced by *Corynebacterium insidiosum*. Phytopathology 60: 1309.

Spencer, J. F. T. & P. A. J. Gorin, 1961. The occurrence in the host plant of physiologically active gums produced by *Corynebacterium sepedonicum* and *Corynebacterium insidiosum*. Can. J. Microbiol. 7: 185–188.

Strobel, G. A., 1967. Purification and properties of a phytotoxic polysaccharide produced by *Corynebacterium sepedonicum*. Pl. Physiol. 42: 1433–1441.

Strobel, G. A. & W. Hess, 1968. Biological activity of a phytotoxic glycopeptide produced by *Corynebacterium sepedonicum*. Pl. Physiol. 43: 1673–1688.

Strobel, G. A., 1970. A phytotoxic glycopeptide from potato plants infected with *Corynebacterium sepedonicum*. J. biol. Chem. 245: 32–38.

Strobel, G. A. & P. V. Rai, 1968. A rapid serodiagnostic test for potato ring rot. Pl. Dis. Reptr 52: 502–504.